Preface

Welcome to Automotive Master Technician

This book has been written because rapid growth in technology used in the production of cars has highlighted the need for a different approach to vehicle diagnosis and repair.

While this technology has improved the comfort, safety, convenience and reliability of vehicles, it has also created an issue with established methods of maintenance and repair. As many of the control systems operate beyond our natural capabilities, diagnostic tooling is required to undertake most of the fault finding duties traditionally conducted by vehicle technicians. Also, the sophisticated nature of advanced system faults will often lead to diagnostic requirements for which there is no prescribed method.

One of the fundamental roles of a Master Technician will be the diagnosis and repair of these complex and advanced system faults, for which diagnostic approaches need to be developed that can provide logical strategies to reduce overall diagnostic time. An effective diagnostic routine should always begin with a logical assessment of symptoms and then uses reasoning to reduce the possible number of options, before following a systematic approach to finding and fixing the root cause.

The chapters will introduce you to health and safety, electrical principles and the psychology of diagnosis, followed by technical chapters on the operation of advanced vehicle systems including:

Advanced Internal Combustion Engine Technology

Advanced Vehicle Driveline and Chassis Technology

Advanced Vehicle Body Electrics

Alternative Fuel Vehicles

It will also include chapters on the non-technical skills required from a Master Technician including:

Providing Technical Support and Advice to Colleagues in Motor Vehicle Environments

Liaising with Vehicle Product Manufacturers and Suppliers on Technical Matters

Diagnostic Consultations with Customers in Motor Vehicle Environments

Fundamental Management Principles in the Automotive Industry

It also lays out key terms, points of interest, safety and diagnostic tips in order to support the information provided within the text.

Chapters:

Advanced Light Vehicle Technology

This book offers:

Ideal support for learners and tutors undertaking automotive qualifications.

Information to help cover the knowledge requirements for Level 4 Master Technicians.

A large number of illustrations to support knowledge and understanding.

Text © Graham Stoakes 2015

Original illustrations © Graham Stoakes 2015

The rights of Graham Stoakes to be identified as author of this work have been asserted by them in accordance with the Copyright, Designs and Patents Act 1988.

Copyright notice ©

Acknowledgements

Graham Stoakes would like to thank Anita and Holly Stoakes for their support during this project.

Thank you to alerrandre for the cover design.

The author and publisher would also like to thank the following individuals and organisations for permission to reproduce photographs:

Shutterstock.com: wavebreakmedia 272, Iakov Filmonov 271, 273, Goodluz 257, 279, Air Images 264, Denis Rozhnovsky 287, Bacho 318, Ensuper 299, bikeriderlondon 319, Andresr 261, 262, Hurst Photo 284, 286, Vereshchagin Dmitry 256.

Cover image: Shutterstock.com - Ensuper

Author

Graham Stoakes AAE MIMI QTLS is a lecturer and author of college textbooks in automotive engineering for light vehicles and motorcycles.

With his background as a qualified Master Technician, senior automotive manager and specialist diagnostic trainer, he brings 30 years of technical industry experience to this title.

Cover design - fiver.com/alerrandre

Published by Graham Stoakes

First published 2015

First edition

ISBN 978-0-9929492-2-8

Automotive Master Technician

Chapter 1 Master Technician and the Psychology of Diagnosis

This chapter will give you an introduction to the personal qualities displayed by a Master Technician. It will cover a brief overview of health and safety legislation for which you may be responsible in your form of managerial role. It will also assist you in developing strategies and approaches to complex diagnostic faults for which there are no prescribed methods. This is done by giving examples and methodologies which can be used or adapted to ensure a systematic approach and reduce overall diagnostic time. The chapter will also explain the setup and use of diagnostic equipment that can be utilised during fault investigation. It will support you with knowledge that will aid you when undertaking both theory and practical assessments.

Contents

There have been over 8000 injuries and 24 deaths in the motor vehicle repair (MVR) industry over the last 5 years (source: Health and Safety Executive).

Your workshop should have in place health and safety policies. These will have been developed by your employer to make sure that government legislation is observed. It is important that you are aware of the **legislation** and your rights and responsibilities, as well as those of your employer. It is your right to expect your employer to fulfil their responsibilities and it is your employer's right to expect you to fulfil yours. Legislation is the law and, if you do not observe it, you are committing an offence. As a Master Technician a certain amount of extra responsibility may be placed on you. One of your roles may involve you managing the general work area, and your colleagues; this could include being a health and safety representative. For this reason, it is important that you are aware of the legislation that affects your workplace and what your roles and responsibilities are.

Legislation – the laws which have been passed by government and which are enforced by the police and other bodies.

The Health and Safety Executive (HSE)

The Health and Safety Executive (HSE) is the national watchdog for work-related health, safety and illness. They are an independent regulator and act in the public interest to reduce work-related death and serious injury across all workplaces in the UK. Health and safety laws relating to your organisation will be enforced by inspectors from the HSE or a health and safety enforcement officer from your local council. They are aware of the special risks of motor vehicle repair work and will give you help and advice on how to comply with the law.

They are legally empowered to:

- Visit workplaces without notice, but you are entitled to see their identification before they come in.
- Investigate an accident or complaint, or inspect the safety, health and welfare aspects of your business.
- Talk to employees and safety representatives, take photographs and samples, and even in certain cases impound dangerous equipment.
- Receive co-operation and answers to questions.

HSE inspectors have powers to issue **improvement notices** and **prohibition notices** if they believe that there are any poor health and safety practices in a workplace they are inspecting.
If an employer does not comply with an improvement or prohibition notice, they can be fined or even imprisoned. If employers do not properly look after the safety of their employees, they can be prosecuted.

This could result in:

- Charges of corporate manslaughter
- Personal fines
- Corporate fines
- Imprisonment

- Loss of production
- Loss of income
- Bad publicity in the media
- Being served with a HSE prohibition notice

Improvement notice – notification that the employer must eliminate a risk, for example a bad working practice such as draining petrol into open tanks. The improvement notice gives the employer a specific period of time to eliminate the risk.

Prohibition notice – notification that the employer has to immediately stop all work until the safety risk is eliminated.

Some of the laws that are relevant to the automotive industry are covered in the next section:

The Health and Safety at Work Act 1974 (HASAWA)

The health and safety of everyone in the workplace is protected by the Health and Safety at Work Act (HASAWA). This law protects you, your employer and all employees while at work. It also protects your customers and the general public when they are visiting your workplace.

If you have five or more employees, you must have a written health and safety policy, it should clearly say who does what, when and how. The policy does not need to be complicated or time-consuming. To help you, there are templates that you can download from the Health and Safety Executive (HSE) website and complete.

Automotive Master Technician

Personal Protective Equipment (PPE) at Work Regulations 1992

This regulation requires that employers provide appropriate personal protective clothing and equipment for their employees.

Figure 1.1 Personal Protective Equipment PPE

When selecting PPE, make sure that the equipment:

- Is the right PPE for the job – ask for advice if you are not sure.

- Fits correctly – it needs to be adjustable so it fits you properly.

- Is properly looked after.

- Prevents or controls the risk for the job you are doing.

- Does not create a new risk, e.g. Overheating.

- Is comfortable enough to wear for the length of time you need it.

- Does not impair your sight, communication or movement.

- Is compatible with other PPE worn.

- Does not interfere with the job you are doing.

The CE mark found on PPE confirms that it has met the safety requirements of the Personal Protective Equipment at Work Regulations 1992. All PPE should have this mark.

Figure 1.2 The CE mark

Provision and Use of Work Equipment Regulations 1998 (PUWER)

The equipment used in your workshop needs to be:

- Safe to use.

- Maintained correctly.

- Inspected regularly.

- Only used by people who have received appropriate training.

The Provision and Use of Work Equipment Regulations 1998 (PUWER) place the responsibility for the safety of workplace equipment on anyone who has control over the use of work equipment, including your employer, you and your colleagues.

Control of Substances Hazardous to Health Regulations 2002 (COSHH)

The legislation which you and your employer must observe when using hazardous substances in the workshop, is the Control of Substances Hazardous to Health Regulations 2002 (COSHH).

There are eight steps that employers must take to protect employees from hazardous substances. These are shown in the next section:

Step 1
- Find out what hazardous substances are used in the workplace and the risks these substances pose to people's health.

Step 2
- Decide what precautions are needed before any work starts with hazardous substances.

Step 3
- Prevent people being exposed to hazardous substances or, where this is not reasonably practicable, control the exposure.

Step 4
- Make sure control measures are used and maintained properly and that safety procedures are followed.

Step 5
- If required, monitor exposure of employees to hazardous substances.

Step 6
- Carry out health surveillance where assessment has shown that this is necessary or where COSHH makes specific requirements.

Step 7
- If required, prepare plans and procedures to deal with accidents, incidents and emergencies.

Step 8
- Make sure employees are properly informed, trained and supervised.

Figure 1.3 Warning labels

Other health and safety regulations

As well as the legislation listed previously, the following regulations apply across the full range of workplaces:

1 The Management of Health and Safety at Work Regulations 1999: require
employers to carry out risk assessments, put in place measures to minimise risks,
appoint competent people and arrange for appropriate information and training for their staff.

2 Workplace (Health, Safety and Welfare) Regulations 1992: cover a wide range of basic health, safety and welfare issues such as ventilation, heating, lighting, workstations, seating and welfare facilities.

3 Health and Safety (Display Screen Equipment) Regulations 1992: set out requirements for work with computers and visual display units (VDUs).

4 Manual Handling Operations Regulations 1992: cover the moving of objects by hand or bodily force.

5 Health and Safety (First Aid) Regulations 1981: cover the requirements for first aid, including the number of trained first aiders required in the workplace.

6 The Health and Safety Information for Employees Regulations 1989: require employers to display posters telling employees what they need to know about health and safety.

7 Employers' Liability (Compulsory Insurance) Act 1969: require employers to take out insurance against work-related accidents and ill health involving employees and visitors to the premises.

9 Noise at Work Regulations 1989: require employers to take action to protect employees from hearing damage.

10 The Pressure Safety Systems Regulations 2000: users and owners of pressure systems are required to demonstrate that they know the safe operating limits, principally pressure and temperature, of their pressure systems, and that the systems are safe under those conditions. This will include compressed air systems, but does not include pressurised systems used for vehicle propulsion.

11 Reporting of Injuries, Diseases and Dangerous Occurrences Regulations 1995 (RIDDOR): require employers to report accidents, near misses and ill health to the HSE and to keep records of these events.

12 Electricity at Work Regulations 1989: require people in control of electrical systems to ensure they are safe to use and maintained in a safe condition.

Accident prevention

Working in the motor industry and with high voltage vehicle electrical systems or fuel used in alternative propulsion is extremely hazardous. There are many dangers that could result in an accident causing injury or even death. The management of these **hazards** is key to reducing the risks involved while working on these systems. Not all hazards can be removed, but they can be identified and measures put in place to reduce the dangers that they pose; this is the purpose of a risk assessment.

A risk assessment is an important step in protecting workers and businesses, and is necessary in order to comply with The Management of Health and Safety at Work Regulations 1999. It is designed to focus on the **risks** that really matter in the workplace – the ones with the potential to cause real harm. In many cases, straightforward measures can control risks.

The law does not expect you to eliminate all risk, but you are required to protect people as far as is **reasonably practicable.**

Reasonably practicable – can be carried out without incurring excessive effort or expense.

Hazard – something that has the potential to cause harm or damage.

Risk – the likelihood of the harm or damage actually happening.

A risk assessment is simply a careful examination of what, in your work, could cause harm to people. It allows you to weigh up whether you have taken enough precautions or should do more to prevent harm.

There are five main steps to risk assessment:

Step 1	Step 2	Step 3	Step 4	Step 5
• Identify the hazards – conduct an inspection of your workplace and make a list of all of the hazards you find.	• Decide who might be harmed and how – remember to include all those who may be at risk, for example: • staff/colleagues • contractors • delivery operatives • customers • general public • young people • people with disabilities	• Evaluate the risks and decide on precautions – for example: • Can you get rid of the hazard completely? If not, what needs to be done to control the risk of harm? • Is there a less risky option? • Can the hazard be guarded or access prevented? • Is PPE required? (PPE should only be used when other methods of reducing the risk are not practical).	• Record your findings and implement them – keep a written record of your risk assessment and what you have done to control the hazards.	• Review your assessment and update if necessary – working situations change, so make sure that you regularly check that your assessment still covers all hazards.

Environmental protection

Damage to the environment can be caused by contaminating the atmosphere, water supply or drainage system. Under the Environmental Protection Act 1990 (EPA), it is an offence to treat, keep or dispose of controlled waste in a way that is likely to pollute the environment (ecotoxic) or harm people. People who produce waste must make sure that it is passed only to an authorised person who can transport, recycle or dispose of it safely. You should have procedures in place for working with and disposing of any material which has potential to harm the environment.

Figure 1.4 Ecotoxic label

Automotive Master Technician

Ecotoxic – a substance that is harmful to the environment.

Controlled waste – any waste which cannot be disposed of to landfill, including liquids, asbestos, tyres and waste that has been decontaminated. There are three types of controlled waste listed under the environmental protection Controlled Waste (England and Wales) Regulations 2012: household, industrial and commercial waste. Depending on the classification and type of waste produced, a charge can be made for its collection and disposal.

Environmental Protection (Duty of Care) Regulations 1991

These regulations describe the actions which anyone who produces, imports, keeps, stores, transports, treats, recycles or disposes of controlled waste must take. These people must:

- Store the waste safely so that it does not cause pollution or harm anyone.

- Transfer it only to someone who is authorised to take it (such as someone who holds a waste management licence or is a registered waste carrier).

- When passing it on to someone else, provide a written description of the waste and fill in a transfer note. (From 2011, the waste transfer note must also include a declaration that you have applied the waste management hierarchy, which means you must consider reusing or recycling your waste before deciding to dispose of it).

- Keep these records for two years and provide a copy to the environment agency if they ask for one.

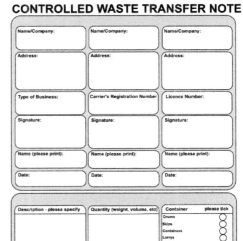

Figure 1.5 Waste transfer note

Safety signs

To assist with health and safety in the workplace, signs are often used to communicate instructions or to give warnings. These signs will normally contain images to convey their meaning.

They use the following formats:

Red signs

Red signs indicate a **PROHIBITION** (something you must not do). They often have a red diagonal line over the symbol on the sign.

Advanced Light Vehicle Technology

Blue signs

Blue signs indicate a **MANDATORY** instruction (something that you must do).

Yellow signs

Yellow signs indicate a WARNING (hazard or danger).

Green signs

Green signs indicate a SAFE condition (useful information). These are often used for signs indicating where to go in emergency situations.

Prohibition – something that you must not do.

Mandatory – something that you must do.

Fire safety

Careful consideration should be given to fire safety in the workplace.
A separate risk assessment should be carried out for fire hazards and safety measures must be put in place.

Fire extinguishers

In case of fire in the workplace, fire extinguishers should be provided and maintained by the company, for safety. The primary function of a fire extinguisher is to enable you to create an escape route. In any other circumstance you should only attempt to tackle a fire if it is safe to do so and if you have had adequate training.

A number of different fire extinguishers are available depending on the type of fire to be tackled.

The type of fire is normally classified:

- Class A: solids, such as paper, wood, plastic etc.

- Class B: flammable liquids, such as petrol, Diesel, oil etc.

- Class C: flammable gases such, as propane, butane, methane etc.

- Class D: metals, such as aluminium, magnesium, titanium etc.

- Class E: fires involving electrical apparatus.

- Class F: waste vegetable oil (WVO), and fat etc.

Every fire extinguisher has a colour-coded label with a description of its contents and a list of the types of fire it is designed to be used on. Extreme caution should be taken when selecting a suitable extinguisher to use on a vehicle electrical fire. Liquid based fire extinguishers will conduct electricity and this may lead to electrocution.

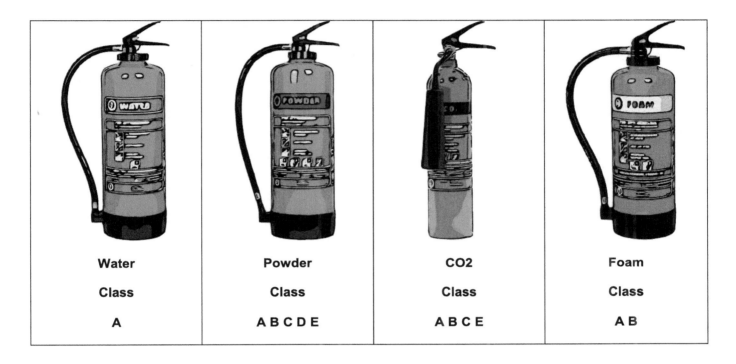

Water	Powder	CO2	Foam
Class	Class	Class	Class
A	A B C D E	A B C E	A B

Figure 1.6 Fire extinguisher types

Basic first aid

An automotive workshop is a high risk environment, and no matter what precautions are taken, there is always the possibility of accidents occurring which may lead to personal injury. The following advice is not a substitute for first aid training, and will only give you an overview of the action you may need to take. You should take care when you attempt to administer first aid that you do not place yourself in danger. Be very careful about what you do, because the wrong action can cause more harm to the casualty.

Good first aid always involves summoning appropriate help; many companies will have a trained first aider on site and must have a suitably stocked first aid box.

First aid box

The minimum level of first aid equipment in a suitably stocked first aid box should include:

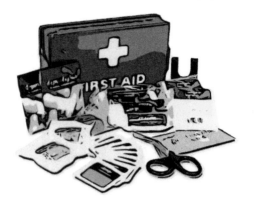

Figure 1.7 First aid box

- A guidance leaflet.

- 2 sterile eye pads.

- 6 triangular bandages.

- 6 safety pins.

- 3 extra-large, 2 large and 6 medium-sized sterile unmedicated wound dressings.

- 20 sterile adhesive dressings (assorted sizes).

- 1 pair of disposable gloves (as required under HSE guidance).

It is important to ensure that the contents of the first aid box are in date and are sufficient, based on the assessment of the workplace's first aid needs.

The law does not state how often the contents of a first aid box should be replaced, but most items, in particular sterile ones, are marked with expiry dates.

Other equipment such as eye wash stations must also be available if the work being carried out requires it.

Getting help

If you need to call for assistance, the main emergency services can be contacted by calling 999 free of charge from any landline or mobile phone.

When calling the emergency services, make sure you give the following information:

- Your telephone number.

- The location of the incident.

- The type of incident.

- The gender and age of the casualty.

- Details of any injuries observed.

- Any information you have observed about hazards, for example high voltage systems, chemical spills, gas or fuel leaks.

The recovery position

When dealing with health emergencies, you may need to place someone in the recovery position. In this position a casualty has the best chance of keeping a clear airway, not inhaling vomit and remaining as safe as possible until help arrives. You should not attempt to put someone in the recovery position if you think they might have back or neck injuries, and it may not be possible if any limbs are fractured.

Automotive Master Technician

Putting a casualty in the recovery position

1. Kneel at one side of the casualty, at about waist level.

2. Tilt the head back – this opens the airway. With the casualty on their back, make sure that their limbs are straight.

3. Bend the casualty's near arm so that it is at right angles to the body. Pull the arm on the far side over the chest and place the back of the hand against the opposite cheek.

4. Use your other hand to roll the casualty towards you by pulling gently on the far leg, just above the knee. This will bring the casualty onto their side.

5. Once the casualty is rolled over, bend the leg at right angles to the body. Make sure the head is tilted well back to keep the airway open.

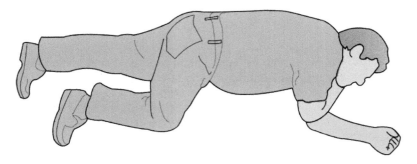

Figure 1.8 The recovery position

To find out more about first aid at work, visit the first aid section of the HSE website – http://www.hse.gov.uk/firstaid/index.htm

Diagnostic approaches (The psychology of diagnosis)

One of the fundamental roles of a Master Technician will be the diagnosis and repair of complex and advanced system faults for which there is no prescribed method. Much of the overall repair time is often taken up with diagnosis, whereas once the root cause of the fault has been correctly identified, it can often be fixed in a relatively short period.

The main issue here will be the unusual nature of the fault, meaning you are often unable to rely on common diagnostic or repair routines developed by other people.

The following section will describe some diagnostic approaches that may be employed during your fault finding routines and can provide logical strategies to reduce overall diagnostic time.

An effective diagnostic routine should always begin with a logical assessment of symptoms and then uses reasoning to reduce the possible number of options, before following a systematic approach to finding and fixing the root cause.

Advanced Light Vehicle Technology

Know your enemy

Knowledge is power. Before you are able to investigate any diagnostic issue, it is vital that you understand the components, function and operation of the system.

Time spent researching information before you start is not time wasted. Information sources may be wide and varied, and examples of these can be found at the start of each technical chapter. Unfortunately, researching information does not always translate into understanding.

A method that can be used to enhance your understanding is by explaining the operation or fault to someone else, especially if the other person is non-technical or has a limited knowledge. (This works particularly well with apprentices, as they may ask questions that you had not initially considered and they will also improve their own learning and performance). In order for you to describe the components, or why the symptoms are occurring, you will have to simplify your explanations. By doing this you will often find that you now understand the system better yourself.

An Internet search for information regarding a specific fault can be very useful in helping you devise a diagnostic strategy, and discussion forums will often explain how a particular issue was overcome. Be aware that a large proportion of information on the Internet may have been written by enthusiastic amateurs and therefore some of the content may be erroneous. Always ensure that you use the Internet to supplement your diagnostic routine, not replace it.

A trip to the library

Technical data and diagnostic information is a vital resource. It is important that you gather as much information as possible before you start and take it to the vehicle with you, correctly interpret electronic circuit diagrams and testing procedures. This will reduce time spent searching for information later on which can interrupt your diagnostic routine.
Interruptions can break your chain of thought, disrupting your routine and upsetting your systematic approach. The diagnostic information could be hard copy (books etc.), electronic (computer based) or notes that you have previously made (*see Dear Diary*). Time spent on researching information before you start is time well spent and will reduce frustration and misdiagnosis.

Interrogation room

Before you begin any diagnostic routine, you should start by carefully questioning the driver about the symptoms. The driver will be used to their own vehicle and have first-hand experience of the fault and the symptoms it produces. Notice I have said 'driver' and not 'customer', as these are sometimes not the same thing. The person that has brought the car to the garage may not be the main driver and any information that you get from them may be second-hand, reducing its overall usefulness.

It is important to use 'open' questions when speaking to the driver about the symptoms produced by a fault. An open question is one that does not have a simple yes or no answer.

Far more diagnostic information can be gained from asking the right type of open question and examples should include:

- Who is the main driver of the vehicle?
- What appears to be the problem/why have you brought the vehicle to the garage?
- When did you first notice the problem?

Automotive Master Technician

- How often does it occur?
- What symptoms are being produced?
- Under what conditions can the problem be recreated?
- How has the vehicle been used/modified lately?

The biggest issue with questioning the driver, is that you may not always be fortunate enough to do so. Someone else may have delivered the car for them, or another member of staff might be the point of first contact and the information you receive may be second or third-hand. In these situations it is worthwhile training other members of staff to also ask the questions listed above, and record the answers given. This way, if you need to speak to the diver directly, you already have a record of what was discussed when the vehicle was dropped off.
It is also advisable to design a pre-diagnostic questionnaire that could be used by any member of staff when receiving a car for repair. It should be relatively generic so that it could be used for many different types of problem, but simple enough so that it can be completed relatively quickly. By using a thesaurus to look up a series of describing words, check boxes can be designed that will further inform your diagnosis. An example of a pre-diagnostic questionnaire is shown in Figure 1.9.

Figure 1.9 Pre-diagnostic questionnaire

Advanced Light Vehicle Technology

A visit to the doctor

When you are unwell and visit the doctor, it is their job to diagnose what is wrong. Before they begin any patient examination, they will have developed a probing questioning routine that will allow them to conduct the most appropriate diagnosis. It will often begin with, 'what appears to be the problem?'.

This is their starting point and you can also use this technique to begin your vehicle fault diagnosis; in the case of a car you would use a scan tool to extract diagnostic trouble codes (DTC's).

Notice the term 'diagnostic trouble code' and not 'fault code' is used here. This is because just like the doctor asking what 'appears' to be the problem, the patient rarely knows the root cause and will only be able to give their best guess. It is then up to the doctor to use this as the starting point of their investigation and examine the patient thoroughly to ensure they have found the problem. The same applies to the technician when diagnosing a faulty car. DTC's should only be used as the starting point for any examination as this is the ECU's best guess as to what may be wrong, based on information it receives from the circuit such as voltages and resistance. Also when scanning a car, remember that no DTC does not necessarily mean no fault.

The diver is also a very important part of the consultation. This can be likened to a relative that has come along to the doctors for support and is able to describe the symptoms displayed. All of this extra information is vital to ensuring that the right diagnosis is made and the correct course of action is taken to find a cure that will not just mask the symptoms.

Another factor that a doctor will consider is patient history. A doctor will have access to the patients' medical notes, and this is similar to a vehicles service or repair history. These should be studied carefully to see if there are any ongoing problems that may have led to this current situation, but this should also be supplemented by further questioning.

Environmental factors may have a part to play, and a doctor will often ask, 'have you been abroad lately?' The same will apply to the use of a vehicle, and it's worthwhile remembering that any recent change in the use of the car may have an impact on its current condition.

At every step of a consultation, the doctor will make notes and so should you.

You can't handle the truth

It is worthwhile remembering that information given by both the driver and the vehicle used for a diagnostic routine might not be entirely accurate. It may not be that they are lying, but can sometimes leave out vital information due to lack of knowledge, or in the case of the driver/owner, embarrassment about something they may have done. Where possible, the driver should be questioned carefully (*see the interrogation room*) and asked for information in such a way that they will feel more inclined to provide a truthful answer (by competing a generic pre-diagnosis questionnaire for example).

When it comes to diagnostic trouble codes produced and stored in an ECU, they might not point towards the actual fault. The system of cause and effect will come into play here. Many faults, particularly those that might be caused by a mechanical issue, can generate spurious trouble codes that can distract you from the true root cause.

Remember, the saying 'Don't get lead up the garden path!' meaning, if you are deceived, or given you false information it will cause you to waste your time.

Differential diagnosis

First of all, differential diagnosis has nothing to do with final drive systems or axles. Differential diagnosis is derived from a medical procedure that produces a systematic method for diagnosing health disorders that lack unique symptoms or signs. This procedure can be used to help identify the likely cause of a fault where multiple alternatives are possible.

Automotive Master Technician

This method is essentially a process of elimination or at least a way of obtaining information that shrinks the 'probabilities' of the root cause to negligible levels. The 'probabilities' at issue, are imaginary parameters in the mind of the diagnostic technician and can be ranked in an order of probability from 0 to 100% as to whether this may or may not be the actual fault.

Differential diagnosis can help eliminate certain issues completely and reduce the number of possibilities to a list which will enable a systematic and more efficient routine to be devised.

There are various methods of performing a differential diagnostic procedure, but in general, it is based on the idea that it begins by considering the most common diagnosis to start off; which means look for the simplest, most common explanation first. Only after the simplest diagnosis has been ruled out should the technician consider more complex or exotic diagnoses.

Differential diagnosis has four steps:

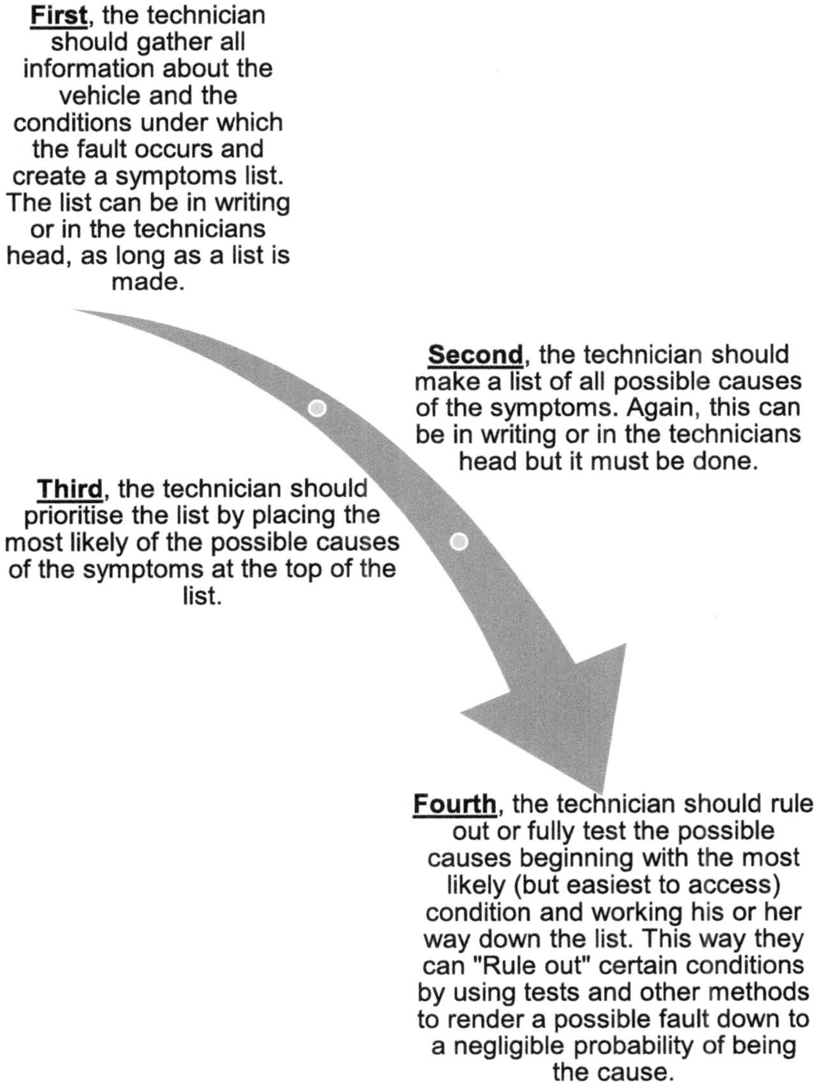

First, the technician should gather all information about the vehicle and the conditions under which the fault occurs and create a symptoms list. The list can be in writing or in the technicians head, as long as a list is made.

Second, the technician should make a list of all possible causes of the symptoms. Again, this can be in writing or in the technicians head but it must be done.

Third, the technician should prioritise the list by placing the most likely of the possible causes of the symptoms at the top of the list.

Fourth, the technician should rule out or fully test the possible causes beginning with the most likely (but easiest to access) condition and working his or her way down the list. This way they can "Rule out" certain conditions by using tests and other methods to render a possible fault down to a negligible probability of being the cause.

Figure 1.10 Differential diagnosis

In some cases, there will remain no diagnosis; this suggests the technician may have made an error, or that the true diagnosis is unknown (i.e. no prescribed method). Removing diagnoses from the list is done by making observations and using tests that should have different results, depending on which diagnosis is correct.

This method of ranking possible faults from high to low can also be greatly improved if it is conducted as a group activity, with the help of other technicians who may be able to provide a different viewpoint. Gathering advice and ideas from others before you start your diagnosis will enable a wider range of faults to be considered, leading to a more thorough and comprehensive fault finding routine.

Connections

Another principle that can be employed alongside differential diagnosis is to link connecting factors. Many faults are able to produce spurious symptoms across a number of different systems.
A good method for reviewing these common factors is to use a Venn diagram.
A Venn diagram or set diagram is a visual representation that shows all possible logical relations between a collection of items (**aggregation** of things).
Venn diagrams normally comprise overlapping circles. The interior of the circle symbolically represents the elements of the set, while the exterior represents elements that are not members of the set.
Once a list of symptoms and possible faults has been created using differential diagnosis, a Venn diagram can be constructed showing how the overlapping factors are represented in each system or component. By laying your diagnosis out in this manner, it can often help you understand contributing factors (things that lead to the fault occurring in the first place) and as a result, ensure that any repairs include preventative measures that help stop the same fault happening again. An example of a diagnostic Venn diagram is shown in Figure 1.11.

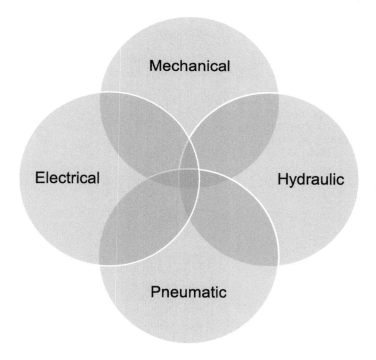

Figure 1.11 Venn diagram

Automotive Master Technician

Aggregation - several things grouped together or considered as a whole.

Brute force and experience (with SMART targets)

There are many faults that are common to particular vehicle types, and it is not unusual for similar issues to arise on previously unrelated makes and models. The brute force and experience method can sometimes locate, or at least rule out, certain faults in a short period of time.

Through many years of experience, you will have developed a catalogue of known vehicle faults (this doesn't necessarily mean a physical catalogue, but a working memory relating to common issues).

This experience can be put to good use by:

- Selecting the vehicle area and system in which the fault is occurring.
- Making a mental list of common faults that occur on other makes and models with this particular system.
- Devise a systematic strategy for your diagnosis which is SMART.

- **S** - Specific (concentrates on a specific area or component which is known to cause issues on other makes and models).
- **M** - Measurable (allows you to compare its condition or operation to known parameters).
- **A** - Achievable (can be done without too much effort or stripping down).
- **R** - Realistic (is a likely fault and not beyond the realms of possibility for this type of vehicle).
- **T** - Time limited (will not take too much time to access or assess).

Investigating to see if a similar fault has occurred (without the replacement of any parts or disturbing the system in such a way that it creates additional issues).

This method can then be used to try out as many 'quick fix' solutions as possible within a self-imposed time frame (*see the 10 minute rule*).

10 minute rule

Many common problems can be reoccurring within a certain vehicle range. It may be that after the driver has described the symptoms, you have a good idea of what the fault is before you start work on the car. Knowing about these reoccurring problems within a vehicle range can lead you to a quick fix, but you should not assume that they are always the correct diagnosis.

It is good practice to set yourself a 10 minute rule.

Allow 10 minutes for checking the easy things first, but don't get drawn deeper and deeper into a false diagnostic routine by frustration when the quick fix doesn't work.

Having used up your 10 minutes, step back from the task and reassess your diagnostic routine. If complex diagnostic equipment needs to be used, contact the customer and inform them of the situation so that they may be prepared for any extra time and cost involved. Some garages have a different labour rate for complex diagnosis as this will help reimburse the investment they have made in equipment and training. (*See gamblers anonymous*).

Keep it simple stupid KISS

The acronym KISS is attributed to first being used by Kelly Johnson, a chief engineer at the Lockheed organisation which made/makes US spy planes.

Advanced Light Vehicle Technology

The use of this terminology should in no way imply that the engineer or technician is stupid, in fact it means just the opposite. Expertise and experience may sometimes cloud your judgement and it then becomes very easy to over complicate a diagnosis or repair. Once this happens you may waste time and become frustrated. As frustration grows, the tendency is for your diagnostic routine to collapse, and then guesswork leads to misdiagnosis.

When presented with a set of symptoms:

- Do not rush into a diagnosis.
- Make sure that you have a good overview of how the system works and what the components do.
- Decide what diagnostic tooling is available to use during your investigations and how it can best be applied to make your task easier.
- Always check the easy things to access first (around the component that is actually displaying the symptoms).
- Don't assume that because it's an unusual fault on an advanced system, that it is the most complicated and expensive part that requires replacement.

Example
Remember that you may not need to test an ECU, and you should be able to use a process of elimination to decide if it is working correctly. Start at the part of the system that is displaying symptoms (the actuator for example) and work backwards through the circuit.

| If the output signals from the ECU are correct, the fault lies with the actuator. | If the sensor signals coming into the ECU are all correct, but the output signals to the actuator are incorrect the fault lies with the ECU. | Check the signals coming in to the ECU, if any are incorrect, the fault lies with the sensor itself. |

Figure 1.12 Electronic control process

The keep it simple stupid principle is best explained by the story of Kelly Johnson giving a team of design engineers a handful of tools. He then challenged them to design a jet aircraft that must be repairable by an average mechanic, in the field under combat conditions with only these tools. The 'stupid' in the acronym KISS refers to the relationship between the way things break and the sophistication available/required to fix them.

Animal vegetable or mineral (a taxonomy approach)

The party game 'Twenty Questions', encourages deductive reasoning and creativity. The premise behind this simple and straightforward activity can be used as a very powerful tool which cuts down on overall diagnostic time and helps focus in on the vehicle/system area which may be causing an issue.
In the traditional game, one player is chosen to be the answerer. That person chooses a subject (object) but does not reveal it to the others, (this can be likened to the vehicle fault).
The other player is the questioner, (this can be likened to the diagnostic technician).

Automotive Master Technician

The player (diagnostic technician) asks a question about the object (fault/symptoms) which can be answered with a simple "Yes" or "No." (If more information is available than just a simple yes or no, it can greatly speed up the procedure).
As the questions continue, the outcome will be narrowed down into a specific area.

If this process is used for a diagnostic procedure, and the questioning is carefully structured, it can help the technician rule out, or at least discount some unnecessary avenues of investigation.
A popular variation of the game is known as 'animal, vegetable or mineral'. This allows the questioning to be broken down into smaller categories known as a taxonomy. This taxonomy approach can be very useful when planning your automotive diagnostic routine. Think about the vehicle symptoms and ask yourself:

Which system:
Engine - transmission - chassis – body

Which subsystem:
Engine: intake - exhaust - fuel - ignition - cooling - lubrication
Transmission: clutch/torque converter - gearbox - final drive/differential - driveline
Chassis: steering - suspension - brakes - wheels/tyres
Body: interior - exterior - driver controls - comfort/convenience – safety

Which type of fault:
Mechanical - electrical - hydraulic - pneumatic

These categories can produce odd technicalities which have crossover within the taxonomy (i.e. a fuelling fault could be electrical or mechanical for example) but they will still help narrow down the search area and help you plan a systematic approach to your diagnosis. Also any commonality can help locate a root cause (*see Venn diagrams*).

Taxonomy - the technique of identification and classification.

When using a taxonomy approach to diagnosis, don't give yourself too many choices at each stage. Try to break your questioning/categories down into only 3 or 4 sets at each stage.

Crime scene investigation CSI and keyhole surgery

The problem with rushing into diagnosis is that there is a tendency to start to strip down the vehicle to carry out a visual inspection or try known fixes and repairs. When attempting to locate a complex system fault, you should try and take a forensic approach.
A faulty vehicle should be treated like a crime scene. Where feasible, try to conduct as much of the diagnosis as you can before stripping anything down, and disturb as little as possible. This will often provide you with more clues about the origin of the problem which can often be overlooked or missed once the system has been dismantled. There are many diagnostic tools produced by specialist equipment manufacturers that can help you with this process.
Once the root cause of a fault has been identified, you should conduct any repairs fully, but avoid unnecessary dismantling of associated systems.

This will reduce the possibility of accidentally disturbing other components which may then develop problems of their own. This works on the same principle as keyhole surgery; a patient will often recover better if the procedure/repair is localised.

Changing rooms

Where possible you should never conduct diagnosis by substitution. This is the process of trying an alternative component in the place of one that is suspected of being faulty. The main problem with this form of diagnosis is that it can actually contribute to system operating issues. Many electronic control units are able to re-programme themselves during operation to reduce the effects on drive-ability caused by faulty components or wear and tear. This reprogramming is often referred to as 'adaptions'. This means that by substituting components, not only do you receive very little diagnostic information that can help you with your investigation of the faults, but you will also be reprogramming the ECU and possibly adding to the symptoms.

Simulation is not the same as diagnosis by substitution. Some equipment manufacturers produce tools that are able to simulate the correct operation of sensors and components. When connected to the circuit suspected of producing an issue, the simulator can be used to produce a known good operating condition and when the system is then scanned or tested, certain components can be ruled out of the investigation. Remember that following any simulation, the ECU will need to be reset by clearing adaptions, so that they do not contribute to any operating issues.

Emergency health check

If someone is admitted to hospital in an emergency, they are immediately tested for their current state of health. The most common checks will be:

- Temperature
- Blood pressure
- ECG

Similar tests can also be conducted on a vehicle, delivering worthwhile information and results.
If faulty, a component will often get hot. A temperature measurement using a non-contact thermometer will give an indication if a component or system is operating within acceptable limits.
Some examples of systems that could be tested using this method are:

- Cooling systems (checking for localised hot or cold spots).
- Engine mechanical components (that appear to be hotter than those in the surrounding area).
- Exhaust manifolds (a branch running cooler than the others may indicate a cylinder misfire).
- Catalytic converters (running hotter than expected may indicate a misfire, rich or blocked operation. Running cooler than expected may indicate inefficient engine combustion/operation).
- Brakes (when compared with each other, individual hot or cold brakes may indicate binding or seizure).
- Tyres (individual tyres that are hotter than the others may indicate overloading or under inflation).
- Electrical components (that generate heat with the circuit switched off may indicate a parasitic drain).

High or low pressures can also give an indication of incorrect system operation. Pressure is created by a restriction to flow, and therefore how efficiently something is functioning.
Pressures that could be checked during a diagnosis include:

- Engine oil pressure
- Hydraulic power assisted steering
- Air conditioning refrigerant
- Cooling system pressure
- Automatic transmission oil pressure
- Turbo/supercharger pressure
- Exhaust back pressure
- Intake manifold pressure
- Compression pressure
- Cylinder leakdown pressure
- Electrical voltage (pressure)

Automotive Master Technician

ECG (electrocardiogram) is a test that measures the electrical activity of the heart. The heart is a muscular organ that beats in rhythm to pump the blood through the body.

In an ECG test, the electrical impulses made while the heart is beating are recorded and usually shown on a piece of paper or an oscilloscope.

The electrocardiogram, measures and records any problems with the heart's rhythm, and gives an indication that it might be affected by an underlying disease.

An oscilloscope can be used as the vehicles' heart monitor.

After a simple set-up (*see how to use an oscilloscope*) using the signal probe, components can be quickly accessed to see if they react when an operation is simulated.

At this stage it is not absolutely necessary to analyse the actual waveform, unless your planned diagnostic routine calls for it.

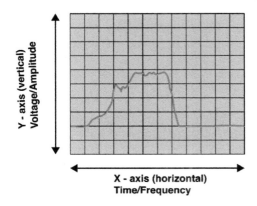

Figure 1.13 Oscilloscope screen

This quick test using an oscilloscope will often help you rule out certain areas from your diagnostic routine, helping to focus your investigation and reduce overall time spent.

Dead on arrival DOA

Quick health checks can often be used to reduce diagnostic time, for example:

If a car arrives at your workshop in a non-running condition (DOA, cranks over but does not start), the following procedure could be used to help focus your diagnosis.

To operate, an engine needs, induction (air and fuel), compression, ignition/combustion and exhaust.

Place your hand over the end of the exhaust pipe, while someone else cranks the engine for you. If you have pressure created on your hand at the tail pipe, you should have air being inducted, compressed and exhausted. If no pressure exists, this could be caused by a broken camshaft drive belt, mechanical damage to the pistons or a blocked exhaust. Your diagnosis should then be directed towards an engine mechanical issue.

Without any stripping down, using an inductive ignition sensor (tachometer, spark plug tester or timing light for example) see if the ignition system is sparking while the engine is cranked. If no indication of a spark is produced, your diagnosis should be directed towards the ignition system.

With the engine being cranked, place the probe from an exhaust gas analyser in the tail pipe of the car. If no Hydrocarbons (HC) are being produced, concentrate your diagnosis on the fuel injection system.

Do not disturb

Complex diagnosis is difficult at the best of times, but when a workshop is busy, this can just serve as a distraction. You may not always have the choice of when a vehicle needs to be diagnosed/repaired, but if possible pick a time when the workshop is quiet or empty so you can concentrate. This can sometimes be achieved by working outside of normal opening times when you can have the workshop to yourself and then taking time off in its place.

If you use this method, ensure that others know what you are doing and procedures are in place in case of an emergency.

Gamblers anonymous

The complex nature of advanced vehicle diagnostics means that the technician cannot afford to rely on luck and guesswork. Experience and a good memory can take you a long way, but unless you test the system or component that you suspect of causing the fault, you are taking a gamble.

If you order a new part to 'try' on a vehicle system, not only are you conducting diagnosis by substation and all the problems that can bring, but many modern vehicle electronic components cannot be returned to the supplier for a refund once fitted. This will increase the customer cost overall and can reduce your professional image.

Advanced Light Vehicle Technology

Having the right tool for the job to enable you to conduct an accurate diagnosis is vital; although this doesn't mean that you shouldn't investigate innovative methods for safely using tools and equipment that you already have. The company should set aside a budget that can be used to invest in tooling, as although there may be an initial outlay, it will increase the amount of jobs that can be conducted and reduce the overall time spent on a diagnosis, leading to a greater profit overall.

If the vehicle diagnosis is complex, it is often worthwhile setting a different labour rate from the organisations standard charge. It doesn't have to be much more than the standard charge, but will help to do three things:

- It will enable the company to differentiate in the customers' mind that the work being undertaken is more complex than standard repairs, reflecting the investment that the organisation has put into training and equipment.
- It can partially reduce the impact of the diagnostic research time on the repair costs, which is often considered 'time wasting', but vital in a good diagnostic routine.
- The extra money can sometimes be reinvested to help towards new tools, equipment and training.

It is important that if the diagnosis is complex and the labour rate is changed, that the customer is informed and it is fully explained why the new charge is being levied.

It is worthwhile developing a relationship with other garages in the area so that certain pieces of equipment might be shared by prior agreement. Not many organisations have the funds available to invest in all of the equipment that is required. If the companies involved in this scheme pool their resources, a wider range of repairs can be conducted. Even if the organisations lending the tool make a nominal charge for its use, this charge could be passed on to the customer, while still keeping the overall cost down by reducing time, effort and misdiagnosis.

Senses working overtime

When conducting a diagnosis, make sure that you take notice of all your senses, and include any symptoms you may be able to detect in planning your routine.
Sight allows you to perform a visual inspection of systems and components.
Hearing allows you to perform an aural assessment of systems and components.
But smell and taste are quite often left out during a diagnosis.
The main reason for this tends to be the toxic nature of many chemical substances associated with vehicle construction and operation and their harmful effects on the human body. This section is not suggesting that you should start inhaling or ingesting chemical substances associated with vehicles, but merely that you can increase your diagnostic capabilities if you make a conscious effort to be aware of smells and tastes that you might pick up during your investigations.
The olfactory system closely links the senses of taste and smell, and it can often be quite difficult to separate the two. Unusual smells will often also leave a sense of sweet, sour or salt on the tongue. The sense of smell is very good at triggering memories, and if you make a descriptive note of any smells/tastes that you encounter during your diagnosis, after a short while, your brain will unconsciously make connections to problems and faults. This way, during future diagnostic routines, extra information can be gathered without really thinking about it.

Dear diary

During your diagnostic routine, keep a record of what you have done or found during each stage of the investigation. Many diagnostic routines are time consuming and may be interrupted mid flow. Recording the information as you go along will allow you to return to your diagnostic routine at any point. Also, if a similar problem occurs in the future, you can refer to your notes and this might save you diagnostic time and effort.

Automotive Master Technician

Make sure that your diagnostic routine builds up 'the big picture' so you have an overall view of system operation. When part of the big picture is missing, this will normally point you to the root of the fault.

Keep the information you have gathered, and build your own library. The more systems that you test, the more voltage, amperage, resistance readings or trouble codes you diagnose, the greater your reference list will become. You can use this information to aid your memory and prompt your diagnostic routine. Spending quarter of an hour checking through your notes and finding previous results can save hours of wasted time and money. You don't have to remember the entire job just that you had a similar problem before.

If you are fortunate enough to have access to a tablet PC, this can be a very good method for recording and tracking your diagnostic work. Many of these devices have a voice dictation application, so it is possible to make notes without touching the screen; very useful in a workshop situation. Also many tablet PC's have a camera attached, and taking photos is another good method to use when making notes about your diagnosis and routine. The notes may then be saved, and if required at a future date, can quickly be located using a keyword search.

The source of the Nile

The most important part of any diagnosis is ensuring that you have found the root cause of any fault occurring. When conducting a diagnosis, remember that many problems described by the driver or displayed by the vehicle are often symptoms and not the actual fault. What you are actually investigating is cause and effect. A common error made can be that when approaching a repair, it is sometimes possible to overcome or cure a symptom, while still leaving the fault present.

No matter what symptoms are created, there will always be an original source. Make sure that you ask yourself, 'is this cause, or effect?'

Remember that even when you have found the faulty component, external factors may have contributed to its failure. Your investigation should not stop at the repair. You need to make sure that whatever caused the fault in the first place will not reoccur and render your repairs inadequate.

Always fully test a system and component following any repairs to ensure the longevity of your fix.

Double tap

Using assumption or jumping to conclusions is often the biggest waste of time in a diagnostic routine. Time wasting has the knock-on effect of causing frustration, leading to a downward spiral and the eventual collapse of your carefully thought out systematic approach.

It is common, as a Master Technician, to take over a diagnosis from another member of staff, when they are unable to find a fault, having tried all of the most straight forwards tests or repairs. Once all of the easy avenues of investigation and repairs have been exhausted, it will now be down to you to find out what is wrong. Do not assume that any previous checks, tests or repairs conducted by another person have been done correctly and discount them from your own diagnostic routine. You must always go back to basics and double check them for yourself. If not, there is the possibility that you might miss an important clue, making your own job harder
.

Also, if your diagnosis leads you to eventually replace a faulty component, but the problem still remains, do not assume that the new component is working correctly. Just because it is brand new, this doesn't mean that it is fully functioning. Time can once again be wasted looking for another issue, when the problem still lies at the heart of your original diagnosis. Whenever you fit a new component, ensure that you test for correct operation, before moving on; this includes functional tests, measurements etc.

Outpatient

Having completed any work to a vehicle system or component, it is important to evaluate the effectiveness of your repairs. The vehicle should be fully tested in order do this.

This should always involve a road test through a complete vehicle drive cycle, where possible, and a re-check of system and components at the end. The ECU should always be rescanned to ensure that no intermittent faults have occurred and created pending trouble codes.

Your evaluation doesn't have to stop there though. If a major repair has taken place, it is often handy to see if the customer/driver will allow you to conduct some follow-up checks. This could be in the form of a brief telephone call a short time after the repairs, bringing the car back to the garage for a quick check or in some cases, 'data logging' (*see tools and equipment*).

If this procedure is conducted correctly, not only can it help promote customer confidence as you are displaying aftercare, but the gathered data can help inform any of your future diagnostic routines or repairs.

Entropy

The word entropy has several different meanings which can be associated with the definition and diagnosis of a complex system fault. Understanding entropy will help you realise what may be going on with your investigations.

Meaning one: The inevitable and steady deterioration of a system or society. This shows that, no matter how well a vehicle has been maintained, systems and components will slowly wear out and eventually break-down. Accepting this fact will help to open your mind to the possibility that even the most reliable parts are subject to failure.

Meaning two: A measure of the disorder or randomness in a closed system. A diagnostic routine must be carefully thought through to ensure that you have a systematic method of working. Once you have decided the order in which you intend to conduct your diagnosis, try and stick to it, so that entropy doesn't lead your investigation into chaos.

Meaning three: A measure of the loss of information in a transmitted message. This describes a process, meaning that it will be rare for you to receive all of the available diagnostic information (either from the driver or by testing the system, i.e. measurements, serial data or diagnostic trouble codes). If you are aware that some information may be missing, it will help prevent you from jumping to conclusions and ending up at a misdiagnosis.

Old habits die hard

Finally, as a Master Technician, you will have considerable experience in your particular field of expertise. This experience will have served you well over the years and will continue to do so. Unfortunately, problems can occur when you do not adapt to changing situations. An approach to complex diagnosis for which there is no prescribed method, means that you will have to have an open mind about how you conduct your investigations.

The English idiom 'you can't teach an old dog new tricks' means, it is impossible, or almost impossible, to change people's habits, traits or mind-set.

It can be accomplished, but you must be open to the idea of learning something new. Adopting new routines will be a slow process, but trying out different ways of working a little bit at a time, on a regular basis, will help these become second nature and add to your considerable existing skill set.

Understanding electrical and electronic values and data

You need to have a basic understanding of the principles of electricity before you can use many of the advanced vehicle diagnostic tools. This knowledge will allow you to make the right judgments and arrive at a successful conclusion, leading to a first time fix.

In cars, electrical energy is created by a chemical reaction (in a battery for example) or by the disruption of magnetic fields near electrical conductors (in a generator for example). You can measure how the electrical energy is created, moved and used, using the electrical units shown in Table 1.1.

Automotive Master Technician

Table 1.1 Electrical units

Volts	Voltage is electrical pressure. Voltage is the potential force in any part of an electrical circuit. Two main types of voltage occur in electrical circuits: Electromotive force (EMF) is potential pressure, and is usually considered to be the open circuit voltage when all electrical consumers are switched off and no current is flowing. It should be higher than electrical system voltage when current is flowing. Potential difference (Pd) is a circuit voltage measurement when components are switched on and current is able to flow. It is a measurement of voltage drop compared to the EMF at different positions within a circuit.
Amps	Amps are the units used to measure the amount of electricity in any part of an electrical circuit. Amps is measured when electricity is allowed to flow in an electrical circuit – this is known as current. There are two main types of electrical current: Direct current (DC) is electricity that flows in one direction only. Alternating current (AC) is electricity that moves backwards and forwards in an electric circuit. Amperage is the same wherever you measure it in the circuit (at the beginning, in the middle or at the end).
Ohms	Ohms are the units used to measure the resistance to electrical flow. Resistance has a direct effect on the operation of any electrical circuit. As resistance rises in a circuit, current and voltage fall, which can restrict the operation of electrical components. In some electrical circuits, resistance can be used as a method of control for electrical components, but in most circumstances a high resistance is undesirable.
Watts	Watts are the units used to measure electrical power made or consumed. Power is defined as the rate at which work is done. When referring to electrical components, the higher the wattage, the more powerful the component will be and the more electrical energy it will use.

A difference in pressure (voltage) between two points in an electric circuit will create flow in the direction of the lower pressure (voltage). With conventional electrics, it is assumed that electricity flows from positive to negative. True electron flow is actually from negative to positive, and you should remember this if the words 'electron flow' appear in a test or assignment.

Series and parallel circuits

Two main types of electrical circuit are used in the construction of motor vehicles:

- Series
- Parallel

Advanced Light Vehicle Technology

Series circuit

In a **series circuit** the consumers are connected in a line one after another. Because they are all in the same circuit, they share the electricity provided depending on the amount of power that they use. If more than one **consumer** is fitted it will only get part of the voltage available.

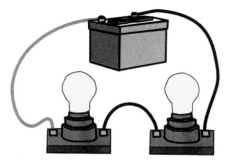

Figure 1.14 A simple series circuit

Series circuit – a circuit with electrical consumers connected in a line, one after another.

Consumer - an item or component that uses up electrical energy, i.e. bulbs, motors etc.

Parallel circuit – a circuit where electrical consumers are connected side by side.

If any one of the consumers fails, the circuit is broken and no electricity can flow. The rest of the consumers stop working. This makes series circuits unsuitable for many systems on cars. For example, if you wired a lighting circuit in series, not only would the bulbs glow dimly, but if one bulb broke, all of the others would go out.

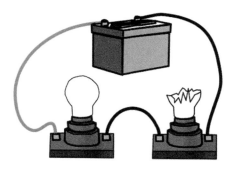

Figure 1.15 A damaged series circuit

Automotive Master Technician

Parallel circuit

In a **parallel circuit** the consumers are connected next to each other. Each has its own power supply and earth return back to the battery.

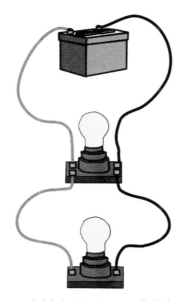

Figure 1.16 A simple parallel circuit

Because each consumer has its own power supply and earth, all the consumers receive the full voltage available, and work at full power.
If one consumer in the circuit fails, the others keep working. For example, in a headlight circuit each bulb has its own 12 volt supply and earth return to the battery. If one bulb breaks, the other will keep working.

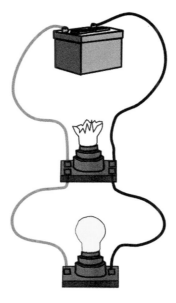

Figure 1.17 A damaged parallel circuit

Advanced Light Vehicle Technology

Ohm's law

If any one of the units within a circuit (volts, amps, ohms or watts) is changed (i.e. increased or decreased), this will affect all the other units. Using a water analogy:

- If the voltage (or pressure) in a water system was increased, more water would flow and the amperage (or quantity) would also increase.
- If the resistance to flow was increased (if a tap was partially closed, for example) then less water would flow and the amperage (or quantity) would also decrease.

This was explained by Georg Ohm with the following mathematical calculations:

amps = volts ÷ resistance
resistance = volts ÷ amps
volts = amps × resistance

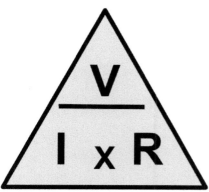

With Ohm's law, if you know two of the electrical measurements, you can calculate the third.
The Ohm's law triangle is a good method for calculating the missing unit. It is laid out as shown in Figure 1.18.

In Figure 1.18:

Figure 1.18 Ohms law triangle

- V = volts (this is sometimes shown as the letter 'E' to represent EMF, but still means volts).
- I = amps (the letter 'I' is used to represent instantaneous current flow).
- R = ohms (the letter 'R' is used for resistance because an 'O' could be confused for a zero).

How to use the triangle
Cover up the unknown unit with your thumb and you are left with the calculation required. For example, amperage is unknown, so cover the 'I' and you are left with V ÷ R (i.e. volts divided by resistance).

Using Ohm's law to help diagnose faults

The relationship between voltage, resistance and amperage can help you to diagnose faults within an electrical circuit. If you take measurements using the different electrical units and then compare them using the Ohm's law calculation, you will be able to work out if the fault is occurring because of:

- **Pressure (volts)**
- If this is lower than expected, component performance is reduced.
- If this is higher than expected, component damage can occur.

- **Quantity (amps)**
- If this is lower than expected, component operation will normally be incorrect.
- If this is higher than expected, component/system operation is being overworked.

- **Resistance (ohms)**
- If this is lower than expected, current may be taking an alternative path to earth (short circuit).
- If this is higher than expected, it will consume electrical energy and reduce system performance.

The power triangle

Watts or power can be calculated in a similar way as:

amps = watts ÷ volts
volts = watts ÷ amps
watts = amps × volts

A power triangle can be used in the same way as Ohm's law. It is laid out as shown in Figure 1.19.

In Figure 1.19:

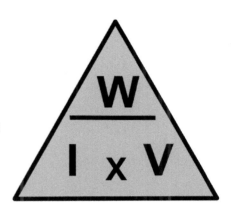

- W = power (in watts – this is sometimes shown as the letter 'P' to represent power, but still means watts).
- V = volts (this is sometimes shown as the letter 'E' to represent EMF, but still means volts).
- I = amps (the letter 'I' is used to represent instantaneous current flow).

Figure 1.19 Power law triangle

How to use the triangle
Cover up the unknown unit with your thumb, and you are left with the calculation required. For example, amperage is unknown, so cover the 'I' and you are left with W ÷ V (i.e. watts divided by volts).

Fault finding

Common electrical faults

Diagnosing electrical faults can be confusing, as the symptoms can be very wide and varied. If you follow a simple approach, the diagnosis can be reduced to four main electrical faults:

- Open circuit
- High resistance (including bad earth)
- Short circuit
- Parasitic drain

Open circuit

In an open circuit, electricity cannot flow. This is normally because there is a physical break in the system. As a potential difference Pd will only occur in a circuit when current can flow, the fault can be diagnosed using a test method known as volts drop.
To diagnose an open circuit, you can use the multimeter as a voltmeter. Once set up correctly and connected to the appropriate circuit, measurements can be taken.
If the circuit is working properly, you should see full voltage all the way up to the consumer, at which point the electrical pressure should be used up. In an open circuit, the voltage will disappear. Using a voltmeter you can see at what point in the circuit this happens.
For example, Figure 1.20 shows using a voltmeter to check a low voltage open circuit in which the bulb does not light up. The voltmeter is connected at various points of the circuit to find out the voltage at these points. Where the voltage is different (between 12 volts and 0 volts), this shows the position of the open circuit. In this example, the open circuit is between points B and C.

Advanced Light Vehicle Technology

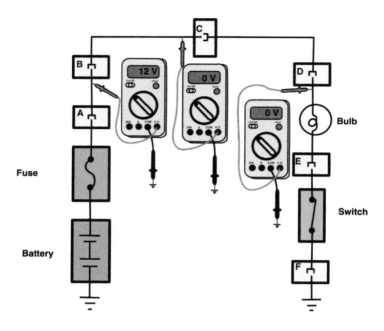

Figure 1.20 Testing for an open circuit

High resistance

In a high resistance circuit, the electricity slows down. This is normally because of a partial restriction in the system. Many high resistance faults are caused by poor, corroded or loose connections. A potential difference Pd will occur in a high resistance circuit, but the total Pd will be shared between the consumer and high resistance. This fault can also be diagnosed using the test method known as volts drop.

The symptoms of high resistance are that the component does not work properly (e.g. a bulb that glows dimly) because the circuit pressure (voltage) is shared with the resistance.

To diagnose this fault, you can use the multimeter as a voltmeter. Once set up correctly and connected to the appropriate circuit, measurements can be taken.

If the circuit is working properly, you should see full voltage all the way up to the consumer, at which point all the electrical pressure should be used up. In a high resistance circuit, the full voltage potential difference is not used up by the consumers. Using a voltmeter you can see at what point in the circuit this happens.

If there is a lower than expected voltage at the consumer, then the fault is in the first half of the circuit. If full voltage is present at the consumer, but some voltage still appears after the component, then the fault is in the second half of the circuit (bad earth).

A high resistance will make circuit current fall and this can sometimes be seen if an inductive amps clamp is connected to the circuit and the value compared with the fuse rating.

Automotive Master Technician

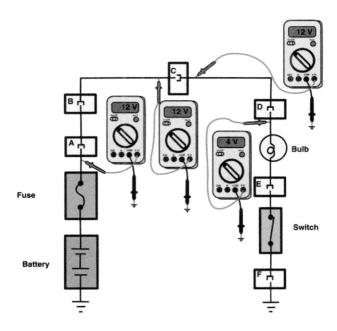

Figure 1.21 Testing for a high resistance

Bad earth

A bad earth is a high resistance after the consumer. If this exists the symptoms will be that the component won't work properly. Sometimes a bad earth can also cause the electrical energy to find an alternative path to the negative side of the battery. If this happens you may see symptoms such as all of the lights in the same unit operating at the same time (e.g. brake lights flashing with the indicators).
To diagnose a bad earth use the same procedure as for high resistance.

Short circuit

Electricity is lazy, and will always take the path of least resistance. (Why travel the full length of the circuit when it can take a shortcut?)
In a short circuit, the electricity doesn't make it all the way to the end. Instead of going through the consumer, the electricity makes its way back to the battery early, and in the process converts its energy to heat.
The sudden discharge of current can cause a lot of damage, so the fuse that is used to protect the system should blow. If this happens the symptoms can make you think that the problem is an open circuit (which it is in a way, as the blown fuse has broken the circuit so that no current can flow).
In this situation, you can test the system with a voltmeter as explained in testing an open circuit. But once you have discovered the blown fuse, you should change your diagnostic routine to look for a short circuit. Any heat damage, including blown fuses, is a good indication that a short circuit might exist.

If a **dead short** to earth exists (e.g. the insulation of a wire has chafed against the metal bodywork of the vehicle), you can use a test lamp to help diagnose this fault (*see Figure 1.22*). It is important to use a test lamp containing a bulb and not an LED, as this could lead to system damage.
Once connected in place of the fuse, if the test lamp illuminates then the electricity is finding an alternative path back to the battery (short circuit). As the bulb is an electrical consumer, it uses up the electrical potential, and shouldn't damage the rest of the circuit. The circuit should then be disconnected systematically from the far end, working back towards the fuse box. When the bulb goes out, you have located the position of the dead short.

Advanced Light Vehicle Technology

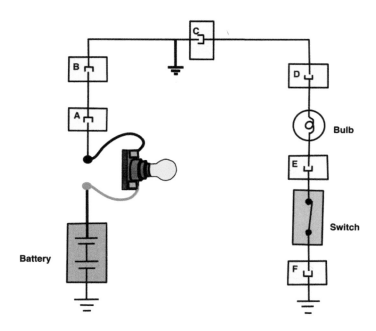

Figure 1.22 Testing for a short circuit

Dead short – an electrical short circuit that goes straight to earth without passing through a consumer.

A blown fuse may also be an indication of excessive current draw. With a test lamp connected as described for a short circuit test, an inductive amp clamp can be attached to the circuit (around the test lamp wire) and real time measurements can be taken by switching components on and off until the one with the high current draw is found.

A blown fuse will initially present the symptoms of an open circuit, which in a way it is. When presented with these symptoms you should conduct your diagnosis for an open circuit as described but as soon as damage to the fuse is seen, then your routine should be changed to a short circuit investigation.

Parasitic drain

A parasitic drain is similar to a short circuit – electricity will continue to flow even if the system is switched off, although this fault may not cause visible system damage. The symptom reported is normally that the battery goes flat if left for a period of time. To help diagnose this fault, you can use the multimeter or an inductive clamp meter as an ammeter. (This acts like a flow gauge to measure the amount of electric current moving in a circuit).

Automotive Master Technician

Checking a parasitic drain

To check for parasitic drain, switch off all electric systems and connect the ammeter. The ammeter must be inserted into the circuit (connected in series) so that it isn't damaged. To do this you may need to disconnect one lead from the battery and use the ammeter to bridge the gap, so that current flows through it.

An inductive clamp can be placed around the battery wiring without disconnecting and inserting in series. This is much safer but is often less accurate than using an ammeter directly.

With everything switched off, there should be no current on the display of the meter. If any current (measured in amps) is shown, then a parasitic drain exists. To help find the parasitic drain, remove the fuses one at a time until **amps draw** falls to zero. This will help you locate the circuit containing the drain. Once you have identified the circuit, disconnect the components in that system until the current draw falls once again. You can now replace the faulty component.

The ammeter function of many multimeters will only provide readings of up to 10 amps before they are damaged. It is far safer where possible to use an inductive ammeter for testing.

Amps draw – the amount of current being used.

Multiplexing and network systems

As the amount of technology on cars has increased, demand for faster computer operation and processing has also risen.

Advances in vehicle management include:

- Engine management
- Body control
- Chassis systems
- Transmission

- Infotainment
- Traction control
- Safety systems

ECU's were becoming bigger to cope with system requirements, and large amounts of wiring were needed to distribute electrical power around the car. These demands also generated a rise in the number of sensors required, leading to complication, extra weight and increases in the cost of manufacture.

ECU's were becoming larger and larger because of the need for more connections and pins where they joined the wiring. There was a limit to how small these connections could be made, and how closely the wires in the loom could be bundled. Multiplexing has permitted a reduction in sensors and wiring, as it allows the ECU's to share information on a network.

To reduce the amount of sensors and wiring needed for system operation, **multiplexing** was introduced. Multiplexing simply means carrying out more than one operation at a time (for example, a multiplex cinema has more than one screen, and is able to show more than one film at once).

Advanced Light Vehicle Technology

Instead of a single large ECU in a vehicle, smaller ECU's were developed that managed individual systems. These single ECU's became known as **nodes**. The nodes are connected to each other by a communication wire, which allows information to be shared in a network. When one of the ECU's receives information from a sensor, it processes the signal and acts if required. It then passes on this information to the communication **network** wire linking the other ECU's, which then use that information if required, and once again pass it on. This means that signals from a single sensor can be shared across a number of different vehicle systems.

Multiplexing – a method of carrying out more than one operation at once.

Nodes – ECU's connected to a computer network (from the Latin word 'nodus', which means knot).

Network – a number of computers connected together so that they are able to communicate with each other.

CAN bus

Controller area network (CAN) was introduced by Robert Bosch in the 1980's. There are a number of different network types and manufacturers available, but the name CAN bus has been adopted by many technicians to describe nearly all network systems.

The word 'bus' is used in various situations. One meaning of bus is a vehicle that collects you from one place and delivers you to another. This is very similar to its meaning within a communication network. Information is picked up at one point on the communication line, it then takes a route around the system and stops at various ECU's (like bus stops).

The nodes are connected by a single communication line, which allows the exchange of multiple pieces of data. The communication line can link these ECU's in the following layouts:

- A large loop, known as a daisy chain.
- A star pattern, known as a server system.
- Connected in parallel to a single bus line.

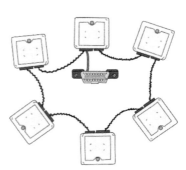
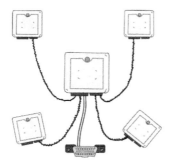
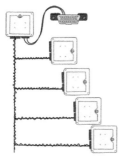

When a daisy chain layout is used, the data sent travels in both directions at once, which gives much greater reliability. If one wire is damaged or broken within the loop, the information can still arrive at the appropriate ECU as it comes from the other direction. The data is not only more reliable but this system also improves malfunction diagnosis.

Automotive Master Technician

Communication data

When an ECU receives a signal from a vehicle sensor, it processes this and places the information on the network bus as a **data packet**.
The data packet is usually made up of the following:

- A header: the equivalent of 'hello, I am transmitting a message'.
- The priority: how important this message is, e.g. vehicle safety information will be more important than a bodywork communication such as a command to open an electric window.
- Data length: this is so the receiver knows it has not lost or 'misheard' any of the information.
- Data type: what type of information is contained, e.g. speed, temperature, etc.
- Data: the actual sensor information itself.
- An error detection code: this says 'has all the information been received?' and is known as a cyclic redundancy check (CRC).
- End of message: 'goodbye'.
- Finally, a request for a response from the receiving ECU: this says 'thank you, I got your message'.

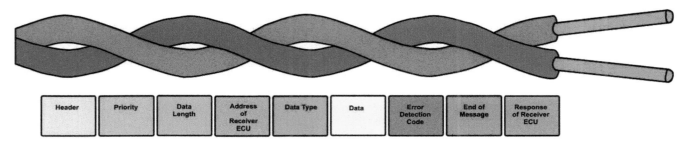

| Header | Priority | Data Length | Address of Receiver ECU | Data Type | Data | Error Detection Code | End of Message | Response of Receiver ECU |

Figure 1.23 A network data packet

Data packet - a bundled information package containing data sent on a communication network line.

To help reduce the possibility of data **corruption** caused by misinterpretation or external electromagnetic interference, a CAN bus system uses two communication wires instead of one, twisted over and over each other in a spiral (known as a 'twisted pair'). The same data is sent on both of these communication wires as an on and off voltage signal. One signal is sent as a positive switch and one is sent as a negative switch, providing a mirror image on each network wire, which are known as CAN high and CAN low. The **potential difference** between the voltages on the two lines produces a digital signal that can be processed into information. The communication wires are exactly the same length, and as the data travels at the same speed, both versions of the data should arrive at the receiver at the same time. These messages can now be compared to help identify data corruption. The opposing voltage of the signals transmitted down the communication wires will also help cancel out electromagnetic interference from other systems.

Corruption – a breakdown of integrity or communication.

Potential difference – the difference between two voltage values.

Advanced Light Vehicle Technology

Bus speeds

There are three main bus speeds:

- Low speed: used for instrumentation, body control and comfort, etc.
- It operates at a rate of 33,000 bits of information per second (33kbps).
- High speed: used for powertrain control and safety critical information, etc.
- It operates at a rate of 500,000 bits of information per second (500 kbps).
- Very high speed: used for high volumes of data transmission in infotainment systems (such as streaming video and music), etc.
- It will operate at a rate of 25,000,000 bits of information per second (25 Mbps).

The data is sent as an on and off signal. It would be seen on the screen of an oscilloscope as a square waveform.

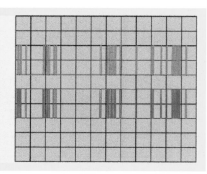

Tip

If the oscilloscope patterns from CAN high and low don't line up, check the channel speeds of your scope before assuming that there is a fault with the system.

System reliability

The CAN bus system is more reliable than a standard wiring system due to the fact that a single open bus wire would not stop communication. Two open bus wires can stop communication, but as more ECU's are used for control, only part of the system may fail.

Short circuits can have a catastrophic effect on network communication. A short to either positive or earth will disrupt the communication on the bus wire, as an on and off signal can no longer be transmitted. If viewed on and oscilloscope screen, this would be a flat line at either 0V or 5V. To avoid total failure of the system, bus-cut relays can be used. These are a type of circuit breaker that isolate part of the network, allowing the rest of the system to continue communicating.

The network will have ECU's known as 'gateways' which connect systems of different speeds. They translate the data packets between the networks and allow information to be shared across all systems.

Automotive Master Technician

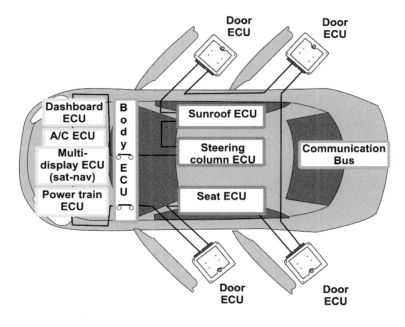

Figure 1.24 A simple network layout with bus cut relays

Multiplex and networked diagnosis

If a critical network failure occurs, such as short to positive or earth, the vehicle may suffer a complete communication loss.

With a networked system, if communication is lost within a certain area, numerous items will not work and several trouble codes may be generated. Having connected a scan tool and retrieved the diagnostic trouble codes, you should look for the code that is the root cause. Communication failures are normally an effect of the original fault. You should ask yourself, 'is this the cause or an effect created by the fault?'

CAN bus systems report communication faults as live data. As a result of this, once you have identified the cause trouble code, you should be able to conduct a diagnosis by disconnecting and isolating components or sections of wiring loom until communication is re-established.

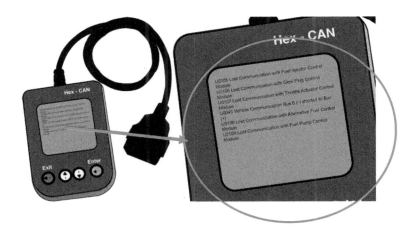

Figure 1.25 CAN Bus diagnostic codes

Advanced Light Vehicle Technology

CAN Bus short circuits can often be diagnosed by isolation. Unplug suspected faulty components completely and re-scan for diagnostic trouble codes. (This can sometimes regain system communication when the faulty component has been isolated, but remember that it can also generate false codes).

Faster than CAN

Although CAN bus systems are very fast, able to receive, process and **disseminate** one million bits of information per second, they are not fast enough to keep up with modern technology demands. Manufacturers are producing faster and faster network systems that are easily able to process over 10 million bits of information per second. These systems work in exactly the same way as a normal CAN bus, just faster.

More information on CAN Bus and multiplexed network systems can be found in Chapter 4.

Disseminate – spread out or disperse widely.

The set up and use of diagnostic tooling

Electricity is invisible and as a result you will need to use specialist electrical diagnostic tooling to enable you to see what is happening in a circuit. When testing high voltage electrical systems, ensure that the test equipment you are using is suitable, and reduces the risk of electric shock.
Some examples of electrical diagnostic tooling are described in the next section.

Test lamps

One of the simplest diagnostic tools you can use is a test lamp. Whether this is a professionally built tool or a bulb and a couple of pieces of wire that you have put together yourself, this tool can be very effective. Its purpose is to check to see if the circuit has power.

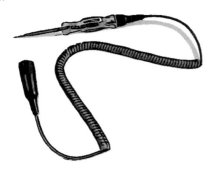

To use a test lamp on a low voltage system:

*   Connect one end of the test lamp to a good earth, such as the vehicle chassis or ground (the negative terminal of the battery is better because this is the end of all electrical circuits on a car).
*   Connect the other end of the test lamp to the part of circuit that needs to be checked.
*   If power exists in the circuit the test lamp will illuminate.

Automotive Master Technician

Many modern test lamps have a point at the end of the probe to enable you to pierce wiring installation. Take care when using this probe as it is very easy to stab your finger. Also, if you pierce insulation, you are opening up the wiring to the effects of **oxidisation** from the air. If the wiring is left open, this oxidisation can lead to a high resistance and electrical problems in the future. It is best, where possible, to **back-probe** an electrical plug so as not to damage the wiring.

If you have to pierce the insulation of the wiring, you should cover it with insulation tape or heat shrink. Wiping a small amount of silicon into the pierced insulation hole is not advisable as it can cause corrosion in the wiring and high resistance.

Oxidisation – the effect of oxygen on metal, which can cause corrosion.

Back-probe – a method of making a test connection at the back of an electrical socket or plug.

Every time you connect another consumer to an electrical supply wire, more electrical current will be drawn from that supply wire until eventually it can take no more. A test lamp contains a bulb and this is a consumer. A standard test lamp has a low resistance, usually around 6 ohms. This means that when testing a low voltage vehicle electrical circuit, 2 to 3 amps of electrical current may be drawn. If a test lamp is used on an electronic circuit, severe damage can be caused as this high amperage moves through the components.

Always take care when using test lamps to diagnose electrical faults on vehicles. They should only be used when it is safe to do so. It is far safer to use an LED test light, if you are likely to be testing near electronic circuits.

Power probe

A power probe is an advanced form of test light, with additional features and capabilities. Power probes are usually fitted with light emitting diodes, LEDs that are able to illuminate in different colours when connected to either a powered circuit (LED glows red) or an earth circuit (LED glows green).

Checking polarity

After a simple connection to the vehicle's low voltage battery, you are able to see quite easily whether a circuit is positive or negative, without having to change **polarity** from one battery terminal to another. The power probe normally comes with two crocodile clips (red and black) – connect these to the appropriate positive and negative battery terminals.

Advanced Light Vehicle Technology

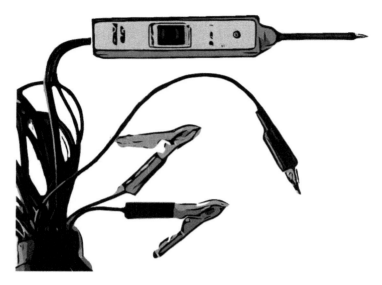

Figure 1.26 A power probe

Polarity – a term used to describe electrical connection to a circuit. It represents the positive and negative connections.

Auxiliary – something that functions in a supporting capacity.

To check that a correct connection has been made, quickly touch the tip of the power probe to each battery terminal in turn:

- The LED should illuminate red when touched to the positive terminal.
- The LED should illuminate green when touched to the negative terminal.

Checking continuity using a power probe

Not only can a power probe check for electrical feed and earth, you can also use it to check for continuity (a continuous or unbroken conductor). You can check continuity on wires or components that have been disconnected from the vehicle's electrical system.

The power probe has an **auxiliary** ground wire – connect this to one end of the conductor, wire or component. Connect the tip of the power probe to the other end. If continuity exists, the LED on the power probe will illuminate.

Always remember to turn off power first before disconnecting a wire or component on the vehicle's electrical circuit.

Automotive Master Technician

Conducting functional tests using a power probe

How to:

You can also use the power probe to undertake functional tests of electrical components.

1. It is recommended that you disconnect the component from the vehicle's electrical system when conducting this test.

2. Connect the auxiliary ground to one terminal of the component and the tip of the power probe should be connected to the other.

3. Check that the LED illuminates to show that the component has continuity.

4. Keeping an eye on the LED, quickly rock the power switch and immediately release.

5. If the LED indicator changed momentarily from green to red, you may proceed with the test.

6. By rocking the power switch forwards and holding it down, electrical potential will be supplied to the component and you can check its operation.

7. If during the initial rocking of the power switch the LED turned off, this normally indicates that the current being drawn by the component is too high for the power probe and the internal circuit breaker has tripped. This may require a manual reset and you will need to check the manufacturer's instructions.

Safety

Power probes are only designed to test components which draw relatively small amounts of current. Never use them to test starter motors, vehicle drive motors, etc.

Multimeters

The multimeter is a piece of electrical test equipment designed to measure a number of different units within an electrical circuit. There are two types of multimeter: analogue and digital.

Analogue multimeters

Analogue multimeters use a needle that moves across a graduated scale to record electrical readings within a circuit. The old-fashioned name for this type of unit was an 'AVO meter', which stood for amps, volts and ohms.

The problem with analogue meters is that they are only as good as the operator. The graduated scale can be difficult to read and so inaccurate readings could be obtained. Depending on the range of the scale provided by the manufacturer, a needle that lies somewhere between two units could be reading any fraction available. Analogue multimeters also have an upper range limit. If the needle flicks all the way to the end of this scale, it is known as full-scale deflection (FSD).

Digital multimeters

Digital multimeters shows digits (numbers) on a liquid crystal display (LCD) screen. These numbers are clearly displayed and are easy to read accurately.

It is quite normal for the last digit on the far right of the screen to continuously change. This is a feature common to most digital multimeters. In a lot of cases, as really high accuracy is not required, this figure can be ignored.

Digital multimeter types

Two types of digital multimeter are common: manually operated and autoranging.
With a manual multimeter, the operator selects the unit and the scale to be measured, normally by turning a dial on the front of the multimeter.

Using a manual multimeter

When you are using a manual multimeter, if you do not know the scale to be used, always follow this procedure:

• For testing volts and amps, first select the highest scale on the dial, then rotate the dial slowly down through the scales until you obtain an accurate reading.
• For testing ohms, first select the lowest scale on the dial and then rotate the dial slowly up until you obtain an accurate reading.

Using an autoranging multimeter

With an autoranging multimeter, the operator selects the unit but the scale of that unit is automatically selected by the multimeter. When using an autoranging multimeter, you must take care that your reading is accurate by taking note of the scale of the unit being displayed. For example, if voltage is measured, the scale might be in:

| • | Millivolts | • | Volts | • | Kilovolts | • | Megavolts |

Using a digital multimeter

You can measure a number of electrical units on a digital multimeter, including volts, amps and ohms, but other measurements can also be taken.

Extra facilities on a digital multimeter may include: temperature, frequency, diode testing, transistor tests and audio continuity testing.

The electrical units of volts and amps are often broken down into two further areas: DC === and AC∿.

• The DC scale is normally shown on the meter as a straight line with a number of dots underneath it ===. This symbol is designed to prevent confusion. If just a single line was used, it might be mistaken for a minus sign and if two lines were used it might be mistaken for an equals sign.

• The AC scale is normally shown on the meter as a wavy line ∿.

Automotive Master Technician

- The ohms scale on a multimeter is normally represented by the Greek letter omega (Ω) because if the letter 'O' was used, it might be confused with zero.

Using a multimeter to check voltage

You can use a multimeter as a voltmeter to measure the pressure difference in an electric circuit between where you place the black probe and where you place the red probe.

How to:

1. Connect the probes to the correct sockets on the front of the multimeter.

- Connect the black lead and test probe to the common socket.

- Connect the red probe and test lead to the voltage socket.

2. Most low voltage systems that you will measure on a light vehicle will use direct current DC, so select the scale with the straight and the dotted lines (===).

3. High voltage systems use a combination of alternating current ∿ (AC), for charging and drive systems, and direct current === (DC) at the battery, so ensure that you select the correct type and scale for the circuit you are testing.

4. Following any high voltage warning instructions connect the voltmeter in parallel.

5. Connect the tip of the black lead to a good source of earth, such as the battery terminal, metal bodywork or engine.

6. Use the tip of the red lead to probe the electrical circuit being tested.

Using a multimeter to check for electrical resistance

You can use a multimeter as an ohmmeter to measure resistance. When checking for electrical resistance, always make sure that the power is switched off first and disconnect the component to be tested from the circuit.

How to:

1. Connect the probes to the correct sockets on the front of the multimeter.

- Connect the black lead and test probe to the common socket.

- Connect the red probe and test lead to the socket marked with the omega symbol (Ω).

2. Before you take any measurements, you need to calibrate the ohmmeter to check that it is accurate.

3. Turn the selector dial to the lowest ohms setting and join the tips of the two probes together.

4. When the leads are connected, the readout should show zero or very nearly zero. (If any figures are shown on the screen you will need to add or subtract them from your final results).

5. When the leads are disconnected you should see OL (meaning off limits) or the number 1, which is used to represent the letter 'I' (meaning infinity).

6. Now connect the ohmmeter in parallel across the components so that you can measure the resistance.

You can also use the ohmmeter to check for continuity.

To check a piece of wire for continuity - Place the red and black probes at each end of the wire. The screen should display a very low resistance reading.

Advanced Light Vehicle Technology

To check a switch for correct operation - Connect the red and black probes across the terminals and operate the switch. In the off position, the display should read OL (off limits) or infinity. In the on position, the reading on the display should be very close to zero.

Using an ohmmeter to check for high resistance in an electrical circuit can be misleading. For example: a 12 volt lighting circuit drawing 10 amps with a bad connection causing a 0.1 Ω resistance will reduce the overall circuit voltage by 1volt (ohms law 0.1 Ω x 10 A = 1 V), but the same 0.1 Ω resistance in a starter motor circuit drawing 100 amps will reduce the overall circuit voltage by 10 volts (ohms law 0.1 Ω x 100 A = 10 V).

This low resistance in two different styles of circuit may have little or no effect in the lighting circuit but cause complete failure of the starter circuit.

It is far better to use a volt drop test when checking for a high resistance as this will give a clearer indication of why a circuit is not working correctly.

Using a multimeter to measure electrical current

When measuring the electrical current in a circuit use the amps setting on the multimeter, so that it is used as ammeter. Take care when using an ammeter because, if it is connected incorrectly, the multimeter can be damaged.

How to:

1. Connect the probes to the correct sockets on the front of the multimeter.

    • Connect the black lead and test probe to the common socket.

    • Connect the red probe and test lead to the socket used for measuring amps. (This socket is normally separate from the one used to measure volts or ohms).

2. Turn the selector dial to amps measurement.

3. You need to break into the circuit being tested, being careful to avoid short circuits.

4. Connect the ammeter in series, turn on the circuit and measure the current.

A good place to connect an ammeter is at the fuse box – remove the fuse completely and replace it with the ammeter.

Never connect an ammeter in parallel (across a circuit). A good ammeter has a very low internal resistance, so if the ammeter is connected in parallel a short circuit is created, causing excessive current flow and the ammeter will be damaged. Also remember that, depending on the quality of your ammeter, the amount of current that you can measure may be restricted to around 10 amps.

Other functions of a multimeter

Many multimeters are capable of other functional tests in addition to checking voltage, amperage and resistance. Some examples of extra functions are described in the next section.

Audible continuity testing

Some multimeters include an audible continuity tester. This means that you can test the continuity of an electrical component without having to look at the screen.

- Connect the test probes to the multimeter: black to the common or ground socket and red to the ohms socket.
- Turn the dial to the audible continuity test setting.
- To calibrate the meter and check correct operation, touch the probes together. You should hear an audible tone.
- As with ohms testing, you must switch off circuit power and remove the component being checked from the circuit.
- Now connect the red and black probes to the terminals of the conductor. If continuity exists, you will hear the audible tone.

Diode testing

Most multimeters include a diode test facility. A diode is a one-way valve for electricity. Conduct the test in a similar manner to the continuity test.

- Connect the test probes: black to the common or ground socket and red to the ohms socket.
- Turn the dial to the diode testing setting.
- To calibrate the meter and check correct operation, touch the probes together. The display should show an ohms reading of zero.
- As with ohms testing, you must switch off circuit power and remove the diode from the circuit. You may need to unsolder the diode to remove it.
- With the diode removed, connect the probes to the terminals. If the diode is operating correctly, the display should show a low ohms reading.
- When the polarity of the probes is swapped over the display should show an off limits or infinity reading.
- If it shows zero in both directions, the diode has become short circuited.
- If it shows off limits or infinity in both directions, the diode has become open circuited.

Figure 1.27 A diode symbol

Frequency testing

Some multimeters have a **frequency** test facility. Frequency is a measurement of how quickly a circuit switches. The reading is normally measured in **Hertz** (Hz). 1Hz is equal to one complete cycle of operation (on and off for example) occurring in one second.

- Connect the test probe leads to the appropriate sockets on the multimeter.
- Turn the dial to the frequency setting.
- Test the component while the circuit is operating.

Frequency – how often something happens.

Hertz – a measurement of frequency.

Transistor – an electronic component which can operate as a switch or amplifier (with no moving parts).

Advanced Light Vehicle Technology

Temperature measurement

Some multimeters have a temperature measurement facility. This normally requires an additional probe to be connected. The temperature probe usually has its own socket for connection. Once you have turned the dial to the appropriate setting, you can measure temperature by placing the end of the probe where the measurement is to be taken. (Temperature measurement can be useful for diagnosing cooling system faults for example).

Transistor testing

Some multimeters have a **transistor** testing facility. This facility is rarely used by automotive technicians. Transistors are small electronic switches with no moving parts. They are normally soldered to an electrical circuit board and have three connections: collector, emitter, and base.
There are two types of transistor in common use: positive negative positive (PNP) and negative positive negative (NPN). If the multimeter has a transistor test facility, a six connector socket will be available, marked PNP or NPN. The transistor must be unsoldered from its circuit and connected to one of these diagnostic sockets. The transistor can now be tested by following the multimeter manufacturer's instructions.

Inductive amps measurement

Using an ammeter to check electric current is intrusive and the circuit must be broken. Also, incorrect connection may cause damage to your ammeter. For these reasons, an alternative method of testing for amperage has been developed: some multimeters come with an inductive amps clamp, or this can be purchased separately as an additional unit.

If the inductive clamp is an additional add-on to a standard multimeter, the wires will be connected to the voltage sockets and the voltage scale selected (but will be used to represent amps displayed on the screen).

The amps clamp uses **electromagnetic interference (EMI)** to measure current flow within a circuit. It does not require connection in series but is simply clamped around the wire to be tested. When the circuit is switched on and current flows, you can read the amperage measurement from the display. (Make sure that you read the manufacturers operating instructions to know how to connect and read the current clamp).
This is not always as accurate as connecting an ammeter in series, but is quicker and should not cause damage if connected incorrectly. It is also able to take much higher amperage readings than a standard multimeter.

Electromagnetic interference (EMI) – a disturbance that affects an electrical circuit due to either electromagnetic conduction or electromagnetic radiation emitted from an external source. It is also called radio frequency interference (RFI).

Automotive Master Technician

There is normally a plus or minus sign on the amps clamp to show which way round it should be connected to an electric circuit.

Due to the nature of high voltage systems used in the operation of hybrid vehicles, you should always conduct current measurement with extreme caution. If you have to take a current measurement, the inductive clamp testing method is recommended as this should not involve disconnecting any high voltage circuits.

By making up a small test lead with an in-line fuse, you can test current draw at the fuse box using an inductive clamp. Simply remove the fuse from the circuit to be tested, and replace it with the test lead (and correct size fuse in-line). The current clamp can now be clipped around the test lead and the circuit operated to obtain the readings.

Inductive amps clamps are quick and easy to use, and very good for testing electrical faults such as the ones shown in Table 1.2.

Table 1.2 Electrical faults that can be tested using an inductive amps clamp

Symptom	Amps clamp use/test
Slow cranking 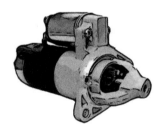	When a customer presents their car because the engine is cranking over slowly, how do you know whether it is the battery, the wiring or the starter motor that is at fault? Connect the inductive amps clamp around the positive or negative battery lead and crank the engine. Initial peak amperage will be shown on the display, indicating the strain that has been put on the electrical system. If the current draw is low, then the starter motor is not struggling and it is the battery that is unable to supply the correct amount of amperage required to turn the engine over at speed. This may indicate that the battery needs replacing. If the current draw is high, then you should suspect the wiring or starter motor and test them. An example of a high current draw on a four-cylinder petrol engine: a peak amperage of around 140 amps should not be exceeded. If you do not know the amount of current that should be drawn on a starting system, connect your inductive amps clamp to another vehicle of similar size and type and compare the two.
Glow plugs	When a customer presents their Diesel engine car and says that it is suffering with poor cold starting, how do you know that the glow plugs are at fault? The normal way of testing the glow plugs would be to leave the car to go cold, sometimes overnight, before the symptoms can be confirmed. If the glow plugs are suspected, they are often removed and tested across a battery. Not only is this time consuming, but it can also be very dangerous. If the tip of a glow plug is contaminated, it may explode when tested. Excessive heat caused by testing

Table 1.2 Electrical faults that can be tested using an inductive amps clamp

a glow plug across a battery may also ignite hydrogen sulphide gas produced within the battery, leading to an explosion.

The Inductive amps clamp allows you to test the glow plugs in a far quicker and safer manner.

Allow the engine to cool for around an hour, connect the inductive amps clamp around the feed wire to the glow plugs and turn on the ignition. As a general rule of thumb, each glow plug will draw approximately 20 amps of current. With a four-cylinder engine this means that you are looking for around 80 amps of current in total if the glow plugs are working correctly. If the inductive amps clamp records a reading significantly lower than this, you can suspect that one or more glow plugs have failed. You will then need to remove and replace all of the glow plugs.

The advantages of this type of test are: the glow plugs do not have to be removed, saving time and cost; there is less danger of personal injury from incorrect test methods; the vehicle does not have to be left overnight to confirm the symptoms.

Alternator output

When testing an alternator for correct operation, amperage output is often overlooked.

Many vehicle technicians only conduct voltage tests when checking alternator operation. A voltmeter is connected across the battery, the engine is started and a load is placed on the system by switching on the headlights. The engine idle is normally raised to around 2000 rpm and, if a voltage of approximately 14.2V is obtained, it is assumed that the alternator is charging correctly.

However, the alternator may still be at fault. Voltage is electrical pressure, and a pressure higher than that which is coming from the battery is required to push electrons back into the battery for charging.

Electrical pressure is not the same as electrical quantity. The pressure or voltage may be high enough to charge the battery, but if the quantity is not available, the electrical components may be using it up quicker than it is going in, resulting in the battery running down over time.

Many modern alternators use a three-phase system, meaning that three coils of wire are connected internally to the charging circuit. This has the advantage of providing three times as much electricity as a normal generator. If one or more of these phase windings fails, correct electrical voltage may still be produced, but the correct quantity might not.

To ensure that an alternator is operating correctly, conduct the following amps test:

With the engine turned off, switch on a number of electrical consumers, such as headlights, heated rear screen, etc. Leave the car for around 10 minutes so that the battery partially discharges. Once the battery has partially discharged, connect the inductive amps clamp to one of the battery cables. Switch off all the consumers and start the engine. Because the battery is now partially empty, the alternator should put out a high quantity of electric current (amps). Check this peak current output to see if it is similar to the one marked on the side of the alternator. If the amperage is well below that indicated on the alternator, the unit should be replaced.

As the battery recharges with current from the alternator, the output from the alternator will fall.

Automotive Master Technician

Table 1.2 Electrical faults that can be tested using an inductive amps clamp

Parasitic drain 	A common fault with a car's electrical system is parasitic drain. The symptoms normally include the battery going flat over a period of time. An inductive amps clamp can be used to assist with diagnosis. The inductive amps clamp should be connected around one of the battery leads. With all electrical consumers switched off, current draw should be almost zero. If a current draw exists, then something is using the electrical energy and this component is considered to be a parasite on the battery. Gain access to the fuse box and then fuses should be systematically removed until the current draw disappears. Once the current draw disappears, the circuit causing the problem can be identified from which fuse has been removed. With the fuse reconnected, and the current draw reinstated, you can now isolate or disconnect consumers within that circuit until the fault has been found.
Lazy fuel pump 	If the customer presents a car with fuelling issues, and you suspect that the fuel pump may be at fault, the normal course of action would be to conduct a fuel pressure test. The current draw on a fuel pump can give an indication of its operation. A high current draw indicates that the pump is struggling, and a low current draw indicates that the pump is finding it too easy. In many cases, you can measure current draw with an inductive amps clamp, without removing the fuel pump. With the rear seat lifted, gain access to the wiring running to the fuel pump in the tank. Place the inductive amps clamp around the feed wire. If there are a number of wires, try each one in turn until you obtain an appropriate reading. Start the engine and record the readings. If the fuel pump is fused individually, you will not need to gain access at the fuel tank. You can remove the fuel pump fuse and use a small loop of wire in its place. (It is recommended that you use an inline fuse in this loop of wire to protect the circuit). Now connect the inductive amps clamp around the loop of wire, start the engine and take readings. A current draw of between 3 and 5 amps should be expected for a standard petrol pump. If you do not know what to expect, try another vehicle of a similar type and compare the readings.

An inductive amps clamp can sometimes help you decide very quickly if a non-working electrical system is being caused by a circuit issue or a physical mechanical fault. Simply attach the clamp to either the positive or negative battery lead and switch on the vehicles ignition. Turn on the inductive amp clamp and wait for the display to settle. Now operate the circuit that appears to be causing the fault. If the current display on the screen alters, the circuit is trying to do something, and this suggests a physical fault is causing the problem.
If the current display on the screen doesn't alter, then the circuit is possibly damaged and will require further electrical testing.

Advanced Light Vehicle Technology

Dedicated battery testing equipment

Tooling manufacturers are now producing specialist battery testers which run fully automated checks that help show the condition of lead-acid batteries and charging circuits. These testers can include an internal charger which brings the battery up to the required state of charge before any checks on the battery are made.

Tests can include:

- Capacity
- Voltage

- Heavy discharge (drop testing)
- Charge system voltage/amperage

- Cold cranking amps (CCA)

The results will often be displayed on a screen as pass or fail. Because the testers use fully automated procedures, they can be operated by people with limited technical knowledge.

Oscilloscopes

An oscilloscope is a piece of electrical test equipment designed to act like a voltmeter or an ammeter. A multimeters measurement readout can't change fast enough to deal with modern electronic systems on motor vehicles – the numbers on the screen can't keep up. The answer to this is to use an oscilloscope.

Unlike a voltmeter, oscilloscopes not only show volts or amps but also time. Instead of a digital readout, the results are shown as a graph of volts or amps against time on a screen (as shown in Figure 1.28).

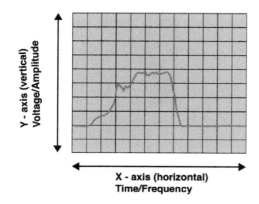

- The graph normally shows voltage or amperage at the side of the screen (on the y-axis) – this axis is often called **amplitude**. Use the scale setting switch in a similar way to the dial on a manual multimeter to choose the amount of volts or amps that are shown on the screen.

Figure 1.28 An oscilloscope screen

- The graph normally shows time across the bottom of the screen (on the x-axis). This axis is often called **frequency**. Use the timescale switch in a similar way to the dial that is used to choose the amount of volts on a multimeter.

An easy way to remember which axis is which on a graph is to say 'X is across' (a cross).

Lots of people are put off using oscilloscopes by the large box containing many wires and connectors. They feel that it will be complicated and time consuming to set up, so they don't bother.

However, to use an oscilloscope for simple electrical testing, you only need two probes – a common and voltage wire – just like a multimeter. To measure amperage, you may need an inductive clamp.

Most of the diagnostic sockets found on oscilloscopes are colour-coded, so after a quick check of the manufacturer's instructions, it should be fairly easy to know where to plug these probes in.

> **Amplitude** – the height of a waveform, measured in volts or amps.
>
> **Frequency** – the time scale of a waveform (how often something happens).

Using an oscilloscope for electrical testing

How to:

Note: The oscilloscope probes may come in different colours, but for the sake of simplicity we will call them red and black here.
1. Connect the tip of the black lead to a good source of earth, such as the battery terminal, metal bodywork or engine. This will then only leave you with the red wire to worry about.
2. Observing any manufacturers safety precautions, connect the red probe to the circuit to be tested. (Some oscilloscopes will require the fitting of a resistor known as an attenuator in order to read high voltages).
3. Adjust the scales until you see an image on the screen.
4. After some practice, you will become familiar with the patterns and waveforms created by different vehicle systems.

If you don't know what voltage or timescale to use on an oscilloscope, find out in the same way as you would with a multimeter. Start with the highest setting available and work downwards until you can see an image on the screen.

Scan tools and fault code readers

Faults with many modern vehicle systems would be difficult to diagnose without the aid of a scan tool. The processes that take place within electrical and electronic circuits mean that these systems are being controlled many thousands of times a second, and faults can occur so quickly that you could miss them.

Since the 1980s, manufacturers have been including on-board diagnostic (OBD) systems as part of their vehicle design. The computers that control the vehicle's electrical systems have a self-diagnosis feature. This allows them to detect certain malfunctions and store a code number. Because these electronic control units (ECU's) are monitoring functions, they are able to record intermittent faults and store them in a keep alive memory (KAM) for retrieval by a diagnostic trouble code (DTC) reader.

It is a common misunderstanding to think that plugging a trouble code reader into the vehicle's OBD system will tell you what the fault is. It actually only points you in the direction of the fault. You must test the system and components to find the actual fault.

To use a scan tool you will need to locate the diagnostic socket. Since the year 2000, the type and position of the diagnostic socket, also known as the data link connector DLC, has been standardised. A 16 pin socket should be located inside the vehicle, within reach of the driver, somewhere between the centre line of the car and the driver's seat.

Advanced Light Vehicle Technology

Not all data link connectors are found on the driver's side of the car. This is often a design issue, particularly if the car was originally intended as a left hand drive. Manufacturer's data is often available to help you locate the diagnostic socket.

The scan tool should be connected to the diagnostic socket and the ignition switched on. The scan tool will then attempt to communicate with the vehicles on-board computer systems.

Once communication has been established, you should follow the on-screen instructions to retrieve information and operate the on-board diagnostic system as required.

If your scan tool says no communication when initially connected to the diagnostic socket, leave connected and switch off ignition, lock car and wait 3 minutes. Unlock and switch ignition back on. This sometimes acts as a re-boot for the diagnostic system and it may now recognise the scan tool.

There are two main types of diagnostic information available on many systems:

- OEM
- E-OBD

OEM

OEM is information from the original equipment manufacturer. To gain access to this information, you will need to enter vehicle specific information such as make, model, engine type and vehicle identification number (VIN). Once this has been done a large amount of manufacturer/vehicle specific data is often available.

Many of the diagnostic trouble codes DTC's have been standardised so that:

- Codes beginning with **P** relate to powertrain faults.
- Codes beginning with **C** relate to chassis system faults.
- Codes beginning with **B** relate to body system faults.
- Codes beginning with **U** relate to network communication system faults.

This standard is not mandatory and as a result, some manufacturers use their own coding system. Also the information from OEM is not generic, meaning that each manufacturer and vehicle type will have its own set of codes.

Automotive Master Technician

When using diagnostic trouble codes, it is good practice to:

Record then clear DTC and fully road test/operate the car/system

⬇

Rescan and concentrate diagnostic routine on any codes that have returned

⬇

Lacate and repair the fault and clear all codes

⬇

Fully road test the car/system - Rescan for any codes that have returned

⬇

Follow up any pending DTC's

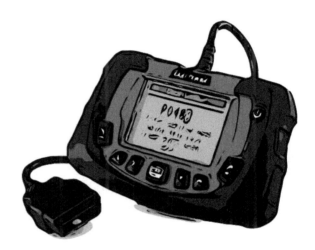

Figure 1.29 Good practice when using scan tools

Features of scan tools

Typical features of scan tools include:

- Retrieval of electronic control unit (ECU) diagnostic trouble codes.
- Erasing of system ECU diagnostic trouble codes.
- Displaying serial data/live data.
- Displaying readiness monitors.
- Resetting of ECU adaptions.
- Displaying freeze frame data.
- Coding of new components, such as fuel injectors.
- Access to information on various vehicle electronic systems.
- Resetting of service reminder lights.
- Vehicle key coding.

Tip

Many diagnostic equipment manufacturers are producing scan tools that use wireless technology to communicate between the serial port (data link connector DLC) and a handheld interface operated by the technician. The unit connected to the serial port is known as the vehicle communication interface (VCI) and will often use a short range Bluetooth wireless connection to transfer two-way data between a PC/dedicated scan tool and the car.

Advanced Light Vehicle Technology

Data logging

Many vehicle electronic management faults are intermittent. This can make the diagnostic process difficult and time consuming. A method that can help with this issue is data logging.
Data logging is a process of 'flight recording' information from the vehicles electronic system. Many scan tools have this capability, meaning that they can be connected to the vehicles diagnostic socket and then the vehicle may be operated, on a road test for example, and serial information can be captured and reviewed after the event.

A number of diagnostic equipment manufacturers have produced specialised tools, made specifically for data logging. These tools are small and compact and are often not much bigger than the size of a matchbox. They can be attached to the data link connector DLC, powered directly from the serial port, and the vehicle can be driven while they unobtrusively gather data. This means that the car can often be returned to the customer to use as normal with the data logger still attached, and if a further fault should occur, the information can be collected from the data logger and reviewed on a computer.

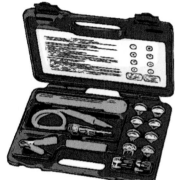

Noid lights

Noid lights are small light emitting diode (LED) testers. They are normally specific to manufacturers, and can be purchased in sets to enable the testing of a wide range of fuel injectors. When connected to the wiring harness at the fuel injectors, with the engine cranking or running, they will flash if an operating signal is being correctly produced.

Cylinder balance testing

A cylinder balance check involves disabling engine cylinders individually and seeing how they affect the performance of the engine. This system uses an engine tachometer to assess the effect that disabling the cylinder has on engine idle.
This dedicated piece of diagnostic equipment is able to shut off individual cylinders without causing over-fuelling or ignition problems which may damage the catalytic converter.

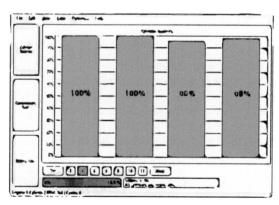

• If you see a large difference in the idle speed when the cylinder is disabled, it is usually an indication that the cylinder is working well.
• If you see a small difference or no difference in the idle speed when the cylinder is disabled, it is usually an indication that the cylinder is at fault.

Notes

An infrared laser thermometer can sometimes be used to help identify a cylinder that is not running very well. If you move the infrared thermometer slowly along the exhaust manifold, you can sometimes see a misfiring cylinder, as it will run cooler than the rest.

Automotive Master Technician

Compression testing

You should always conduct an engine compression test if you suspect that a pressure leakage might have occurred within a cylinder.

How to:

1. Isolate the engine so that it cannot start.
2. Gain access to and remove all of the spark plugs.
3. You can now attach the compression tester to the engine, in place of the plugs.
4. Once the gauge is attached to a cylinder, crank the engine with the throttle held wide open. Continue to crank the engine for a short period of time, until the maximum reading is shown on the gauge. Record the reading and compare the figure with the manufacturer's specifications. Repeat the procedure for the remaining cylinders. (This is known as a dry test).
5. Using an oil can, introduce a small amount of oil to the cylinder down the plug hole and repeat the compression test. Record the results and compare with the original readings. (This is known as a wet test).

During a **wet compression test**, the oil can often sit on top of the piston crown, and form a temporary compression seal around the top piston ring.

- If the wet test shows an increase in compression pressure over a **dry test**, it gives an indication that the leakage is occurring down past the piston and rings.
- If the compression pressure stays low, this gives an indication that the leakage may be coming from the cylinder head or valve area.

Wet test – where oil is used during a compression test.

Dry test – where no oil is used during a compression test.

Leakdown testing

A cylinder leakdown tester is a tool that you can use to help diagnose compression leakage within a cylinder. It consists of a pressure gauge, regulator and adapter which is screwed into the spark plug or injector hole in the engine.

How to:

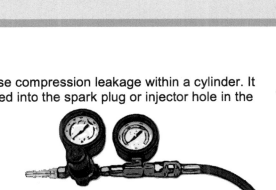

1. Strip out and remove the spark plugs or injectors.
2. Rotate the engine until the cylinder to be tested is on its compression stroke, and the inlet and exhaust valves are both closed.
3. Lock the crankshaft to prevent it turning. You can do this physically, or by placing the car in gear and applying the handbrake.
4. Screw the leakdown tester into a spark plug hole and connect the cylinder leakdown tester to a source of compressed air.
5. Introduce a regulated air pressure into the cylinder.

The gauge can indicate if a significant difference in pressure exists in the cylinder when compared with the pressure introduced by the compressed air. If a pressure difference exists, then cylinder leakage may have occurred.

Advanced Light Vehicle Technology

To help diagnose the location of the cylinder leakage you can:

- Listen for air leaking from the exhaust pipe: this is an indication that the exhaust valve is leaking.
- Listen for air leaking from the intake manifold (remove the air filter housing): this is an indication that the inlet valve is leaking.
- Listen for air leaking from the dipstick tube or the rocker cover: this is an indication that pressure is leaking down past the piston.
- Remove the cooling system pressure cap, and check for bubbles in the engine coolant: this is an indication that pressure is leaking past the head gasket or cylinder head.

Smoke

A smoke generator is a tool that you can use to help diagnose engine system leakage. A chemical or oil is heated in the smoke generator which then produces pressurised smoke from a pipe that can be connected to various engine systems.

• To check for an inlet system leak, remove a vacuum pipe from the inlet manifold and connect the smoke generator. Then block the air inlet using special bungs. Once the smoke has been introduced to the inlet system, you can then look for any signs of leakage which may to cause engine running problems.
• To check for an exhaust system leak, the pipe from the smoke generator is placed in the exhaust tailpipe and sealed. Once the smoke has been introduced to the exhaust system, you can then look for any signs of leakage which may result in an engine emissions fault.

Notes

A smoke generator is not only useful when finding intake and emission leaks. The special adapters supplied with the smoke kit will often allow you to connect it to many other systems and the harmless smoke produced can help identify various leakage issues.

Tip

Induction air leaks around the rocker cover etc. can cause engine management issues (use a smoke generator to check).
Use a smoke generator to check for leaks in turbo piping.

Cylinder block tester

A cylinder block tester is a tool which you can use to help diagnose compression leakage past the cylinder head gasket into the cooling system. You will need to fill a container with a special chemical and place it in the neck of the cooling system radiator or expansion tank. Once the engine has reached its normal operating temperature, draw the fumes given off in the cooling system through the liquid using a hand pump. If the chemical liquid changes colour, it indicates the existence of hydrocarbons from exhaust gases. These hydrocarbons will normally have leaked from the combustion process inside an engine cylinder past the head gasket.

Tip

You can use an exhaust gas analyser to conduct a similar test to a cylinder block tester. With the engine at normal operating temperature, carefully remove the cooling system pressure cap and hold the probe from the exhaust gas analyser above the coolant to sample the fumes. You can then assess the amount of hydrocarbons, measured in parts per million, from the gas analyser screen. **(Be careful not to dip the exhaust probe of the gas analyser into the coolant, as the liquid can be sucked up and damage the machine).**

Safety

Always allow the engine to cool before removing the cooling system pressure cap. If you remove the cap while the engine is hot, a sudden drop in pressure may cause the liquid coolant to boil, resulting in scalding or severe burns.

Cooling system pressure testing

A cooling system pressure test is used to ensure that the cooling system can be pressurised and that there are no leaks. Many testers can also be used to check the correct operation of the cooling system pressure cap (a vital check that is often overlooked by many vehicle technicians).

How to:

1. Allow the engine to cool.
2. Once the engine is cool, slacken the cooling system pressure cap so that no pressure can build-up. Then run the engine until it has reached an operating temperature where the thermostat is open. Carefully remove the cap.
3. Following the tool manufacturer's instructions, attach the cooling system pressure tester in place of the system pressure cap.
4. Without operating the pump of the pressure tester, start the engine. If you see a rapid rise in pressure on the gauge, this can indicate head gasket failure.

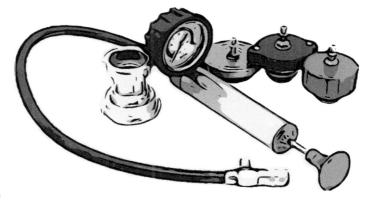

5. Following this, operate the pump on the pressure tester until the gauge reaches the maximum pressure indicated by the vehicle manufacturer. (The maximum cooling system pressure can often be found as a marking on the radiator cap).
6. Check to see if the pressure decays over a short period of time, which may indicate a leak. Now conduct a visual inspection for leaks. This can include checking the inside of the car in case the heater matrix is leaking. (You can sometimes check for this by feeling the carpet area around the driver and passenger foot well. If the heater matrix is leaking, the carpet will often be wet).
7. If there are no visible signs of leakage, the car can be left with the cooling system pressure tester attached for a longer period of time. If the gauge pressure falls significantly during this period, there may be a system leak, which might require further investigation.
8. Once the cooling system has been checked, it is good practice to test the pressure cap. Attach the cap to the tool following manufacturer's instructions and operate the pump. Check that the pressure cap releases at the specified value.

- Too low release pressure may cause overheating.
- Too high release pressure may cause system damage and leaks.

Advanced Light Vehicle Technology

Endoscopy

Endoscopy normally involves inserting a small camera inside an engine to visually inspect for damage before the engine is stripped down. For example, you can put an illuminated probe with the camera attached down a spark plug hole, and use the image shown on the screen to check for physical engine damage.

Exhaust emission testing

Many workshops own an exhaust gas analyser, but don't use it to its full potential. If you have a good understanding of combustion processes and the chemical emissions that are produced, a gas analyser can be a very useful diagnostic tool (*see Chapter 2*).

How to:

1. If you are conducting an analyser test in a workshop, ensure that suitable exhaust extraction is used. Make sure that there is sufficient oil and coolant, and that the engine is in a suitable condition to conduct an exhaust gas test.
2. Start the engine and allow it to warm up to its normal operating temperature.
3. Switch on the exhaust gas analyser, and allow it to warm up and go through its calibration procedure (do not bypass this stage as it could lead to incorrect emission values being obtained). Conduct a leak check on the equipment using the equipment's vacuum decay test.
4. When the engine is warm, raise the engine speed to around 2,000 RPM and hold for approximately 30 seconds to allow the exhaust system to clear any accumulated gas.
5. Return the engine to idle and insert the exhaust probe into the exhaust.
6. Allow the readings on the display to settle (this can take up to 15 seconds) and record all of the results.

7. With the exhaust probe still inserted, raise the engine speed to around 2,000 RPM and hold for approximately 30 seconds. When the readings on the display have stabilised, record all of the results.
8. Return the engine to idle, switch off and remove exhaust gas test equipment/extraction.
9. Examine all of the exhaust gas results and use them to conduct a full assessment of engine operation and performance (*see Chapter 2*).

Automotive Master Technician

Preparing for assessment

The information contained in this book can help you with theory or practical assessments used to identify your skills or competence when undertaking vehicle repairs or a recognised qualification. It is possible that some of the evidence you produce may contribute to more than one qualification. You should ensure that you make best use of all your evidence to maximise the opportunities for cross-referencing between units or qualifications.

You should choose the type of evidence that will be best suited to the type of assessment that you are undertaking (either theory or practical). The types of evidence you could use are listed below:

- Direct observation by a qualified assessor
- Witness testimony
- Computer based
- Audio recording
- Video recording
- Photographic recording

- Professional discussion
- Oral questioning
- Personal statement
- Competence/Skills tests
- Written tests
- Multiple-choice tests
- Assignments/Projects

Before you attempt a written or multiple-choice test, make sure you have reviewed and revised any key terms that relate to the topics in that subject area. Make sure you read all questions carefully. Take time to digest the information so that you are confident about what the question is asking you. With multiple-choice tests, it is very important that you read all of the answers carefully, as it is common for two of the answers to be very similar, which may lead to confusion.

For practical assessments, it is important that you have had enough practice and that you feel that you are capable of passing. It is best to have a plan of action and work method that will help you.

Make sure that you have the correct technical information, in the way of vehicle data, and appropriate tools and equipment. It is also a good idea to check your work at regular intervals. This will help you to be sure that you are working correctly and to avoid problems developing as you work. When undertaking any practical assessment, always make sure that you are working safely throughout the test.

Chapter 2 Advanced Internal Combustion Engine Technology

This chapter will help you develop an understanding of the construction and operation of advanced internal combustion engine (ICE) technology. It will cover mechanical design and computerised electronic technology which enhances vehicle performance, fuel economy and helps to reduce emissions. It will also assist you in developing a systematic approach to the diagnosis of complex system faults for which there is no prescribed method. This is achieved with hints and tips which will help you undertake a logical assessment of symptoms and then uses reasoning to reduce the possible number of options, before following a systematic approach to finding and fixing the root cause.

Contents

Safe working when carrying out complex engine system diagnostic activities

There are many hazards associated with the diagnosis and repair of advanced engine systems for both yourself and the vehicle. You should always assess the risks involved with any diagnostic or repair routine before you begin, and put safety measures in place.

You should always use appropriate personal protective equipment (PPE) and vehicle protective equipment (VPE) when you work on engine systems. Make sure that your selection of PPE will help to protect you from the hazards you may encounter and reduce the possibility of damage to the customer's vehicle.

Information sources

Many of the diagnostic routines that are undertaken by a Master Technician are complex and will often have no prescribed method from which to work. This approach will involve you developing your own diagnostic strategies and requires you to have a good source of technical information and data (*see knowledge is power*). In order to conduct diagnosis on advanced engine systems, you need to gather as much information as possible before you start and take it to the vehicle with you if possible.

Automotive Master Technician

Sources of information may include:

Table 2.1 Possible information sources

Verbal information from the driver (*see the interrogation room*)	Vehicle identification numbers
Service and repair history (*see the interrogation room*)	Warranty information
Vehicle handbook	Technical data manuals
Workshop manuals/Wiring diagrams	Safety recall sheets
Manufacturer specific information	Information bulletins
Technical helplines	Advice from other technicians/colleagues
Internet	Parts suppliers/catalogues
Jobcards	Diagnostic trouble codes
Oscilloscope waveforms	On vehicle warning labels/stickers
On vehicle displays	Temperature readings

Always compare the results of any inspection or testing to suitable sources of data. Remember that no matter which information or data source you use, it is important to evaluate how useful and reliable it will be to your diagnostic and repair routine.

Electronic and electrical safety

Working with any electrical system has its hazards. The main hazard is the possible risk of electric shock, but remember that the incorrect use of electrical diagnostic testing procedures can also cause irreparable damage to vehicle electronic systems. Where possible, isolate electrical systems before conducting the repair or replacement of components.

Always use the correct tools and equipment. Damage to components, tools or personal injury could occur if the wrong tool is used or misused. Check tools and equipment before each use.
If you are using electrical measuring equipment, always check that it is accurate and calibrated before you take any readings.

If you need to replace any electrical or electronic components, always check that the quality meets the original equipment manufacturer (OEM) specifications. (If the vehicle is under warranty, inferior parts or deliberate modification might make the warranty invalid. Also, if parts of an inferior quality are fitted, they might affect vehicle performance and safety). You should only carry out the replacement of electrical components if the parts comply with the legal requirements for road use.

Table 2.2 The operation of electrical and electronic systems

Electrical/Electronic system component	Purpose
ECU	The electronic control unit is designed to monitor the operation of vehicle systems. It processes the information received and operates actuators that control system functions. An ECU may also be known as an ECM (electronic control module).

Advanced Light Vehicle Technology

Table 2.2 The operation of electrical and electronic systems

Sensors 	The sensors are mounted on various system components and they monitor the operation against set parameters. As the vehicle is driven, dynamic operation creates signals in the form of resistance changes or voltage which are sent to the ECU for processing.
Actuators 	When actioned by the ECU, motors, solenoids, valves, etc. help to control the function of vehicle for correct operation.
Digital principles 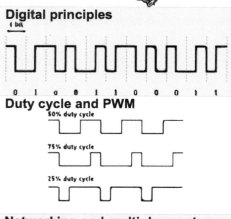	Many vehicle sensors create analogue signals (a rising or falling voltage). The ECU is a computer and needs to have these signals converted into a digital format (on and off) before they can be processed. This can be done using a component called a pulse shaper or Schmitt trigger.
Duty cycle and PWM 	Lots of electrical equipment and electronic actuators can be controlled by duty cycle or pulse width modulation (PWM). These work by switching components on and off very quickly so that they only receive part of the current/voltage available. Depending on the reaction time of the component being switched and how long power is supplied, variable control is achieved. This is more efficient than using resistors to control the current/voltage in a circuit. Resistors waste electrical energy as heat, whereas duty cycle and PWM operate with almost no loss of power.
Networking and multiplex systems	Many modern vehicle systems are controlled using computer networking. In these systems a number of ECU's are linked together and communicate to share information in a standardised format. The most common network system is the Controller Area Network (CAN Bus). *See Chapter One.*

Internal combustion engine design

Most internal combustion engines (ICE) work on the four stroke cycle of induction, compression, ignition/power and exhaust. This design concept is mainly attributed to Nickolaus August Otto, who in 1876 produced the first effective gas powered engine, although a number of earlier patents for four-stroke cycle engines exist before this time.

In nearly a century and a half, the use of the four-stoke cycle has seen designers produce a number of different styles of internal combustion engine, but most are fundamentally the same in their operation.

Figure 2.1 A modern internal combustion engine

Automotive Master Technician

The need to find alternative methods of propulsion for motor vehicles has led to a new resurgence in engine design and fuel types; some of these will be discussed in chapter 5. Having said this, advances in engineering, design and production have enabled the conventional internal combustion engine to develop into a much more efficient and economic power plant.

Performance characteristics

The key goal of most modern engine designs is to enhance performance, particularly in the areas of power and torque, whilst ensuring that overall capacity is kept generally low. This can often be achieved by using methods which will help to improve mechanical or **volumetric efficiency**.

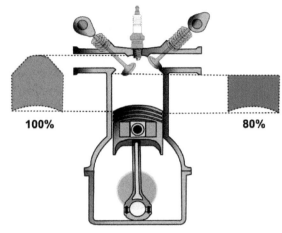

Volumetric efficiency is how well the engine cylinder can be filled with a fresh charge of incoming air and fuel in the correct quantities. A **naturally aspirated** engine relies on atmospheric pressure to force air into the cylinder above the descending piston. As the piston descends, a low pressure, or **depression**, is created above the piston crown, and air is forced into the cylinder by atmospheric pressure through the open inlet valve(s).

Figure 2.2 Volumetric efficiency

Due to the speed of operation, it is very difficult to fill the cylinder completely and, as a result, historically only about 80% of the available space is filled. As engine speed increases, this figure often falls and volumetric efficiency is reduced, which affects overall engine performance. Engine manufacturers and designers have come up with a number of methods to help improve volumetric efficiency. These can often include forced air induction such as turbo charging and supercharging, variable valve control, advanced intake manifold and combustion chamber design.

Volumetric efficiency – how well an engine cylinder can be filled with air and fuel.

Naturally aspirated – not turbocharged or supercharged.

Depression – a low or negative pressure.

With effective design and new technologies, power and torque can not only be improved overall, but also varied according to road and driving conditions, giving better control of driver response, fuel economy and lower emissions.

Power and torque

Power and torque are measurements that describe the performance output from an internal combustion engine. They are normally tested using a piece of equipment known as a **dynamometer**, or 'dyno' for short. A dynamometer can be connected directly to an engine during factory testing, but most technicians are familiar with the more common 'rolling road' type which can be used to assess engine and transmission performance during tuning adjustments. A rolling road type dynamometer works in a similar manner to a brake rolling road that is used to assess the efficiency of the vehicles braking system, but in reverse. During brake testing, the wheels of the vehicle are placed in the rollers and a motor is used to rotate the rollers, and therefore the wheels of the car.

Advanced Light Vehicle Technology

The driver will then apply the brakes, slowing down the roller drive motors, and the strain on the system is calculated as braking effort.

With a performance dynamometer, the wheels are placed in the rollers and the vehicle engine and transmission system provide the driving force. An internal brake within the dynamometer, tries to slow the rollers down, and the effort required can be used as an assessment of performance for the engine and transmission and leads to the term 'Brake Horse Power' Bhp.

In order for the dynamometer to calculate engine performance at the road wheel, the vehicle should be accelerated up through the gears, and then slowed down through the gears. This will then allow the software to assess the drag created by the transmission system and exclude this from the final results.

Power and **torque** describe two different elements of performance. Depending on the intended purpose of the vehicle, different combinations of power and torque will be used.

Dynamometer – an instrument for measuring mechanical force or power.

Power – the amount of energy that is efficiently released during the combustion cycle, or the rate at which work is done. This will normally give a vehicle its speed.

Torque – the turning effort produced at the crankshaft as it is rotated. This will normally increase the amount of weight a vehicle can pull.

Power gives a vehicle its speed. You can think of power as being like a sprinter, who delivers a sudden burst of energy to provide speed.

Torque is the effort produced, giving a vehicle its strength. You can think of torque as being like a weightlifter, who has a great deal of strength to lift heavy weights.

The sprinter and weightlifter are both very fit athletes, but they perform two completely different disciplines, and it is the same for power and torque.

When designing a vehicle, the manufacturer must take into account what that vehicle will be used for. For example, a sports car may require a great deal of power, but because it is lightweight it does not necessarily require large amounts of torque. In contrast, a vehicle designed to carry heavy weights, such as a van or people carrier, may require more torque and less power.

The systeme international (SI) unit of power is the watt, and most engine calculations of performance are measured in Kilo-watts (Kw). There are 746 watts to 1 horse power (United Kingdom), and 1.34 horse power (United Kingdom) equals 1Kw.

Automotive Master Technician

A number of diagnostic tool manufacturers now give the opportunity to conduct a simulated form of dynamometer testing via the vehicles serial output from the data link connector (DLC). Having connected a scan tool to the DLC and input the required technical information, the vehicle may now be driven and the software will calculate measurements of engine power and torque from the engine management serial data.

It is worthwhile remembering that this is simulated data, and may not be a fully reliable source of power and torque, although it can be a very handy tool when evaluating repairs or performance enhancements (*see outpatient*). Also it will not be possible to conduct this test on a public highway, as it will often require that the vehicle is driven in a manner that contravenes driving regulations and road laws. This test should only be conducted on a closed road or track.

Understanding power and torque graphs

The readout of a performance dynamometer is normally displayed as a graph with power and torque laid out as two separate curved lines on the same chart, *see Figure 2.3*. The 'x' axis of the graph displays a timescale usually relating to engine speed or road speed. The 'y' axis displays the amplitude for both power (measured in watts W) and torque (measured in Newton meters Nm). This will give an indication of overall function of the engine and transmission as well as a general impression of mechanical efficiency.

- The amplitude of the power curve will often rise as a relatively smooth line and peak around three-quarters of the way up the rev range before it begins to tail off or fall. This is because many engines are designed to give peak power performance before maximum revs, otherwise it would have to be run 'flat-out' in order to achieve full power.

- The amplitude of torque curve will often begin relatively high and rise to its maximum depending on engine design and fuel type. (A Diesel engine will often have its peak torque at a lower rev range than petrol due to the nature of its combustion process). No matter which fuel or engine design is used, vehicle manufactures must match transmission gear ratios to this torque delivery, in order to extract the maximum amount of turning effort from the engine.

Both power and torque curves on a dynamometer graph will often show small dips as the transmission changes from one gear to the next.

Many dynamometers are able to calculate the mean (average) effective pressures generated inside the engine by adding peak power for each gear ratio and dividing them by the number of gear changes.

If the dynamometer is able to be connected directly to the engine crankshaft, then is often able to display the pressure rise inside the combustion chamber in direct relation to crankshaft rotation. (i.e. crank rotation in degrees on the 'x' axis and cylinder pressure in Bar on the 'y' axis). This will help engine designers calculate the thermal efficiency of an engine, and extract the maximum amount of energy from the combustion process.

Some scan tools are able to simulate virtual dynamometers. This is a very useful function for checking the effectiveness of repairs that affect vehicle/engine performance.

Advanced Light Vehicle Technology

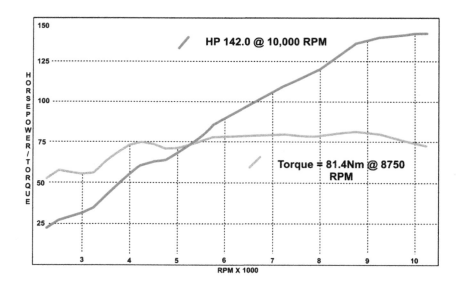

Figure 2.3 Dynamometer power and torque graph

Once engine designers have calculated the power and torque from a particular engine design and fuel source, they are able to map engine management systems to the mechanical output and control fuelling and air/fuel ratios which give improved economy, and lower emissions while maintaining a high level of power output for different driving conditions.

Ideal air/fuel ratios produce a balanced chemical reaction where the amount of fuel is accurately matched to the amount of oxygen in the inducted air so that when it is combusted, the least amount of chemical components remain, improving efficiency and lowering emissions. This is known as a stoichiometric value and is often referred to as a ratio of 14.7:1 of air to fuel by mass.

With careful engine management, this ideal can be cleverly manipulated when outright performance is not required by making it run weak (more air, less fuel) for example, and further enhance fuel economy and reduce exhaust emissions.

Direct and indirect injection engine design

An internal combustion engine operating on petrol or Diesel will mainly deliver its fuel by injection. This can be indirect injection, into a manifold or pre-combustion chamber or direct injection straight into the cylinder.
Indirect injection often provides favourable production costs, but direct injection gives accurate fuelling advantages, which when coupled with advanced engine management will provide gains in performance, economy and the reduction of emissions.

Petrol injection

Currently, the most common type of petrol fuel injection is multipoint indirect, but many vehicle designers are now including the option of direct injection for a proportion of their engine types.

Automotive Master Technician

Multipoint indirect injection petrol

In multipoint petrol fuel injection, each cylinder has its own fuel injector mounted just before the intake valve(s). The air intake tract only draws air, normally from a **plenum chamber** in the manifold, and fuel is injected at the last moment before it enters into the cylinder. In this way, the metering of fuel and its accuracy can be adequately controlled and reduces the possibility of **intake robbery**.

Petrol is injected at a relatively low pressure of around 3Bar, directly onto the back of the inlet valve(s) during the induction stroke which helps to atomise the fuel and mix it with the incoming air charge.

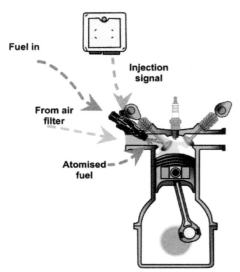

Figure 2.4 Multipoint petrol fuel injection

Three main systems of operation can be used with multipoint indirect petrol injection:

- **Simultaneous** - this is where the injectors are not timed to any particular cylinder and all injectors operate at the same time, controlled by a common driver circuit.
- **Grouped** - this is where the injectors are grouped in combinations (normally pairs) and the grouped injectors are operated at the same time by a common electronic driver circuit for each pair.
- **Sequential** - this is where each individual injector works independently of the others and is timed to operate with the cylinder induction stroke in the same patterns as the engine firing order.

The amount of time that each injector is held open, its pulse width, will be varied in order to provide the correct amount of fuel to achieve an ideal air/fuel ratio of 14.7:1 by mass.

The shape of the engine's combustion chamber in this design of engine will affect its thermal efficiency. An engine that can extract more heat energy from a quantity of fuel will give better performance, be more economical and produce lower emissions. The shape of a multipoint petrol fuel injection combustion chamber will be based around four main designs:

- Bath tub
- Wedge
- Hemispherical (hemi)
- Penthouse roof

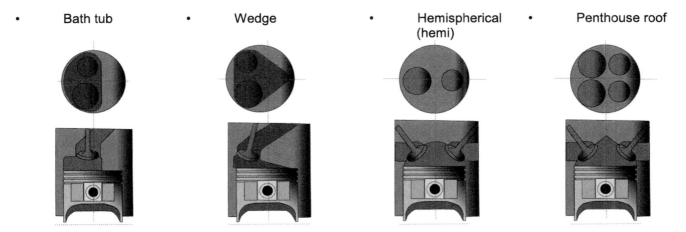

Figure 2.5 Combustion chamber designs

The bath tub and wedge shapes are easy to design and cheap to produce, so overall production costs can be kept down, resulting in a cheaper engine. But their **thermal efficiency** is low so they do not perform as well.

Plenum chamber – an air reservoir normally formed in the intake manifold design.

Intake robbery - where the intake stroke of adjacent cylinders steal air/fuel in the intake manifold from each other due to proximity.

Thermal efficiency – how well an engine extracts heat energy from the fuel. The more efficient the engine is, the more heat energy it will extract from the fuel.

Hemispherical and penthouse roof designs improve performance but are more costly to design and build. Inlet and exhaust valves are normally positioned on either side of the combustion chamber. So with these shapes, two operating camshafts are often required.

The bigger the inlet and exhaust vales, the greater the performance gained from the engine. To make best use of the surface area available in the combustion chamber, many of these designs use multiple valve arrangements. Instead of having a single inlet and a single exhaust valve, it is common to have four valves per cylinder (two inlet and two exhaust valves). In this way a larger opening can be formed for the inlet and exhaust gases, improving **volumetric efficiency**. Also this multi-valve arrangement can give designers the opportunity to add variable valve control, further improving performance, economy and emissions.

Volumetric efficiency – how well the cylinder can be filled with air and fuel.

Direct injection – Petrol

Advances in fuel delivery processes have allowed manufacturers to design engines with direct petrol injection into the combustion chamber. (This is sometimes referred to as gasoline direct injection or GDI).
In this system, petrol is delivered to a common fuel rail from an engine driven high pressure pump at approximately 35 to 200Bar. Instead of being mounted in the intake manifold, as with indirect injection, the fuel injectors are located in the cylinder head with the tips inside the combustion chamber. When the engine is operating, only air is drawn through the intake valves and the petrol is injected in varying quantities directly into the combustion chamber. The direct injection of petrol gives the opportunity to create 'lean burn' technology for economy and low emissions. Unlike multipoint indirect injection, the shape of the combustion chamber and piston have been specifically designed to create turbulence, mixing fuel evenly during **homogenous** injection, or concentrated areas of fuel during **stratified** injection.

Homogenous - a uniform composition throughout (evenly mixed).

Stratified - to form or arrange in layers.

Automotive Master Technician

Modes of operation

Homogeneous operation: When a smooth even delivery of power is required, the air and the fuel are evenly mixed at the ideal air/fuel ratio of 14.7:1 by mass. This is achieved by injecting the fuel during the intake stroke, and the turbulence created by the shape of the combustion chamber and descending piston will thoroughly combine the air and fuel for a standard operation.

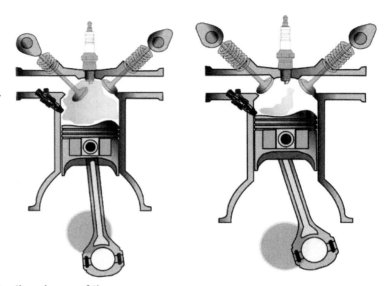

Stratified operation: when outright power is not required, the GDI engine is able to operate in a lean burn mode which helps to lower fuel consumption and reduce some exhaust emissions. This is achieved by injecting only small quantities of petrol, using a reduced pulse width, during the compression stroke when the intake valve has closed. The turbulence created by the shape of the

Figure 2.6 Direct injection petrol

combustion chamber and rising piston, will help direct the injected fuel into a concentrated area around the spark plug at a lean ratio of approximately 40:1.

Lean burn technology can create large quantities of the exhaust pollutant oxides of nitrogen (NOx). To help monitor oxides of nitrogen and regulate injection, GDI systems include a NOx sensor after the catalytic converter in a similar position to a post-cat lambda sensor.

With stratified fuel injection vehicles, damage to NOx sensors can be caused if the wrong fuel grade is used. Check with the customer that they are aware of which fuel grade they should be using.

Indirect injection – Diesel

This form of engine is slowly being phased out as it is old fashioned technology. With an indirect injection Diesel engine a fuel injection pump is used to pressurise Diesel in the system until it reaches approximately 120Bar, a point where it can overcome the force of a spring inside a mechanical injector. When this happens, a needle in the injector lifts off of its seat, spraying (atomising) fuel into the combustion chamber. Instead of Diesel fuel being atomised directly into the combustion chamber, the injector is mounted in a small pre-combustion or swirl chamber. As the fuel starts to burn, it is contained within the pre-combustion chamber and the flame spreads out in a relatively even manner to provide the pressure inside the cylinder to operate the piston on its power stroke.

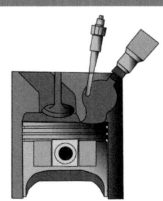

Figure 2.7 Indirect injection Diesel

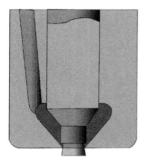

An indirect injection engine will usually use an injector of the 'pintle' type (*see Figure xx*) and will be much quieter in operation than a standard direct injection Diesel engine, but is less thermally efficient. Cold starting with indirect injection will also require the use of glow plugs to pre-heat the air in the swirl chamber.

Advantages of standard indirect injection Diesel engines include:
Small, lightweight engine designs can be produced.
Slightly lower combustion pressures and a more even burn lead to quieter engine operation.
Simple engine designs can be used because the direction mounting of injectors is less important.

Figure 2.8 Pintle fuel injector

These engines are more suited to running on alternative fuel types such as waste vegetable oil (see Chapter 5).

Direct injection – Diesel

The combustion chamber of a direct injection compression ignition engine is much smaller than in a spark ignition engine. This provides the very high pressures required to heat the air, so that when fuel is injected it will spontaneously combust.

The combustion chamber can be manufactured in the cylinder head or in the piston crown. The shape of the combustion chamber will be specifically designed to try and produce a swirling action of the incoming air (*see Figure 2.9*). This swirling action will try and mix the air and fuel, as it is injected, in an even manner. If the air and fuel are not evenly mixed, as the fuel is injected, droplets of Diesel will start to ignite in a number of different positions at the same time. As the fuel burns and the flame spreads, a collision of flame fronts produces a distinctive noise known as 'Diesel knock'.

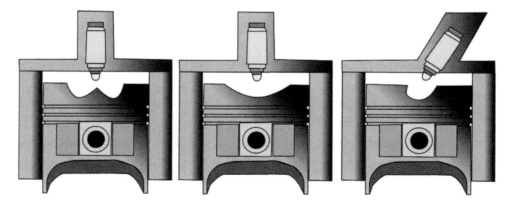

Figure 2.9 Diesel combustion chamber designs

A major disadvantage of standard Diesel fuel injection for both direct and indirect types is that engine speed has an effect on the pressure build-up in the pump, leading to inaccurate fuelling/injection at slow speeds. Some of these inaccuracies can be reduced by the use of electronic Diesel control (EDC).

Automotive Master Technician

Electronic Diesel control (EDC)

Advances in electronic control have allowed standard Diesel engines to be managed more accurately. Fuel injectors have been developed with needle motion sensors and 'drive by wire' systems are used, where a throttle potentiometer controls the injection pump via an ECU, instead of a throttle cable.

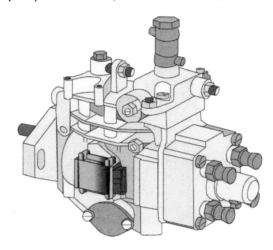

Figure 2.10 Electronic Diesel control

EDC has a number of advantages over a purely mechanical system:

- Better control over the amount of fuel injected.
- Stable idle speed control.
- The easy addition of cruise control.
- Electronic data gathering.
- EGR control.

Common rail direct injection Diesel (CRD)

Modern Diesel engines have moved away from mechanical pumps and injection systems and now operate in a very similar manner to the electronic fuel injection processes in a petrol engine. A low pressure electric pump is often used to transfer fuel from the tank to the engine through a fuel filter. An engine-driven high pressure pump is then used to raise the pressure of the fuel to around 1800 bar, at which point it is stored in a common pressure accumulator or fuel rail that feeds all fuel injectors. As a result, this type of system no longer relies on engine speed to control injection pressures. (An early system with a mechanical pump provided poor pressure at low speeds or tickover).

Due to the extremely high pressures involved in common rail Diesel injection systems, precautions need to be taken when working on these components. If the fuel system is not correctly depressurised before work is started, Diesel can be released at high pressure causing severe injury or death. Always follow manufacturer's instructions and recommendations.

Advanced Light Vehicle Technology

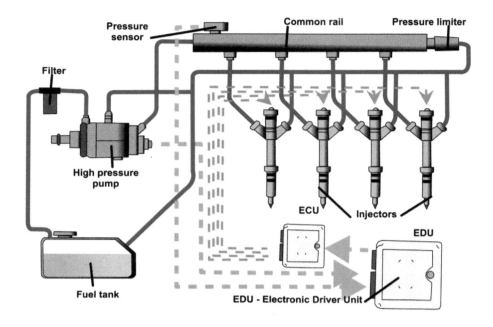

Figure 2.11 Common rail direct injection Diesel CRD

A number of sensors are used to monitor engine requirements, including load, speed, and temperature. This information is then processed by the ECU and used to control electronic solenoid or piezoelectric fuel injectors, which atomise Diesel directly into the superheated compressed air of the combustion chamber.

Fuel supply

The main components making up the fuel supply circuit are:

- Fuel tank.
- Low pressure fuel pump (depending on manufacturer).
- Fuel Filter.
- Engine driven high pressure fuel pump.

- Fuel rail or accumulator.
- Fuel pressure regulator and return system.
- Fuel injectors.

The fuel pressures in a common rail direct injection Diesel are so high that as it is injected the fuel is 'pulverised' reducing the amount of droplets created during atomisation. This helps to reduce the possibility of droplets igniting in different points within the cylinder, and therefore lessens the collision of flame fronts, creating a much quieter operation.

Fuel tank

The fuel tank is the main reservoir used for supply of clean fuel for the common rail injection system.

Low pressure fuel pump

A common rail fuel injection system needs a method of transferring fuel from the tank to the engine. This is the job of the low pressure fuel pump.

Many fuel pumps are of the roller cell type and operate in a similar manner to those found on petrol fuel injection systems.

Fuel filters

Because of the very small tolerances found within a common rail fuel injection system, tiny amounts of dirt would easily block fuel injectors or damage high pressure pumps. As a result a fuel filter is fitted between the low pressure pump and the high pressure pump.

High pressure fuel pump

A common rail fuel injection system needs extremely high fuel pressures in order to correctly atomise/pulverise the fuel from the injectors in the combustion chamber. A mechanically driven pump is operated by the engine which is able to raise system pressures to around 1800 bar, regardless of engine speed.

Fuel rail and pressure regulator

The fuel rail is a common reservoir of Diesel, held at a constant high pressure from which all the injectors are fed. The pressure in the rail is monitored from an electronic sensor and a solenoid valve is able to allow any excess pressure and fuel to return to the tank.

Fuel injectors

There are two main types of fuel injector used with common rail systems:

- Solenoid type
- Piezoelectric

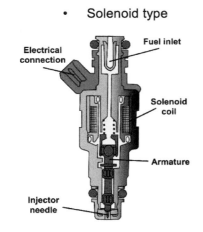

Figure 2.12 Solenoid type injector

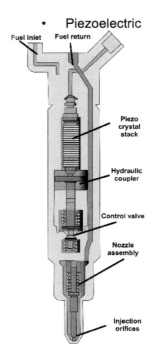

Figure 2.13 Piezoelectric type injector

Advanced Light Vehicle Technology

Solenoid fuel injectors

In a solenoid type fuel injector, a small coil of wire is wound around a movable armature. When current is applied to this coil, a magnetic field is created drawing the armature through the middle of the winding. When current is removed from the coil, the magnetic field collapses, and a strong return spring is used to move the armature back into its original position.

The armature forms a needle valve at one end and fuel pressure is supplied to the needle valve. When the winding of the solenoid is supplied with current and the needle of the armature lifts from its seat, fuel can be sprayed under extremely high pressure into the combustion chamber.

Piezoelectric injectors

To improve the speed and accuracy of fuel injectors, many engine manufacturers now use **piezoelectric** injectors. Instead of using solenoids to open the injectors, a series of piezoelectric crystals are stacked above the needle. When supplied with an electric current from the ECU, the crystals expand in the electric field. As the crystals expand, the injector is opened extremely fast and with accurate control. As the amount of expansion of the crystals in each fuel injector may be slightly different due to its construction, each injector will often have to be calibrated. To do this a method known as injector coding is used. Each injector will have a serial number marked on the body, which must be programmed into the fuel injection systems ECU using a scan tool or manufacturer specific device. The serial number contains information about the function of the individual injector and allows the engine management system operate it accurately.

Piezoelectric - A material that generates an electric charge when mechanically deformed. Conversely, when an external electric field is applied to piezoelectric materials they mechanically deform.

The advanced nature of common rail systems allows Diesel fuel to be introduced by the injectors in distinct steps known as phasing:

- **Pilot injection:** A small amount of fuel is injected prior to the main injection. The purpose of this is to initiate combustion and reduce **the delay period** normally associated with compression ignition. As a small amount of combustion is already underway when the main injection takes place, the flame spread is much more controlled. Detonation is reduced, performance is improved and Diesel knock/noise is lessened.

- **Main injection:** Most of the fuel is atomised in the combustion chamber for power delivery. As the fuel is initially injected, the combustion chamber temperature and pressure falls slightly, leading to a pause before the fuel begins to burn. This is known as the delay period and could be seen on an engine dynamometer as a small dip in the amplitude of pressure rise during the compression stroke. As the fuel ignites, pressure and temperature rapidly rise as the flame spreads out in the combustion chamber, forcing the piston down on its power stroke. As the piston moves down, pressure drops and the combustion process ends with a complete burn.

- **Post injection**: A small amount of fuel is injected into the combustion chamber after the main injection. The purpose of this is to help complete the combustion process and reduce the amount of hydrocarbon emissions. If the emission control system also incorporates a Diesel oxidation catalyst and a catalysed soot filter DPF, the additional fuel injected into the cylinder is used for the purpose of increasing the DPF inlet temperature during the soot regeneration.

Automotive Master Technician

Sensors

The common rail direct injection system uses many of the same sensors found in petrol electronic fuel injection including:

- Mass air flow sensor.
- Throttle/pedal position sensor.
- Intake air temperature sensor.

- Engine coolant temperature sensor.
- Camshaft position sensor.
- Crankshaft position sensor

The monitoring of air mass in a Diesel engine is often used by the engine management system to control the amount of exhaust gas recirculation EGR for performance and emissions purposes.

As with standard Diesel engines, glow plugs are often used to pre-heat the air in the combustion chamber to assist with cold starting. Many plugs are left on for a short period after the engine has started and this is normally referred to as 'post glow'. They run at a reduced current to help prevent them from burning out, and the additional heat supplied helps to lower start-up emissions.

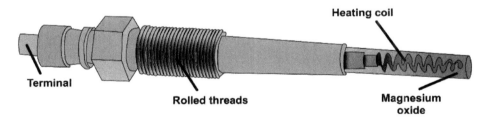

Figure 2.14 Diesel glow plug

Combustion chamber design

As with standard direct injection Diesel engines, the very small combustion chamber may be formed in the cylinder head, the piston crown or a combination of the two. Often the shape of the combustion chamber creates less turbulence than those found on equivalent engines as the extremely high pressures used in common rail direct injection systems produce a very fine atomised spray of Diesel known as pulverisation. This pulverisation, coupled with injector phasing, produces a very smooth combustion with good thermal efficiency, giving improved performance, fuel economy, low emissions and quiet operation.

Advanced Light Vehicle Technology

Fuel system operating characteristics for petrol and Diesel under different driving conditions
Petrol

The speed of a petrol engine is directly related to regulating the amount of air that is allowed to enter the cylinder during the induction stroke. Traditionally, air has been regulated by the use of a throttle butterfly. Some engine designers are now removing the throttle butterfly completely, as this is a very large restriction to the air flow in the intake manifold, reducing overall performance, and instead are using variable intake valve lift to regulate the air on the induction stroke (*see variable valve control*).

During most engine running conditions, an indirect petrol injection engine management system tries to maintain an ideal air/fuel ratio of 14.7:1 by mass. This is achieved by taking measurements from various engine sensors and calculating the precise injector pulse width to allow the correct quantity of fuel to be injected during all running conditions. To ensure that the ideal air/fuel ratio is maintained, lambda sensors in the exhaust monitor the oxygen content of the exhaust gasses and inform the engine management ECU, which is then able to adjust the short term fuel trim to compensate for load, environmental conditions and system wear.

Although stoichiometric values give a balanced chemical reaction for most engine operating conditions, there are times when these values are adjusted.

When the engine is started from cold, a slightly richer mixture is required to provide a smooth running operation. During this period, the engine management system runs open-loop and ignores any readings from the lambda sensor and operates on pre-programmed values from its stored memory.

When the vehicle is cruising at speed under light load, the engine management system is often able to run the fuel injection in a slightly lean mode to reduce fuel consumption and exhaust emissions.

A direct injection petrol engine (GDI) will have two main modes of operation, depending on whether performance or economy/low emissions is required.

Fuel trim - the ability to alter the pulse width of the injector to reduce or increase the amount of fuel injected in relation to the measured air/fuel ratio.

Open loop - an operating state when readings from a sensor are ignored as it has not yet reached its required operating condition.

Closed loop - an operating state where sensor readings are used to influence the operation of an actuator or system.

Diesel

The speed of a compression ignition engine is directly related to regulating the amount of fuel injected during the compression stroke. Traditionally the fuel was regulated by controlling the injection period inside a mechanical high pressure pump. This standard control could lead to many emission issues because, as a Diesel engine draws the same amount of air on every induction stroke, mixing different quantities of fuel with the air gives extremely weak mixtures at low engine speeds, and even during acceleration would find it difficult not to run weak. It is almost impossible to achieve an ideal air/fuel ratio of 14.7:1 by mass.

The introduction of electronic Diesel control (EDC) and common rail Diesel systems (CRD) combined with **closed-loop** feedback EGR, has managed to overcome some of the air/fuel ratio issues. During operation, air flow is accurately monitored for various engine speed and load conditions. A regulated amount of exhaust gas from the EGR system is then introduced into the intake and will occupy some of the space in the combustion chamber, leaving less room for the induced fresh air charge. The amount of fresh air charge can be then closely matched to the quantity of fuel injected and get closer to an ideal air/fuel ratio. Although the exhaust gas does not contribute to the overall combustion process/chemistry it is still a compressible gas which will raise combustion chamber pressures and temperature, allowing the fuel to ignite as it normally would.

Air/fuel ratios of a standard Diesel engine with no closed-loop feedback EGR are approximately:
- 50:1 to 145:1 by mass at idle or under light load
- 17:1 to 29:1 by mass under load

Advanced engine design features

Although the internal combustion engine has been in use for around 150 years, its overall design has changed very little in that time. Most modern engines are still based on a standard reciprocating piston design, with the main advances in engine technology being related to the materials used in construction, the engineering processes and adaptions to the ancillaries which provide better volumetric efficiency.

Advances in engineering have allowed designers to reduce the overall size of many engines. The use of lightweight materials such as aluminium alloys, titanium etc., have reduced weight; although both of these have cost implications. By making the engines generally smaller and lighter, power to weight ratios have greatly improved overall general performance.

Advances in design of ancillaries, such as camshaft, crankshaft and pistons have increased the overall thermal efficiency while cutting pollution through noise and exhaust emissions.

Table 2.3 Current sound emission levels under Noise directive (70/157/EEC)

Vehicle categories	Limit values (dB)
Vehicles used for the carriage of passengers and capable of having not more than nine seats, including the driver's seat.	74 (dB)
Vehicles used for the carriage of passengers having more than nine seats, including the driver's seat, and a maximum authorised mass of more than 3.5 tonnes.	
With an engine power less than 150 kW (ECE)	78 (dB)
With an engine power of 150 kW (ECE) or above	80 (dB)
Vehicles used for the carriage of passengers having more than nine seats, including the driver's seat; vehicles used for the carriage of goods with a maximum authorised mass not exceeding 2 tonnes.	76 (dB)
With a maximum authorised mass greater than 2 tonnes but not exceeding 3.5 tonnes	77 (dB)
Vehicles used for the transport of goods with a maximum authorised mass exceeding 3.5 tonnes.	77 (dB)
With an engine power less than 75 kW (ECE).	
With an engine power of 75 kW (ECE) or above but less than 150 kW (ECE).	78 (dB)
With an engine power of 150 kW (ECE) or above.	80 (dB)

NB: there are exceptions to the regulations listed in the Table above, for more information see Regulation No 51 of the Economic Commission for Europe of the United Nations (UN/ECE) — Uniform provisions concerning the approval of motor vehicles having at least four wheels with regard to their noise emissions.

Advanced Light Vehicle Technology

Notes

On 9 December 2011, the European Commission published a proposal for a Regulation on the sound levels of motor vehicles. If adopted, it will replace the existing Vehicle Noise directive (70/157/EEC) and provide slightly tighter noise emission limits for cars, vans, lorries and buses.

Initially the new standards would only apply to entirely new types of vehicles, so have no effect on current models. The standards would only affect which vehicles can be sold from 2019. The new limits would be 68 decibels for cars, 70 decibels for vans and 78 decibels for lorries. The regulation will also introduce a new noise test method for vehicles.

Exhaust emission standards

Exhaust emission standards are continually being updated and revised, with Euro 4 being the most common standard currently. The stricter Euro 5 regulations are now compulsory for all new cars currently on sale in the UK.

The European emission standards are set out in Table 2.4. They show the limits set in grams per kilometre (g/km).

Table 2.4 European emission standards

Standard	Commencing	Carbon monoxide limits	Total Hydrocarbons	Non-Methane Hydrocarbons	Oxides of Nitrogen	Hydrocarbons plus Oxides of nitrogen	Particulate matter (soot)
Limits for Diesel engine cars							
Euro 1	July 1992	2.72 (3.16 COP)	N/A	N/A	N/A	0.97 (1.13 COP)	0.14 (0.18 COP)
Euro 2	January 1996	1.0	N/A	N/A	N/A	0.7	0.08
Euro 3	January 2000	0.64	N/A	N/A	0.50	0.56	0.05
Euro 4	January 2005	0.50	N/A	N/A	0.25	0.30	0.025
Euro 5	September 2009	0.500	N/A	N/A	0.180	0.230	0.005
Euro 6	September 2014	0.500	N/A	N/A	0.080	0.170	0.0025

Automotive Master Technician

Table 2.4 European emission standards

Standard	Commencing	Carbon monoxide limits	Total Hydrocarbons	Non-Methane Hydrocarbons	Oxides of Nitrogen	Hydrocarbons plus Oxides of nitrogen	Particulate matter (soot)
Limits for petrol engine cars							
Euro 1	July 1992	2.72 (3.16 COP)	N/A	N/A	N/A	0.97 (1.13)	N/A
Euro 2	January 1996	2.2	N/A	N/A	N/A	0.5	N/A
Euro 3	January 2000	2.3	0.20	N/A	0.15	N/A	N/A
Euro 4	January 2005	1.0	0.10	N/A	0.08	N/A	N/A
Euro 5	September 2009	1.000	0.100	0.068	0.060	N/A	0.005
Euro 6	September 2014	1.000	0.100	0.068	0.060	N/A	0.005
COP = Conformity of Production							

Exhaust gas emission analysis

Since the introduction of engine management, the re-tuning of engines has stopped. It was once common that at every major service, an exhaust gas analyser was used to set-up carburettors to ensure that they performed correctly and operated within required limits. As manual adjustments are no longer possible, the exhaust gas analyser is now mainly used for MOT emission testing. This is a powerful diagnostic tool, but the skills required to interpret the results of an emission test are being lost, due to lack of use.

Emissions are tightly controlled by regulations and, as a result, they tend to be regarded as more important than fuel economy and performance. The main emissions produced by a spark ignition engine are:

- Carbon monoxide (CO).
- Hydrocarbons (HC).
- Oxygen (O2).
- Carbon dioxide (CO2).
- Oxides of nitrogen (NOx).

If an engine is accurately controlled, and a **stoichiometric** combustion process is produced, many of these emissions can be reduced considerably. As the air/fuel ratio of a petrol engine approaches the ideal of 14.7 to 1 by mass, carbon monoxide and hydrocarbons fall considerably, but unfortunately carbon dioxide and oxides of nitrogen rise.

If you burn a fossil fuel, one by-product is carbon dioxide. Although carbon dioxide is not considered a poisonous gas, it has an environmental impact as it is believed to contribute to global warming. The better the combustion process, the higher the output of carbon dioxide.

Oxides of nitrogen also rise as the ideal air/fuel ratio is reached. Oxides of nitrogen are produced by the high combustion temperatures present and must be dealt with in another way (for example, by exhaust gas recirculation EGR).

Advanced Light Vehicle Technology

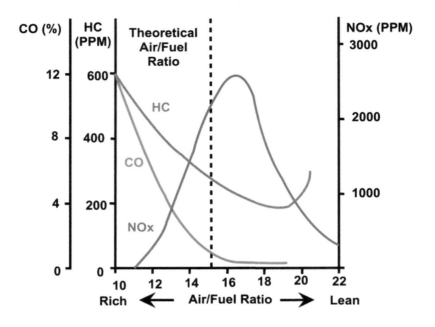

Figure 2.15 Exhaust emissions graph

Stoichiometric – a balanced chemical reaction often used to describe the ideal air/fuel ratio.

Emissions diagnosis

The correct use of an exhaust gas analyser can provide some very useful diagnostic information. Table 2.5 shows some examples of exhaust gas diagnostic readings and their possible faults.

Table 2.5 Exhaust gas analysis

Abbreviation	Exhaust gas (full name)	Approximate acceptable values	Diagnostic information that can be acquired
CO	Carbon monoxide	Less than 0.3%	Too high – running rich (lack of sufficient oxygen for full combustion). Fuel extinguished by quenching zones (colder cylinder walls, valves etc.) poor thermal efficiency.

Automotive Master Technician

Table 2.5 Exhaust gas analysis

CO2	Carbon dioxide	More than 14%	Too low – poor engine efficiency (unless air injection is used to dilute CO2, but this would also show a high oxygen concentration).
HC	Hydrocarbons	Less than 200 PPM (as low as possible)	Too high – misfire, raw fuel left over due to very poor mechanical efficiency (similar to quenching zones).
O2	Oxygen	Between 0.5% and 0.3% approximately	Too high – running weak, induction or exhaust air leak, air injection (coupled with a lower than expected CO2). Too low – running rich (coupled with a high CO).
NOx	Oxides of Nitrogen	As low as possible (can only be tested when car is running and under load)	Excessive – very high engine temperatures, running weak, over advanced ignition timing, faulty EGR or reduction catalyst.
λ	Lambda	Lambda 1.000 – is equivalent to an AFR of 14.7:1 by mass, but as this is calculated value it gives a representative reading before or after the catalytic converter	Too high (above 1.020) – weak running. Too low (below 0.980) – rich running. Incorrect when combined with other emission readings – faulty stoichiometry

Exhaust gas analysers and Lambda

Lambda λ is the eighth letter of the Greek alphabet, and is often used to represent the exhaust gas oxygen sensor of a car. Lambda is also used in physics to indicate the wavelength of any wave, especially in electronics engineering, and mathematics. Anyone who has looked at the output from a standard exhaust gas oxygen sensor will have noticed that it shows a rising and falling waveform.

Tip

When using an exhaust gas analyser, multiply the lambda reading by 14.7 to get the engines operating air/fuel ratio.

Notes

A wideband oxygen sensor, as the name suggests, will also produce a waveform, but due to the nature of its construction and operation, the amplitude curve is often much flatter and its wavelength longer (wider). This gives a much more stable reading than other exhaust oxygen sensors, providing more accurate monitoring of exhaust oxygen content and therefore air/fuel ratio. It is however more sensitive to combustibles in the exhaust stream, or exhaust gas temperature EGT, which is a combination of flame temperature and volume (power).

Advanced Light Vehicle Technology

This use of the term Lambda, when seen on an exhaust gas analyser machine, will sometimes lead to the misinterpretation that it is measuring or doing the same job as an exhaust gas oxygen sensor EGO. This misunderstanding can often lead to the incorrect diagnosis of faulty catalytic converters, or exhaust oxygen sensors. The reading of Lambda on an exhaust gas analyser is actually based on a mathematical calculation developed by Dr Johannes Brettschneider. The Brettschneider equation is often the standard method used to calculate the normalised air/fuel ratio (Lambda) for many pieces of emission diagnostic equipment. It does this by comparing the ratio of oxygen molecules to carbon and hydrogen molecules in the exhaust. The equation is a little complex, but is relatively easily calculated from the measured values of CO, CO2, unburned HC, and unconsumed O2 in the exhaust (in other words, it takes the measured values from a four gas analyser and puts them in an equation to work out Lambda).

Where:

$$\lambda = \frac{[CO_2] + \left[\dfrac{CO}{2}\right] + [O_2] + \left[\dfrac{NO}{2}\right] + \left(\left(\dfrac{H_{cv}}{4} \times \dfrac{3.5}{3.5 + \dfrac{[CO]}{[CO_2]}}\right) - \dfrac{O_{cv}}{2}\right) \times ([CO_2] + [CO])}{\left(1 + \dfrac{H_{cv}}{4} - \dfrac{O_{cv}}{2}\right) \times ([CO_2] + [CO] + (C_{factor} \times [HC]))}$$

Figure 2.16 The Brettschneider equation

[XX]= Gas Concentration in % Volume.
Hcv = Atomic ratio of Hydrogen to Carbon in the fuel.
Ocv = Atomic ratio of Oxygen to Carbon in the fuel.
Cfactor = Number of Carbon atoms in each of the HC molecules being measured.

The equation compares all of the oxygen in the numerator (the bit above the line), and all of the sources of carbon and hydrogen in the denominator (the bit below the line). The result of the Brettschneider equation is the value of 'Lambda' (λ) shown on the exhaust gas analyser, and relates nicely to the stoichiometric value of air to fuel.

- At the stoichiometric point, Lambda = 1.000
- A Lambda value of 1.050 is 5.0% lean
- A Lambda value of 0.950 is 5.0% rich

Once Lambda is calculated, A/F ratio can be easily determined by simply multiplying this result by the stoichiometric A/F ratio for the fuel used - e.g. 14.7:1 for petrol. Because the calculation takes into account all of the values recorded by an exhaust gas analyser, it will provide a very accurate measurement of air fuel ratio regardless of whether the measurement is taken before or after the catalytic converter. This makes it a very useful diagnostic indicator as the catalyst itself has very little effect on the value of Lambda shown on the display.

Automotive Master Technician

It is important to understand the effect that air leaks or secondary air injection may have on lambda calculation. The percentage of extra air in the exhaust gases will result in the same percentage error in the Lambda calculation, i.e. a 5% air leak will not only dilute (lower) the CO, HC, CO2 and NOx gas readings by 5%, but will increase the Oxygen reading by about 1.00% (5% of 21%) and will result in the calculated Lambda being 5% leaner than it should. That means that a perfect Lambda of 1.000 will be reported as 1.050 if there is 5% air leak or air injection. This significant error can occur relatively easily, so any air leaks should be dealt with and corrected before any lambda calculations using measured gases are attempted. Also secondary air injection should be disabled for testing.

General rules for exhaust emissions analysis

If carbon monoxide goes up, oxygen goes down, and conversely if oxygen goes up then carbon monoxide goes down. - (carbon monoxide is an indicator of a rich running engine and oxygen is an indication of a lean running engine).

If hydrocarbons increase due to a lean running misfire, the amount of oxygen in the exhaust gas will also increase. Carbon dioxide will fall for either of the above due to an air/fuel imbalance or misfire.

An increase in carbon monoxide doesn't necessarily mean there will be an increase in hydrocarbons. (HC will only be created when the air/fuel ratio begins to reach a rich misfire, i.e. 3% to 4% CO).

High hydrocarbons, oxygen and low carbon monoxide at the same time indicates a misfire due to lean mixture or an EGR dilution.

High hydrocarbons, with normal to marginally low carbon monoxide and high oxygen, indicates a misfire due to ignition or engine mechanical fault.

Normal to marginally high hydrocarbons, with normal to marginally low carbon monoxide and high oxygen indicates a misfire due to lean mixture or false air.

Table 2.6 explains some common terms associated with fuelling, combustion and exhaust emissions.

Table 2.6 Fuelling, combustion and exhaust emission terms

Term	Meaning
Flame travel	This is the way that the flame spreads in the burning air/fuel mixture within the combustion chamber. To provide good performance and efficient combustion, the engine is designed to ignite the fuel at a particular position inside the combustion chamber. This is normally around the tip of a spark plug or fuel injector. The flame created by the burning mixture should then spread out evenly away from this point, giving a smooth but rapid pressure rise within the cylinder. This pressure will react correctly against the piston crown and obtain maximum effort to turn the crankshaft.
Pre-ignition	Pre-ignition normally occurs on petrol engine vehicles. This is a condition where the air/fuel mixture ignites before the operation of the spark plug. Pre-ignition is normally a result of a combustion chamber deposit (carbon build-up) which is burning/glowing red hot and will act as an alternative source of ignition.

Advanced Light Vehicle Technology

Table 2.6 Fuelling, combustion and exhaust emission terms

Detonation	Detonation is a condition that exists when the fuel in the combustion chamber spontaneously ignites due to a rapid rise in heat, normally created by compression pressures. This can occur in both petrol and Diesel engines. Fuel droplets begin to burn in a number of different places at once, and the collision of flame fronts as they spread out, creates a distinctive knocking noise which is particularly apparent in Diesel engines.
Pinking	Pinking is a condition that occurs when the ignition of the air/fuel mixture happens too early in the combustion cycle. The most common reason for this is over-advanced ignition timing on petrol engines or over advanced injection on Diesel engines. This results in the pressure build-up occurring before the piston and the connecting rod have reached the required position after top dead centre. If the piston is at top dead centre and the connecting rod is vertical, turning effort to the crankshaft is wasted and performance is lost. The condition can sometimes be identified by a light pinging or metallic tapping noise from the engine when accelerating or putting the engine under load.
Octane rating	The octane rating of a fuel refers to petrol. It is a measurement of petrol's resistance to detonation i.e. how stable the fuel is under compression pressures. The higher the octane rating, the more stable the fuel is and therefore the more suitable it is for high performance or high compression engines. Fuel is classified with a research octane number (RON) which helps to identify its properties.
Cetane rating	The cetane rating of a fuel refers to Diesel. It is a measurement of Diesel's delay period i.e. how quickly it will ignite and burn after injection. The higher the cetane rating, the shorter the delay period and the faster the ignition of the fuel. A Diesel fuel with a high cetane rating is desirable for good combustion and performance, as it gives a smooth and even flame spread within the combustion chamber.
Volatility	Volatility refers to a fuel's ability to vaporise at certain temperatures. Petrol is a volatile fuel which turns into a vapour readily at room temperature. Diesel is less volatile and needs its temperature to be raised before it will vaporise. The volatility of a fuel is important as it is the vapour of a fuel that burns, not the liquid.
Calorific value	The calorific value of a fuel is a measurement of the heating power it contains. It refers to the amount of energy released when the fuel is burned under specified conditions. The amount of energy stored in a fuel depends on its composition and, as a result, different fuels have different values.
Flash point	Flash point is the lowest temperature at which a fuel will vaporise to form an ignitable mixture in air. Because of the volatility of petrol, it has a lower flash point than Diesel.
Hydrocarbon content	Whether a fuel is petrol or Diesel depends on how much hydrogen and carbon there are in the fuel. This is based on the number of carbon molecules in the hydrocarbon chain. The amounts can differ slightly depending on how the fuel is refined, but as a general rule: When the chain has between five and nine carbons, the hydrocarbon is petrol. When the chain has about twelve carbons, the hydrocarbon is Diesel. Depending on how Diesel and petrol are refined, approximate percentages of carbon and hydrogen are 84% carbon and 14% hydrogen (plus 2% other elements).
Composition of air	During the induction process, air is blended with fuel to form an ignitable mixture – this is because oxygen is required for combustion. At sea level, air is composed of approximately 21% oxygen and 78% nitrogen (plus 1% other gases).
Air/fuel ratio	Air/fuel ratio refers to the quantities of air and fuel in the mixture created for combustion. An ideal air/fuel ratio for correct combustion is 14.7:1 by mass (weight). This means that the amount of air in the combustion mixture will weigh 14.7 times more than the amount of fuel.

Table 2.6 Fuelling, combustion and exhaust emission terms

Stoichiometric ratio	Stoichiometric ratio refers to a balanced chemical reaction where all the component elements are used up during a reaction process. This is often used to describe the correct air/fuel ratio of 14.7:1 for ideal combustion.
Lambda window	The term lambda window is sometimes used to describe when exhaust emissions fall within a desired tolerance for correct engine operation. Engine management is designed to use various sensors to monitor and control engine functions and maintain this lambda window.
Carbon monoxide	If during the combustion process, the burning fuel goes out, perhaps due to lack of oxygen or rapid cooling, carbon monoxide (CO) is produced. Carbon monoxide is a product of incomplete combustion. Carbon monoxide is harmful to health. It is colourless, odourless and tasteless, but if inhaled it is poisonous. It replaces oxygen in the blood and starves the organs of required oxygen. Measurements of carbon monoxide in exhaust gases can give an indication of air/fuel ratios. Modern engines operate with approximately 3% CO. A Figure higher than this indicates a rich mixture. A Figure lower than this may indicate a weak mixture. (To get an accurate measurement of CO, you need to take a sample before the catalytic converter).
Carbon dioxide	With effective combustion, the chemical elements carbon and oxygen combine to form carbon dioxide (CO_2). This should normally be higher than 14% by volume. Carbon dioxide can be used as an effective diagnostic gas and can give a good indication of engine mechanical operation and condition. If its percentage by volume is low, you should assess the engine condition and function. Carbon dioxide is a greenhouse gas and is considered an environmental pollutant.
Hydrocarbons	If fuel passes through the combustion process with no chemical change, hydrocarbons (HC) are given off. These should be kept as low as possible, usually under 200 parts per million (PPM) at idle. High values of hydrocarbons normally give an indication of an engine misfire. Hydrocarbons are considered harmful to health and may cause lung damage or cancer.
Oxides of nitrogen	A large proportion of the combustion process takes place at temperatures between 2000 and 2500°C. In this extreme heat, the oxygen and nitrogen in the incoming air are combined to produce oxides of nitrogen (NOx), which is a pollutant. Unfortunately, the better the combustion process, the more oxides of nitrogen are produced. Because of this, specific methods of control are required, such as exhaust gas recirculation (EGR) and reduction catalysts. Oxides of nitrogen can cause health issues for the lungs and may also damage plant life and reduce visibility. Most four gas exhaust analysers are unable to measure oxides of nitrogen.
Oxygen	The oxygen found in exhaust gas is not considered harmful to health or an environmental pollutant. If the stoichiometric ratio is correct, nearly all of the available oxygen in the inducted air (21%) will be used up during the combustion process, leaving around 0.5%. High levels of oxygen in the exhaust gases can indicate an air leak in either the induction or exhaust system. Low levels of oxygen in the exhaust gases may indicate an over-rich mixture caused by excessive fuelling.
Lambda	Lambda is often used to represent the ideal air/fuel ratio of 14.7:1 by mass. Most exhaust gas analysers are unable to measure true lambda, but instead use a mathematical calculation created from the measured gases (CO, CO_2, HC and O_2) to work out the air/fuel ratio. Any discrepancy in one of the measured gases can lead to an incorrect value of lambda being displayed on the analyser. If lambda values are incorrect, your diagnostic routine should focus on any of the four measured gases, which may be out of tolerance.

Advanced Light Vehicle Technology

Bore diameter, stroke length and con-rod to crank ratio

When designing an engine a number of factors should be considered, this will vary from manufacturer to manufacturer, and the intended use/vehicle type.
The internal capacity of an engine is determined by its **bore** and **stroke** and the way it performs will be a combination of these two factors.
In general:
The larger the bore diameter when compared to the length of the stroke, the higher revving the engine will be and the more power it can produce.
The longer the stroke when compared to the bore diameter, the lower revving the engine will be, but it will tend to produce more torque.

There is also a relationship between the **throw** of the crankshaft and the length of the connecting rod. This is known as the con-rod to crankshaft ratio and is calculated by dividing the length of the con-rod (from the centre of the small end to the centre of the big end) by the throw of the crankshaft. These ratio values will normally fall between approximately 1.7:1 and 2.3:1 and in general:
The lower the ratio (short con-rod), the lower down the peak rev band occurs.
The higher the ratio (long con-rod), the higher up the peak rev band occurs.

Low con-rod to crankshaft ratios can produce high pressures on the piston and cylinder walls, increasing wear on the thrust side of the cylinder and can sometimes lead to premature failure of the piston and rings.

Bore - the area of an engine's cylinder.

Stroke- the distance travelled by the piston from bottom dead centre BDC to top dead centre TDC.

Throw- the length of the offset of the crankshaft from the centre of its main bearing journal to the centre of the crank pin journal (the throw will determine the length of the stroke and therefore have a direct effect on the amount of torque produced).

An Atkinson cycle engine is able to vary the length of its stroke depending on its position in the four stroke cycle. The piston and connecting rod are linked to the crankshaft using a pivoting intermediate arm. As the crankshaft rotates through one complete revolution, the intermediate arm will provide a different angle of connection to the big end of the con rod during 180 degrees of rotation. This can give the engine a different length induction and compression stroke from its power and exhaust stroke. By reducing the length of the induction and compression stroke and maintaining the length of the power and exhaust stoke, the engine is able to make effective use of the thermal energy contained in the air/fuel mixture at the expense of power. This engine type produces very good fuel economy but would not be suitable for use in driving situations demanding a high power output.

Automotive Master Technician

Compression ratios

The compression ratio of an engine is the difference between how much air is drawn into the engine, and the space into which it is squashed. This is illustrated in Figure 2.17.

To calculate the compression ratio of an engine, you need to find how many times the clearance volume will fit into the swept volume and add 1. Therefore, swept volume divided by clearance volume plus one equals compression ratio. You can use this formula:

$$\text{Compression ratio} = \frac{\text{Swept volume}}{\text{Clearance volume}} + 1$$

For example:

To calculate the compression ratio of an engine with a cylinder swept volume of 234cc and a clearance volume of 26cc:

$234 \div 26 = 9$

$9 + 1 = 10$

This gives compression ratio of 10:1

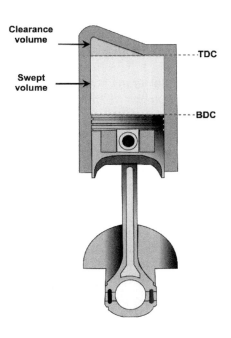

Figure 2.17 Compression ratio

The higher the compression ratio, the more performance or power is gained from the engine.

In a petrol engine, as the compression ratio increases so does the performance, but there is a point at which the compression ratio will make the air/fuel mixture become unstable and possibly auto-ignite (self-combust).

If the fuel ignites by itself, instead of waiting for the spark plug to begin the combustion process, it is acting like Diesel. This unstable ignition can lead to a misfire and engine damage.

Typical compression ratio values are:

- Petrol engines 8:1 to 12:1
- Diesel engines 16:1 to 20:1

Number and arrangement of cylinders (including rotary engines)

When it comes to the arrangement and number of cylinders vehicle manufacturers use in the design of their engines, they must consider how the vehicle will be used and any cost or space restrictions. Another consideration will be how any dynamic vibrations caused by the reciprocating pistons can be reduced. When a piston moves up and down or side to side it will be subject to **inertia** forces. During operation, the piston and con-rod assembly must be accelerated in one direction, and as it reaches the end of its stroke, the mass must be slowed down and stopped and then accelerated in the opposite direction. The piston and connecting rod actually moves further in the upper 90° of its stroke than it does in the bottom 90°, so most of the vibrations occur at top dead centre TDC and when the piston reaches the mid-point of its stroke. This can be explained by using the formula shown in the next example:

Advanced Light Vehicle Technology

For example:

The following formula is for a 2.5- litre, four-cylinder engine:

- Total length = rod + throw (stroke/2)
- 209mm = 157mm + 52mm
- Piston travel during the first 90 degrees of crankshaft rotation:
- 209mm − 157mm² + 52mm² = 209mm − 21945mm = 148mm
- Total − difference = piston travel to mid-point
- 209mm − 148mm = **61mm (travel for first 90 degrees of crankshaft rotation)**
- Bottom half of travel = distance left − rod length + throw
- 148mm − 157mm + 52mm = **43mm (travel for last 90 degrees of crankshaft rotation)**

Smaller engines may use a balance shaft to counteract these vibrations. A shaft with balance weights is timed with the crankshaft and rotates at twice engine speed in the opposite direction. This way it will act as a counterbalance twice during the stroke of the piston and con-rod so that it can react not only at top and bottom dead centre, but also in the middle of its stroke.

Inertia - the tendency for a body to resist acceleration.

Another method that is employed by engine manufacturers to counteract engine vibrations, and also to fit design constrictions of size and weight, is to use multiple cylinders and arrange them in different layouts/configurations. Some examples of common configurations are described below:

- **In-line:** An in-line engine is one where the pistons are arranged in a line next to each other. This is probably the most common design, but it can take up a great deal of space because the length of the cylinder block needed to hold the pistons.

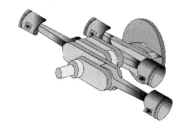

- **Flat:** A flat engine is also sometimes known as a 'boxer' engine or 'horizontally opposed' engine. In this type of engine, the pistons are laid out flat on each side of the crankshaft. This means that if you have a four-cylinder horizontally opposed engine, there would be two pistons on each side of crankshaft. A flat engine provides a low centre of gravity, and the crankshaft can be kept relatively short because there are only two pistons side by side. This makes the engine compact.

- **Vee:** A vee engine has its cylinders laid out in the shape of a letter V. In a similar way to the flat engine, the crankshaft can be made shorter and more compact. When the firing impulses on each bank occur, they help to balance each other out, providing a smooth delivery of power.

- **W:** A W engine has its cylinders laid out in the shape of a letter W and is a variation on the theme of a vee engine design but with three banks of cylinders. When the firing impulses on each bank occur, they help to balance each other out, providing a smooth delivery of power and a compact engine design for the number of cylinders that can be housed within the block.

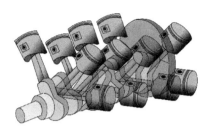

Single and multi-cylinder engines

In a standard four-stroke engine, there is only one combustion operation for every two revolutions of the crankshaft (720°). The large gap between **firing impulses** can cause vibrations, which makes engine operation harsh and noisy. To overcome this problem manufacturers use multi-cylinder engines with pistons operating at different phases of the four-stroke cycle. This way, as one piston finishes its power stroke, another is just beginning. This produces a much smoother delivery of power and reduces the need for very large **flywheels**.

Firing impulse – another term used to describe an engine power stroke.

Flywheel – a heavy metal disc bolted to the crankshaft which stores kinetic (movement) energy.

- If a two-cylinder engine is used, there is a firing impulse every 360°.
- If a four-cylinder engine is used, there is a firing impulse every 180°.
- If a six-cylinder engine is used, there is a firing impulse every 120°.
- If an eight-cylinder engine is used, there is a firing impulse every 90°.

Notes

Some advanced engine designs are beginning to offset crank pins at different angles and mix firing orders to enable a different delivery of power. A cross-plane crankshaft for example is sometimes used in an inline engine design, but instead of the crank pins being set 180° apart, they are manufactured at 90° intervals. The crankshaft and cylinder layout design, is quite often unique to the individual manufacturer.

Tip

If engine mechanical failure seems premature due to low mileage, check for tow bars etc. that may have caused excessive engine stress and wear.

The operating cycle of a rotary engines

Rotary engines are a form of spark ignition (SI) petrol engine that do not use a conventional piston.

The type of rotary engine used in many modern cars is based on a design made by a German mechanical engineer, Felix Wankel, in the 1950s.

Advanced Light Vehicle Technology

A rotary engine is different from a standard reciprocating (up and down) engine because it does not have a normal piston. Instead of a piston, it has a three-sided rotor sitting inside a squashed oval-shaped cylinder called an 'epitrochoid'.

Because the rotor has three sides, it is effectively three pistons in one. This means that a different **phase** of the Otto cycle (induction, compression, power or exhaust) will be happening on each side of the rotor at the same time.

Unlike a standard four-stroke engine, the rotary engine doesn't have valves. Instead, it has ports in the cylinder wall that are opened and closed as the rotor turns.

> **Rotor** – a triangular component used instead of a piston in the rotary engine.
>
> **Phase** – a term used instead of 'stroke' when describing the operation of a rotary engine.

The operation on one side of the rotor is as follows:

Induction

As the rotor turns, the tip uncovers the inlet port. An expanding chamber is produced which creates the low pressure (or depression). Atmospheric pressure or a turbo charger forces air/fuel mixture through the open inlet port and tries to fill the cylinder. As the rotor reaches the end of the induction phase, the tip covers the inlet port, sealing the combustion chamber.

Compression

As the rotor continues to turn, the sealed chamber starts to reduce in size, compressing the air/fuel mixture into a smaller and smaller space, which raises its pressure. This rise in pressure performs two functions:

- Firstly, as the pressure rises, the temperature also increases. This rise in temperature helps to vaporise the fuel. This is important because it is the fumes or vapour of the petrol that actually burn, not the liquid.
- Secondly, when a fuel is rapidly burnt in a confined space, it releases its energy in a much more powerful manner.

Power

At the point of highest compression, a spark plug ignites the air/fuel mixture. Because the gases are sealed in a rotor chamber, a rapid pressure rise occurs, forcing the rotor round on a power phase.

The energy from the power phase is transferred through the rotor to the crankshaft, which is turned.

The rotation of the crankshaft on the power phase is the only active operation within the Otto cycle, but as the rotor has three sides, it is quickly followed by another power phase.

This means that a smaller, lighter flywheel can be used than in a standard reciprocating engine. The crank produces a lower amount of torque than a standard piston engine, but can give a high power output.

Exhaust

At the end of the power phase, the rotor tip uncovers the exhaust port. The chamber continues to decrease in size, forcing the burnt exhaust gases out through the manifold and exhaust system. At this point the process repeats itself.

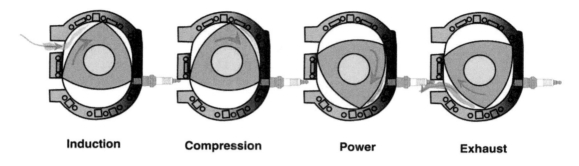

Induction   Compression   Power   Exhaust

Figure 2.18 Rotary engine operating cycle

Crankshaft rotation in the rotary engine cycle

Because the rotary engine has a rotor with three sides, there is a power phase every 120° of revolution. Many rotary engines combine two rotor chambers and will provide a power phase every 60°.

Intake and exhaust

Modern engines are designed to meet very strict manufacturing and performance guidelines. In many cases engine capacity has reduced in order to improve fuel economy and lower exhaust emissions. Designers and manufacturers are continually improving their systems in order to keep up with technology and legislation.

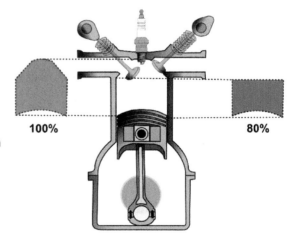

Figure 2.19 Volumetric efficiency

The performance of an engine is affected by many factors, but one of the most important is **volumetric efficiency**. Volumetric efficiency is how well the engine cylinder can be filled with a fresh charge of incoming air and fuel in the correct quantities. A **naturally aspirated** engine relies on atmospheric pressure to force air into the cylinder above the descending piston. As the piston descends, a low pressure called a **depression** is created above the piston crown, and air is pushed into the cylinder by atmospheric pressure through the open inlet valve(s). A naturally aspirated design can have cost and engineering advantages over a comparable pressure charged engine.

Due to the speed of operation, it is very difficult to fill the cylinder completely and, as a result, only about 80% of the available space will be filled. As engine speed increases, this Figure often falls and volumetric efficiency is reduced, which affects overall engine performance. Engine manufacturers and designers have come up with a number of methods to help improve volumetric efficiency.

Advanced Light Vehicle Technology

These include:

- Advanced intake and exhaust designs.
- Variable valve timing and lift.
- Forced air induction.

Volumetric efficiency – how well an engine cylinder can be filled with air and fuel.

Naturally aspirated – not turbocharged or supercharged.

Depression – a low or negative pressure.

Intake robbery - an undesirable condition where neighbouring cylinders steal air and sometimes fuel during induction.

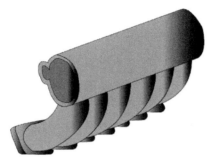

Figure 2.20 Plenum chamber

It is vitally important in a naturally aspirated engine that any restriction to incoming air and outgoing exhaust gas is kept to a minimum.

The length of the air inlet tract (from intake near the air filter, to the inlet valve) will change from manufacturer to manufacturer. The ideal length of the air inlet tract depends on engine load and speed demands. This means that performance will be affected if the inlet is too long or too short. To help overcome this, some manufacturers use a plenum chamber in the design of their intake manifolds. This is a chamber that can be used as a reservoir of air. The cylinders can draw from this, which means that the intake air does not have to travel the entire length of the intake tract. In addition to a plenum chamber, some engine designers use a set of butterfly valves in the manifold which help control air flow through the plenum chamber. At low revs, the butterfly valves remain closed, and the air has to take a long path through the plenum chamber, but as engine revs rise the butterfly valves are opened and the air can now take a short-cut. This design gives a variable length intake and can improve volumetric efficiency in two distinct rev bands and also help to reduce **intake robbery**. This design is known as a pulse tuned intake.

Exhaust systems should also allow the flow of exhaust gas from an engine with the least amount of restriction or back-pressure. The exhaust manifold and down pipe can be designed to assist the removal of exhaust gasses from the combustion chamber during the scavenging process. The shape and connection of the manifold tubes can be joined in such a way, that the exhaust gas leaving one cylinder, creates a depression which helps draw the exhaust gasses from a companion cylinder. This process is similar to intake robbery but in reverse.

Variable valve control

Another method of reducing the restriction caused by the inlet and exhaust valves is variable valve control. Variable valve control is able to produce a number of different operating conditions, including:

- **Camshaft phasing:** The operation of the camshaft is **advanced** or **retarded** in order to open the valves earlier or close them later. By altering the operation of valve timing during different engine running conditions, enhanced performance is achieved.

- **Variable valve lift:** The amount that the valve is opened is changed during different engine running conditions. If valves are only opened a small amount during slow speed operation, this achieves smooth running, fuel economy and low emissions. If the valves are opened a large amount during high speed operation, this achieves enhanced performance. Some manufacturers are now using complex systems of variable valve lift, such as valvetronic and multi-air to fully control intake, meaning that the throttle butterfly is no longer required.

- **Valve operation speed**: The speed of the camshaft is varied during different engine running conditions. If the camshaft is sped up during engine slow speed operation, the valve is only open for a short period of time which gives smooth running, fuel economy and lower emissions. If the camshaft is slowed down during engine high speed operation, the valves are held open for longer, allowing more air and fuel to enter the engine which produces enhanced performance.

Advanced – ahead of time (early).

Retarded – lagging behind (late).

Variable valve control mechanisms

Different manufacturers use distinct types of valve drive mechanisms on their engine designs. Some examples of variable valve control are described in the next section:

VVT-I

The VVT-I system is a method used to manage the phasing of the inlet camshaft(s). A controller mechanism is mounted on one end of the camshaft. This controller connects the camshaft to the timing chain. The VVT-I mechanism allows a small amount of rotational movement between the camshaft and the timing chain drive gear, which means that the timing and operation of the valves can be varied according to engine speed and load.

The VVT-I mechanism is held in a standard timing position by spring pressure. Engine oil is then directed to chambers inside the VVT-I unit. This provides a hydraulic pressure, which rotates the camshaft slightly in relation to the timing chain drive gear. Depending on which pressure chamber engine oil is directed to, camshaft timing can be advanced or retarded. If oil pressure is lost, or when the engine is first started, the spring-loaded mechanism inside the VVT-I unit returns the camshaft to a standard timing position.

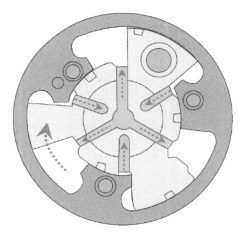

Figure 2.21 VVTI controller

VANOS

The VANOS system is a method used to control the phasing of the inlet camshaft(s). A VANOS mechanism is mounted at one end of the camshaft. The controller section connects the camshaft to the timing chain.

The VANOS controller uses a small **intermediate** gear with a spiral **helix** cut into one surface and **splines** cut into the other to join the camshaft with the timing chain drive mechanism. The intermediate gear is able to slide along the camshaft on the splines. As it does this, the spiral helix acts against the timing chain drive mechanism, causing the camshaft to rotate slightly. The small amount of rotational movement between the camshaft and the timing chain means that the valves can be advanced or retarded according to engine speed and load.

Advanced Light Vehicle Technology

Hydraulic oil pressure from the engine is directed to one side or the other of the VANOS controller unit, which moves the intermediate gear backwards or forwards along the camshaft splines to alter the valve timing when required.

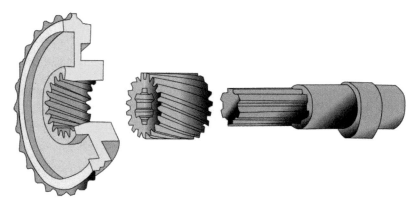

Figure 2.22 VANOS controller

Intermediate – in-between.

Helix – shaped like the spiral of a coil spring.

Splines – grooves machined along the length of a shaft.

Profile – the outline of a shape.

Performance VTEC

The Performance VTEC system is a method used to control how far the inlet valves open. Different cam **profiles** are machined on the same camshaft, and rocker arms are used to transfer the movement from the different cam profiles to the inlet valves when required.

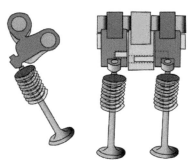

A shallow cam profile will only open the inlet valve a short distance, whereas an aggressive cam profile will open the valve fully. At low engine speed and load, a shallow cam profile is used to provide smooth running, fuel economy and lower emissions. At high engine speed and load, an aggressive cam profile can be used to provide performance.

To switch between the different cam profiles, hydraulic oil pressure is used to move a locking pin between two rocker arms. At slow engine speed, the shallow cam profile operates the low lift rocker arm to open the valve(s); the high lift rocker arm moves freely against a return spring (idles) with no effect on engine operation.

Figure 2.23 VTEC controller

Automotive Master Technician

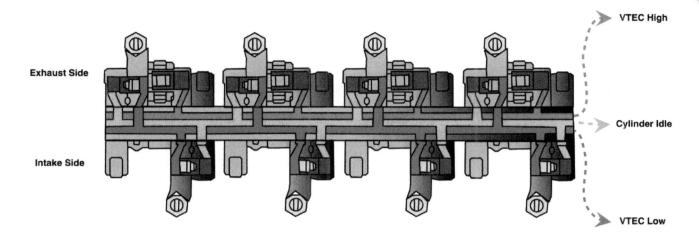

Figure 2.24 VTEC rocker system

As engine speed increases, hydraulic oil pressure locks the high lift rocker arm to the low lift rocker arm. The aggressive cam profile now takes over, opening the inlet valve fully. As engine speed falls, hydraulic oil pressure is directed away from the locking pin and a return spring is used to unlock the two rocker arms and the low lift cam takes over. Due to the nature and operation of the VTEC system, the change in performance can often be felt by the driver as they accelerate.

An economy VTEC system is also available which works to stall the valve operation of both the inlet and exhaust valves, meaning that they can be independently controlled to remain closed. This is particularly important when incorporating this design into a hybrid drive vehicle, as it can help reduce engine pumping losses during overrun and improve regenerative braking. (*See Chapter 5*)

VVC

The VVC system is a method used to control the speed of operation of the inlet camshaft. If the speed of the camshaft can be varied according to engine performance requirements, the inlet valve can be held open for longer or shorter periods of time.

At slow engine speed and load, the inlet valve can be held open for a short period of time, giving smooth running, fuel economy and lower emissions. At high engine speed and load, the inlet valve can be held open for a longer period of time, allowing more air and fuel to enter the engine and improve performance.

In order to keep the valve timing correct, the camshaft must still rotate at half crankshaft speed. This means the speeding up and slowing down of the inlet camshaft must be completed in the same period of time that it would normally take to make one rotation, and independently of the crankshaft. Because of this, the variation of speed is controlled so that during half a rotation of the camshaft, it is moving fast and during the other half the rotation of the camshaft it is moving slowly (the camshaft still completes one revolution in the same amount of time as it would in a normal system).

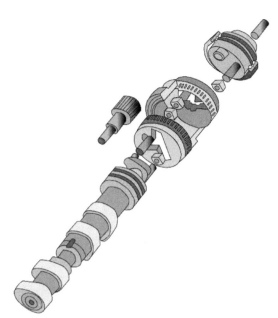

Figure 2.25 VVC controller

The drive from the timing belt to the camshaft is controlled by an **eccentric** drive pin. The drive pin is able to move in a slot to vary the amount of eccentricity created between the timing belt and the camshaft. This variation of eccentric movement is able to speed up and slow down the operation of the inlet camshaft depending on engine performance requirements.

Because the operation of the eccentric drive would affect all cylinders at the same time, a method is needed of controlling each set of inlet valves separately. In a four-cylinder engine, four separate inlet camshafts are used. Two timing belts (one at the front of the engine and one at the back of the engine) each drive a pair of separate camshafts. Each pair of camshafts has one solid camshaft and one hollow camshaft, with the solid camshaft being driven through the middle of the hollow camshaft. An ECU determines engine speed and load and directs hydraulic oil pressure to the eccentric drive pin mechanism.

In this way, as engine speed and performance requirements increase, the amount of eccentricity on the drive can be controlled so that, as the cam lobe rotates and opens the valve, it is slowed down (so the valve is held open for longer) and, as the cam lobe rotates away from the valve, it is speeded up and still completes one revolution in the same amount of time it normally would.

Eccentric – off centre, not in the middle.

Variable valve control can be compromised by poor oil pressure/quality. Before attempting a mechanical strip down and overhaul you should always check oil level, pressure and condition.

As the engine wears it is common for oil to leak past the valve stem oil seals and get burnt in the combustion chamber. This will often lead to blue smoke emitting from the exhaust system. Oil that is leaking past the valve stem oil seals will be most apparent on start-up or if the vehicle is left at idle for a period. (Blue smoke during acceleration is often due to worn piston rings or turbocharger seals).
To help diagnose oil smoke issues, the vehicle should be road tested under various driving conditions, and the amount of smoke and when it occurs needs to be assessed.

Automotive Master Technician

Valvetronic

An advancement in the use of variable valve control is valvetronic. In this system, the vehicle manufacturer has removed the need for the throttle butterfly valve used to control the amount of air entering the engine. Instead, an electronic system which controls the amount that the inlet valves open is used to regulate the amount of air entering the engine.

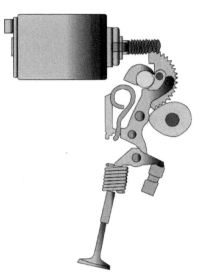

An electric motor mounted near the inlet camshaft drives an eccentric gear mechanism. The eccentric gear mechanism is connected to an intermediate shaft which can be moved towards or away from the inlet camshaft. The intermediate shaft is fitted with a roller which follows the opening face of the camshaft profile. The lower edge of the intermediate shaft operates the rocker mechanism connected to the inlet valves.

When the engine is running at slow speeds, the motor and eccentric gear move the intermediate shaft away from the camshaft so that only a small amount of movement is transferred to the rocker and inlet valve mechanism. To increase the engine speed, the driver presses the accelerator pedal and a voltage from the throttle position sensor sends a signal to the ECU which controls the electric motor. The motor turns the eccentric gear, which moves the intermediate shaft towards the camshaft – a larger amount of movement is transferred to the rocker, opening the valve further and allowing more air to enter the engine which increases its speed.

Figure 2.26 Valvetronic controller

Multiair

Multiair is an electro-hydraulic variable valve lift technology, controlling the intake air without a throttle butterfly in a similar manner to a valvetronic system. Instead of using an eccentric gear mechanism to vary the lift of the intake valve, the cam profile operates the valve(s) via a hydraulic chamber (tappet) that can be pressurised with engine oil in varying quantities. When controlled by an ECU actuated solenoid valve, the hydraulic valve tappet chamber can open the inlet valve with varying degrees of lift, manipulating the amount of air entering the engine and therefore regulating the engine speed. This gives accurate dynamic control for various different load conditions.

Figure 2.27 Multiair controller

Notes

Valvetronic and Multiair are compatible with both naturally aspirated and pressure charged engine designs.

Advanced Light Vehicle Technology

Other methods of valve control

Three-dimensional camshaft profiles: The cam lobe is tapered so that if the camshaft is moved **longitudinally** a different profile is able to operate on the valve mechanism controlling lift.

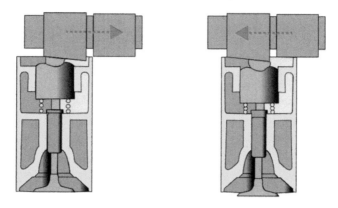

Figure 2.28 Three-dimensional camshaft profile

Longitudinally - running lengthwise rather than across.

Composite camshafts: Separate cam lobes are mounted on a driveshaft that allows slight rotational or axial movement.

* When actuated, rotational cam lobes are able to advance or retard the valve timing.
* When actuated, axial cam lobes are able to swap between different profiles and vary the amount of valve lift.

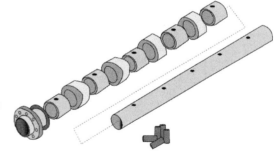

Figure 2.29 Composite camshaft

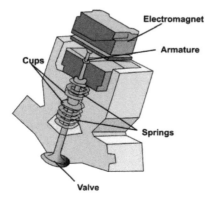

Solenoid operated valves: Electronically controlled solenoid motors are used to open the inlet and exhaust valves instead of a camshaft. Using solenoid motors gives engine designers complete control over both timing and lift for any engine operating condition – adaptions can be made to compensate for most driving situations. As this system has no camshaft or valve train to drive, loads placed on the engine are reduced, which leads to improved performance, greater fuel economy and lower emissions.

Figure 2.30 Solenoid operated valves

Automotive Master Technician

Camshaft/crankshaft sensor range/implausible signals may be caused by loose timing belt.

Pressure charged systems

Turbocharging

To improve the performance of internal combustion engines, air is sometimes forced into the cylinder above the piston. (The more oxygen there is contained in the cylinder, the more fuel can be burned, and therefore the more energy can be released). One method that can be employed to raise the pressure of the incoming air is to use a turbocharger.

A turbine driven by exhaust gas rotates a set of compressor vanes and force air through the intake system, and into the combustion chamber.

The advantages of using a turbocharger in vehicle design include:

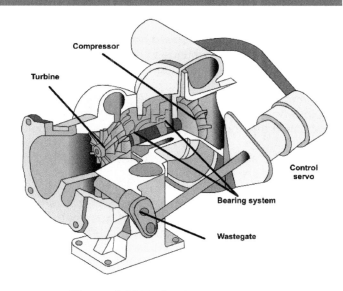

Figure 2.31 Turbocharger

- Improved volumetric efficiency of above 100%.
- An increase in engine performance.
- Turbochargers harness and recycle the energy produced by engines, transforming more of the fuel energy consumed into power, giving greater fuel economy.
- Smaller capacity engines can be used in vehicle design but output performance is similar to that of larger engines.

- No loss of performance at altitude where air pressure is lower.
- More responsive drive produced giving safer overtaking.
- The turbo increases the amount of air entering the engine, which can help reduce exhaust emissions produced.
- The increased performance improves the enjoyment of driving.

A turbocharger has four main disadvantages which can sometimes be overcome through design features, as described in the next section:

Lag

Disadvantage: The turbine has to be spinning at considerable speed before it produces any usable boost in performance. This leads to a condition called turbo lag, where during the initial acceleration, no noticeable performance increase is felt. As the turbine speed increases, a sudden surge of power is introduced, which is called boost.

Solution(s): Multiple turbochargers (Bi-turbo and Tri-turbo system design) can be fitted to an engine. They are manufactured in different sizes, which help to give a smooth delivery of boost pressure.

A small turbine will **spool up** to speed quickly but only produce a small amount of boost pressure. If this is combined with a larger turbocharger, as the small turbocharger runs out of boost, the larger turbine is up to speed and can take over.

Advanced Light Vehicle Technology

A turbocharger can be designed with variable vane geometry. A small set of moveable blades or vanes are fitted around the outer edge of the compressor turbine. The angle of these blades or vanes can be varied to allow for different amounts of boost at different engine speeds. The variable vane geometry can be controlled by intake system pressure or by ECU actuation.

Overboost

Disadvantage: As engine speed rises, the exiting exhaust gas turns the turbine faster and faster. This means that boost pressures may continue to rise above safe limits. Excessive boost pressure can affect engine performance and may lead to engine damage.

In a petrol engine, the overcharging of a cylinder with compressed air can raise the effective **compression ratio** to a point where the petrol begins to auto-ignite. With auto-ignition, the fuel no longer requires a spark plug to initiate combustion and therefore acts like a compression ignition Diesel engine.

If the boost pressure is not limited, the forces acting on mechanical engine components can be so great that premature failure may occur.

Solution(s): To help prevent the turbocharger overboosting, causing detonation of the fuel and engine damage, an exhaust wastegate is often used. The wastegate is a mechanically controlled valve which opens to allow some exhaust gases to bypass the turbine. The wastegate is connected to a **servo** valve, which is acted on by pressure or vacuum created in the inlet manifold.

As the engine speeds up and boost increases, the servo is able to operate at a pre-set value, wasting some of the turbine energy from the exhaust. The actuator arm from the servo is often adjustable to allow the technician to set the wastegate operating pressure.

Many modern engines incorporate an electrically operated solenoid valve into the controller unit of the wastegate servo. When actuated by the ECU through a duty cycle, this solenoid valve is able to accurately control the operation of the wastegate servo, and therefore the amount of boost pressure that is available to the engine at any time.

Temperature

Disadvantage: The temperature of the induction air is raised due to the heat of the turbocharger and the action of compressing the incoming air. This rise in temperature makes the incoming charge less dense, reducing the effect that the oxygen contained has on the combustion process.

Solution(s): To help increase the **density** of oxygen in the compressed air charge from the turbocharger, a method of removing heat is needed. This is normally achieved by using an intercooler.

An intercooler is a small radiator placed after the turbocharger, through which the compressed air charge travels. As it passes through the intercooler, some of the heat energy created in the turbocharger is removed. The cooling of the air charge helps increase the density of the oxygen contained in the incoming air.

To improve the operation of the intercooler still further, some manufacturers spray water onto the outer surface of the intercooler. As the water evaporates from the surface of the intercooler radiator, **latent heat** created by evaporation reduces the temperature still further.

Automotive Master Technician

Back-pressure

Disadvantage: Back-pressures created in the inlet tract have a tendency to try to slow the turbine down and reduce its performance. This is particularly noticeable during gear change.

As the driver closes the throttle in order to change gear, the butterfly valve restricts airflow in the intake manifold, creating pressures which slow down the turbine. When the driver resumes acceleration, the turbine has to speed back up before boost is once again usable. This creates lag during gear change and reduces performance.

Solution(s): To ensure that the turbocharger does not lose performance during gear change, some vehicle manufacturers and **aftermarket suppliers** incorporate a component called a **dump valve**.

During operation, as the driver lifts their foot off of the accelerator to change gear, a vacuum pipe connected to the inlet manifold operates a servo controller in the boost pressure side of the turbocharger. A valve opens and dumps the boost pressure from the inlet manifold side of the turbocharger, often accompanied by a loud whoosh as the air is allowed to escape.

Because boost pressure has now been dumped, back-pressure in the manifold has not slowed the turbine down. Once the gear change has taken place, the driver puts their foot back on the accelerator and a spring closes the servo in the dump valve and boost is returned. As the turbine is still spinning at speed, the boost is almost immediate and turbo lag is reduced.

Notes

Some turbo manufactures are using recycle or recirculation valves instead of dump valves in their designs. These operate in a similar manner to a dump valve, however they return any boost during gear change back into the inlet of the compressor rather than releasing it to atmosphere.

Terms

Spool up – wind up to speed.

Compression ratio – the difference in volume above the piston between when it is at the bottom of its stroke and when it is at the top of its stroke.

Servo – a control system that converts mechanical motion into one requiring more power.

Density – how closely packed together the molecules of a substance are.

Latent heat – the temperature at which a substance changes from a liquid to a vapour or gas.

Aftermarket suppliers – companies that manufacture components that can be fitted as alternatives or upgrades to the original equipment manufacturer (OEM).

Dump valve – a component used to get rid of turbo boost pressure during gear change.

Advanced Light Vehicle Technology

Superchargers

A supercharger is a mechanically driven air compressor connected directly or indirectly to the engine's crankshaft. This can be achieved using drive gears, drive chains or drive belts.

Unlike a turbocharger, which is driven from exhaust gases, the boost provided by a supercharger is instant, with no lag. Unfortunately, a supercharger is considered a **parasite** of engine power. This means that some of the turning effort created at the crankshaft is being used to drive the supercharger and, as a result, not all the power created by the supercharger is converted into performance. Some manufacturers have created a variable drive supercharger with a clutch which disengages at lower engine revs. This allows performance, economy and emissions to be kept within reasonable limits.

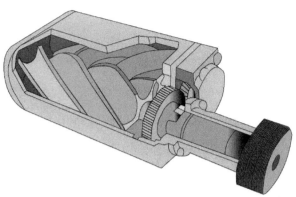

Figure 2.32 Supercharger

A number of different types of supercharger compressor can be used, including:

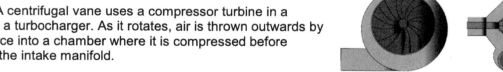

- **Reciprocating** piston: This uses a piston, connecting rod and crankshaft in a similar manner to a standard reciprocating engine.

- Rotary vane: This system uses two or three rotor gears, machined and connected with very close **tolerances**, to compress the air for the intake charge.

- Centrifugal: A centrifugal vane uses a compressor turbine in a similar way to a turbocharger. As it rotates, air is thrown outwards by centrifugal force into a chamber where it is compressed before moving on to the intake manifold.

- Axial: This uses a set of rotor blades in a similar way to a jet engine. These rotor blades force air through smaller and smaller channels until it has been compressed to a level that will create extra boost for the engine.

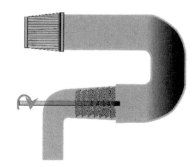

Automotive Master Technician

Parasite – something that draws energy away from the engine without gain or advantage.

Reciprocating – moving backwards and forwards or up and down.

Tolerance – an allowable difference from the optimum measurement.

Engine management systems

Engine management systems have greatly enhanced the operation and control of light vehicle powertrain systems. Engine management improves performance, fuel economy and helps to reduce emissions.

Electronic control principles

To manage an engine, the operation involves:

Input (from sensors) → **Processing (in an ECU)** → **Output (to actuators)**

Figure 2.33 Electronic control principles

The electronic control principles used to manage vehicle engines work in a similar way to how the brain operates different functions in the human body. Information is received, that information is processed, and an action is taken.

Input

The information received form sensors monitoring engine conditions may be analogue, in the form of a rising/falling voltage or resistance, or in a digital format that signals by switching a circuit on or off. A modern engine management system will use many sensors to provide information about engine condition and operation. These may include but are not limited to:

- Engine temperature sensors (ECT).
- Oxygen sensors (sometimes called lambda sensors) (EGO or HEGO).
- Engine oil pressure sensors (EOP).
- Engine oil temperature sensors (EOT).
- Knock sensors (KS).
- Airflow sensors (AFM).
- Mass airflow sensors (MAF).
- Manifold absolute pressure sensors (MAP).
- Throttle position sensors (TPS).
- Pedal position sensors (PPS).

- Air temperature sensors (ATS, AAT or IAT).
- Fuel pressure sensors (FPS).
- Fuel temperature sensors (EFT).
- Camshaft position sensors (CMS).
- Crankshaft position sensors (CPS).
- Barometric/atmospheric pressure sensors (BARO or APS).
- Vehicle speed sensors (VSS).
- Clutch and brake switches (CPP or BPP).
- Power steering pressure sensors.
- Air conditioning pressure sensors.

The information from these sensors is then sent to the electronic control unit (ECU) for processing.

Advanced Light Vehicle Technology

This list of sensors shown is not exhaustive, and new types of sensor are continually being developed by vehicle manufacturers to improve sensing and therefore engine control. In fact many sensors are now being produced with their own small integrated circuits which conduct some of the processing originally handled by an ECU. This reduces the amount of work that has to be processed and speeds up operation as each sensor or actuator only has to handle its own information. If the sensors and actuators are then connected to a multiplexed network, such as CAN Bus, all other systems that need their information have access to it.

Processing

Input signals produced by the sensors are processed within the ECU, which will then send signals to the **actuators** (i.e. fuel injectors, auxiliary air valves and throttle controllers etc.), providing engine control.
An ECU requires digital information (on and off) in order for it to conduct processing. Many signals sent by engine management sensors are analogue (rising and falling voltages or resistances) and must be converted to digital. This is achieved using a pulse shaper or Schmidt trigger (comparator).

Pulse shapers - these are normally **transistors** which have the analogue input joined to the base connector. The analogue signal is then used to switch the collector and emitter on and off converting it to a digital signal.

Schmidt triggers (comparators) - these are normally a **thyristor** or a form of latching transistor which will stay either switched on or off when a threshold current/voltage is reached. The analogue signal is now able to be converted to a digital signal only when certain values are reached.

Transistor - a semi-conductor electronic component which acts like a switch with no moving parts.

Thyristor - a semi-conductor electronic component with four layers of alternating Negative and Positive type material. They act as bistable switches, conducting when their gate connection receives a current trigger, and continue to conduct while they are forward biased (that is, while the voltage across the device is not reversed).

Actuator- a mechanism that puts something into action.

Analogy - The basic operation of an ECU is similar to a calculator. Various engine sensors send information to the ECU, as voltage signals, resistance values or digital frequency. The ECU is programmed to carry out a complex mathematical calculation (algorithm) – to add, subtract, multiply or divide these readings. When a result is obtained, the answer will correspond with a setting to turn one of the actuators of the fuel injection or ignition system on or off. The table of results and settings that are programmed into the ECU could be considered the engine 'map'.
The map is designed to ensure that the engine operates within environmental and mechanical tolerances and will be set by the manufacturer in the factory. As these maps are essentially a computer program, they are often up-dateable, if the correct equipment is used. This is known as 're-mapping'.

Automotive Master Technician

Electronic control is achieved using computers. Different manufacturers call these computers various names:

- electronic control unit (ECU)
- electronic control module (ECM)
- powertrain control module (PCM)

Despite their different names, essentially they all function in a very similar manner.

The performance of vehicle engines is often de-tuned by the manufacturer in order to comply with environmental and legal requirements. This will also ensure that the engine is reliable and has a long lifespan.
Vehicle/engine performance can often be improved by updating the management system. This may be achieved using software updates -'re-mapping' or ECU hardware processor updates - 'chipping'.

Re-mapping or chipping may improve engine/vehicle performance, but if either of these are used, environmental and legal requirements might be breached and engine reliability and lifespan may be considerably reduced.

Once processing is complete, many ECU's operate as **earth switches**, completing the electrical circuit of an actuator and allowing it to operate. Because the ECU completes the electric circuit for many of the engine management actuators, it is important that it has a good earth. Any resistance created where it attaches to the vehicle body can lead to poor performance or running issues.
For all their complexity, most ECU's can be considered a large earth switch connected to a calculator.

Earth switches- a switch on the negative side of an electric circuit.

Engine management power supply

In order to process information coming from engine sensors, ECU's rely on voltage and frequency patterns. Unfortunately, vehicle system voltages are unstable and battery voltage rises and falls in accordance with engine operating conditions. For example:

- When the engine is cranked, the potential difference created at the battery may make system voltage fall to around 8V.
- When the engine starts and the alternator charges, system voltage will rise to around 14.5V. If the vehicle uses a smart charging system, this can sometimes be as high as 18V.

These rising and falling voltages will have a large impact on ECU processing capabilities – basically they confuse the ECU because the input signals are not standardised.

Advanced Light Vehicle Technology

As a result many systems use a resistor to reduce the processing voltages to the sensors to around 5V. This means that, regardless of engine operation, cranking or charging, signal voltages will be stabilised to between 0V and 5V at all times, resulting in correct interpretation under all engine running conditions.

Table 2.7 lists some of the processing operations that take place within an ECU.

Table 2.7 Processing operation inside an ECU

Read only memory (ROM)	The read only memory contains a programme to control the ECU, and therefore engine management. The programme is vehicle generic and not restricted to one vehicle type. The information in this part of the ECU cannot be changed.
Random access memory (RAM)	This is a temporary memory inside the ECU. It is used by the central processing unit to store and retrieve information. With random access memory, information is lost when power is disconnected.
Programmable read only memory/erasable programmable read only memory (PROM/EPROM)	The programmable read only memory contains information which is vehicle specific, and is sometimes known as the engine management map. This information is updateable, sometimes called remapping. Some systems allow this remapping to be done via the serial port or data link connection, where a new program or map is substituted for the original. This information is read only.
Keep alive memory (KAM)	The keep alive memory system of a vehicle's ECU allows it to adapt to different running conditions. This is known as adaptive strategy or adaption. As engine operation changes from that found in the ECU map, due to wear, faults or driving style, changes are substituted to the calculations carried out by the central processing unit. This allows fuel injection and ignition systems to adapt. Because of this, a stable engine operation is achieved, and drivability remains constant regardless of the situation (within reasonable limits).
Limited operating strategy (LOS)	If a fault occurs in a sensor or actuator, default information is supplied and may by-pass incoming and outgoing data to the ECU. The engine is now running on pre-programmed values and maintaining a relatively stable engine operation. This is known as limited operating strategy (LOS) or sometimes 'limp home'.
Open and closed loop operation	For the ECU to accurately measure and control the engine, certain operating conditions must be met. (The engine should be at normal operating temperature, for example). Until these conditions have been met, the ECU controls the engine using pre-programmed values and ignores sensor inputs – this is known as open loop. When the operating conditions have been achieved, sensor information is used to monitor and control engine operation – this is known as closed loop.

It is important when using a scan tool to help carry out the diagnosis of an engine management fault, that the 'readiness monitors' are checked. This step tells the technician that the system is ready to be tested as it has achieved the correct conditions to be running in a closed loop operation.
If this step is skipped or ignored, incorrect results could be obtained leading to misdiagnosis.

With keep alive memory, information is not lost when power is disconnected.

Adaptive strategy within the keep alive memory system of the ECU can cause issues if a technician's diagnostic strategy involves diagnosis by substitution (*see gamblers anonymous*).

For example: if the technician does not fully understand the operation of an engine management system, and an engine is running badly, he or she may substitute components to see if it makes a difference to the running of the engine.

Every time a component is substituted, adaption can occur inside the ECU's keep alive memory, making any running issues worse. The technician is basically reprogramming the ECU.

As the situation becomes worse the technician may resort to substituting the ECU with a new one from a supplier. Once connected, the running problems on the vehicle may reprogram the ECU's adaption strategy. This means that if the ECU is removed from the vehicle and sent back to the supplier, they now have a faulty ECU. If not reset to original values, the next person to use this ECU will suffer engine management operation issues.

Output

Once an ECU has processed the incoming sensor information, it will then complete an electric circuit which switches on or off an actuator. A modern engine management system can use a number of actuators to provide overall control. These may include but are not limited to:

- EVAP vacuum controllers.
- EGR valve.
- Canister purge valve.
- Canister air valve.
- Secondary air injection solenoids.
- Throttle motors.
- Cruise control actuators.
- Idle actuators.
- Fuel pressure regulators.
- Ignition coils.

- Fuel injectors.
- Injector pressure regulators.
- Variable valve control systems.
- Turbo wastegate/boost controllers.
- Malfunction indicator lamps (MIL).
- Smart charging regulators.
- Engine start/stop.
- Passive anti-theft systems.
- Tachometer.

Many of these actuators are fed with system voltage via a relay and then switched on by the ECU by earthing the circuit. This is often regulated by pulse width modulation or duty cycle to maintain overall control.

Ignition control principles

To ensure a spark arrives at the correct spark plug, at precisely the right time, regardless of speed and load, an ignition management system can be used.

- CPS or crankshaft position sensors provide information on engine speed and top dead centre position of cylinder number one and its companion cylinder.
- Distributor pickup or camshaft position sensors provide information for cylinder recognition, so that firing order can be determined.
- Engine coolant temperature sensors (ECT) provide information so that ignition timing can be advanced or retarded in response to engine temperature and fuelling requirements.
- Air measurement sensors such as MAF or MAP provide information on engine load. This information may also be used to advance or retard the ignition timing.
- Battery voltage is often sensed so that the rise time (or dwell period) of the ignition coil can be precisely calculated ensuring that a good spark is available under all running conditions.

The combination of the sensor signals can be processed by the ECU. The ECU will send out signals to the ignition coil, switching off the primary circuit or discharging a capacitor through the primary circuit. The resulting collapse in magnetic field will induce a high voltage in the secondary winding, producing a spark at the plug.

Advanced Light Vehicle Technology

Fuel system control principles

A fuel management system may be used to ensure that the correct quantity of fuel is injected, by the correct fuel injector, at precisely the right time, regardless of speed and load.

- Airflow meters (AFM), mass airflow meters (MAF) or manifold absolute pressure sensors (MAP) can be used to provide information on the volume of airflow in the intake manifold. This means that the correct amount of fuel can be injected for stoichiometric operation.
- Engine coolant temperature sensors (ECT) provide information so that the quantity of fuel injected may be increased or decreased due to engine temperature.
- Throttle position sensors (TPS) provide information to the ECU on throttle butterfly position. When used in conjunction with airflow volume or mass, this can help the ECU calculate engine load.
- Distributor pickup or camshaft position sensors provide information for cylinder recognition, so that **sequential fuel injection** may be used.
- Battery voltage is often sensed so that the pulse width of the fuel injector can be altered in accordance with the rising and falling voltage found when an engine is running.

- If voltage is high, the fuel injector will open early and close late.
- If voltage is low than the fuel injector will open late and close early.

This will have an effect on the amount of fuel injected.

You can sometimes check fuel injection ECU operation by watching pulse width or short term fuel trim (STFT) when disconnecting a component.
Remember to only conduct this test if safe to do so and any diagnostic trouble codes or adaptions caused by this action will require re-setting following your repairs.

Sequential fuel injection – when fuel is injected in the same sequence as engine firing order.

Engine management is designed to bring these two separate systems (ignition and injection) under one umbrella, combining them to provide a system with the lowest emission output, better fuel economy and good all-round performance. Many early engine management systems had one ECU to control both the fuel injection system and the electronic ignition system. With the introduction of network/multiplex systems, many ECU's are able to share information. This means that separate processing systems have been developed to work in harmony with each other in order to provide overall engine control.

Emission control

Advances in engine management, and the control of fuel injection and ignition systems, have made it possible to limit most exhaust emissions, but further methods are needed in order to meet all of the strict limits imposed. Other methods to help control exhaust emissions are shown in Table 2.8.

Automotive Master Technician

Table 2.8 Methods of controlling exhaust emissions

Catalytic converters 	Catalytic converters are fitted to some vehicles to help reduce the amount of harmful pollutants present in exhaust gases. A catalytic converter is normally mounted in the exhaust system very close to the inlet manifold where the exhaust gases exit the cylinder head. (The closer the catalytic converter is to the exiting exhaust gases, the quicker it heats up and begins to operate in an efficient manner). From the outside, a catalytic converter may look similar to an exhaust system silencer. On the inside, it is normally made up of a matrix-style mesh which contains precious metals such as platinum and rhodium. The mesh-style matrix increases the surface area exposed to the exiting exhaust gases. When the gases pass over the precious metals, a chemical/catalytic reaction takes place (a catalyst is anything that makes a situation or substance change). The exhaust pollutants are not removed from the exhaust gases, but are converted to far less harmful pollutants. When the exhaust gas comes into contact with the reduction catalyst of a three-way converter, a chemical reaction takes place which strips the nitrogen molecules from any oxides of nitrogen and retains them in the catalyst, while allowing the oxygen to continue. When the exhaust gases come into contact with the oxidation catalyst, a chemical reaction takes place which recombines hydrocarbons and carbon monoxide with some of the remaining oxygen to produce carbon dioxide and water. This way, the amount of harmful emissions produced the exhaust system are reduced. To work efficiently, catalytic converters require the engine to operate within the ideal stoichiometric values. This means that accurate engine management is essential.
Selective catalytic regeneration (SCR) 	On some three-way catalytic converters, oxides of nitrogen are removed using a chemical reaction. As a result of the chemical reaction, the nitrogen molecules are retained inside the catalytic converter and must periodically be cleaned out. Some engine management systems achieve this by changing their fuel injection programme to make the system run rich – this alters the chemical processes going on inside the catalytic converter, helping to remove the nitrogen molecules. Other systems have a reservoir containing a special chemical, normally based on urea. This is injected into the exhaust gas before the catalytic converter and alters the processes going on inside the catalytic converter, helping to remove the nitrogen molecules. This chemical additive is a service item that must be topped up according to the manufacturer's maintenance schedule.
Secondary air injection 	A secondary air injection system is sometimes fitted to vehicles to help with emission-related issues. A small air pump, controlled by the car's computer (ECU), can blow air into the exhaust manifold close to where it exits the cylinder head. By introducing extra oxygen to the hot exhaust gases, unburnt hydrocarbons can continue the combustion process and be burnt away. As with blowing on the embers of a fire, introduced air helps heat the exhaust gases. If the vehicle is fitted with a catalytic converter, this will assist with the converter's warm-up period, making it operate efficiently sooner.

Table 2.8 Methods of controlling exhaust emissions

	The extra oxygen introduced by a secondary air injection system is also able to dilute the carbon dioxide produced by the combustion process, and this should be taken into account when conducting diagnosis of exhaust emissions (*see Chapter 1*).
Diesel particulate filters (DPF) 	A Diesel particulate filter can be mounted in the exhaust system of a compression ignition powered car to help remove soot and other small contaminate particles that are produced during the combustion cycle. A car fitted with a DPF in good working order should show no visible smoke from the tailpipe when running. Unlike a catalytic converter, a DPF is a filter – the exhaust gases are passed through the filter and particulate matter is retained. This means that a DPF requires maintenance to prevent it from becoming blocked. Some Diesel particulate filters are a replaceable service item, while others can be regenerated in a similar manner to SCR. Pressure and temperature sensors are often mounted in the exhaust system so that the engine management system is able to monitor and maintain correct function of the DPF.
Exhaust gas recirculation (EGR) – petrol 	Exhaust gas recirculation (EGR) can be used as an emission control system to reduce the production of oxides of nitrogen (NOx). EGR is a method for taking some of the burnt exhaust gases and directing them through pipes or galleries back into the inlet manifold. (With variable valve control it is sometimes possible to create small amounts of internal EGR by altering the valve overlap period). By introducing exhaust gas back into the inlet manifold of a petrol engine, the amount of incoming air and fuel that reaches the cylinder is reduced. This means that the performance output of the engine is also reduced, and so engine temperatures will be lower. This helps prevent the production of NOx. It is important that performance is not affected during most of the operating conditions under which a car is driven. A device called an EGR lift valve is fitted so that exhaust gas is only recirculated when outright performance or acceleration are not required. For example, when cruising on a motorway at a steady speed with your foot resting only lightly on the accelerator pedal, exhaust gas can be recirculated and the production of NOx reduced.
Exhaust gas recirculation (EGR) – Diesel 	EGR in Diesel engines reduce the production of oxides of nitrogen in a similar manner to that in a petrol engine, however it will also have a second function. In a Diesel engine, it is very difficult to achieve ideal air/fuel ratios of 14.7:1 because Diesel engines draw the same amount of air every time, and it is the amount of fuel that is injected that dictates how fast the engine will run. This means that when a standard Diesel engine is operating under normal conditions, the air/fuel ratio will be extremely weak at low revs and idle, and this enriches during acceleration. These extremes of air/fuel ratio create emission and pollution problems. By introducing a quantity of burnt exhaust gases back into the intake manifold of a Diesel engine, the amount of fresh air entering the cylinder can be more accurately controlled, which results in better air/fuel ratios. Exhaust gas will still compress and raise the temperature in the cylinder for ignition of the fuel.

Table 2.8 Methods of controlling exhaust emissions

Evaporative emission control (EVAP)	Petrol fumes are a pollutant and are therefore not allowed to escape into the surrounding air. Because of this, fuel tanks are sealed to reduce the possibility of hydrocarbon emission due to evaporation. If the tank has nowhere to store these evaporated fumes, high pressures could build up in the tank, causing leaks or damage. The evaporation of fuel continues as with any other fuel system, but the vapours must be captured, stored and disposed of in an environmentally friendly way. Fuel emission vapours are stored in a charcoal canister. When the engine temperature is above 55° C and the throttle has been opened past the canister purge port, the fumes are vented into the intake manifold via a duty cycle controlled solenoid valve, and burnt along with the rest of the air/fuel mixture. This makes sure that all of the HC vapours go through the combustion process and eliminate HC emissions. This is called an evaporative emission system (EVAP).
Positive crankcase ventilation (PCV) or breather system	Positive crankcase ventilation (PCV) was introduced in the 1960's and marked the beginning of evaporative emission control systems in the cars. The initial purpose of the PCV system was to capture crankcase vapours and prevent them from being vented into the atmosphere through atmospheric breather pipes. Combustion chamber gasses are able to escape into the crankcase through a process known as "blow-by". Blow-by occurs when the compressed fuel/air mixture is forced past the seal created between the piston ring and the cylinder wall. When this happens these gasses become trapped in the crankcase where hydrocarbons may then leak to atmosphere. A PCV valve meters the return of the crankcase vapours to the engine's intake manifold. The vapours then mix with the engine's intake air and/or fuel/air mixture and re-enter the combustion chamber to be burned. This makes sure that all of the HC vapours go through the combustion process and eliminate HC emissions from the crankcase to the atmosphere.

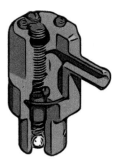

Lambda sensors (pre- and post-catalytic converter)

To make sure that the correct air/fuel ratio is achieved from the engine management system, an oxygen sensor (also called a lambda sensor) is fitted in the exhaust system before the catalytic converter. The lambda sensor measures the oxygen content of the exhaust gases, and instructs the ECU if the engine is running too rich or too weak.

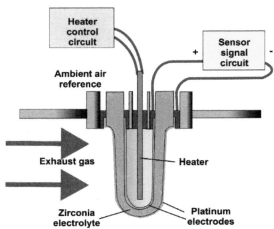

Figure 2.34 Zirconia oxygen sensor

- Too much oxygen in the exhaust gas and the engine is running weak.

- Too little oxygen in the exhaust gas and the engine is running rich.

Many lambda sensors use a zirconia ceramic to measure the oxygen content. One end of the sensor is inserted into the exhaust gas and the other end is exposed to the outside air. The difference in oxygen between these two points causes a corresponding resistance in the zirconia ceramic and signals the ECU, which can now calculate the amount of oxygen in the exhaust gas. The resistance in the sensor will continuously rise and fall, creating an average figure that can be used to maintain an ideal air/fuel ratio.

Advanced Light Vehicle Technology

A lambda sensor will only begin to operate correctly once it has reached an operating temperature above 300°C. Because of this it normally contains an additional heating element to get it up to operating temperature as quickly as possible. This additional heating will also help prolong the lifespan of the sensor.

- During the warm-up period of the catalytic converter and lambda sensor, signals from the oxygen sensor to the ECU are ignored. When this happens, the engine management system is running **open loop**.

- As soon as the catalytic converter and lambda sensor are up to the required operating temperature, the signals produced are used to correct fuel injection for the ideal air/fuel ratio. When this happens, the engine management system is running **closed loop**.

Open loop – when the engine management ECU operates on pre-programmed values from its memory.

Closed loop – when the engine management ECU operates with signals supplied by sensors.

To make sure that the catalytic converter is operating correctly, some manufacturers include a second lambda sensor mounted after the catalytic converter. The signals produced by the two lambda sensors (pre-cat and post-cat) are compared. If they are found to be similar, no chemical reaction is taking place within the converter and therefore the catalyst has failed. The engine malfunction indicator lamp (MIL) will be illuminated and a diagnostic trouble code stored.

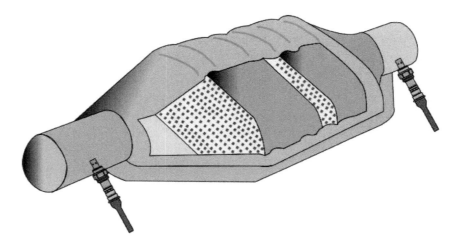

Figure 2.35 Pre and post catalyst lambda sensors

Remember, the use of the term Lambda, when seen on an exhaust gas analyser machine, will sometimes lead to the misinterpretation that it is measuring or doing the same job as an exhaust gas oxygen sensor EGO. This misunderstanding can often lead to the incorrect diagnosis of faulty catalytic converters, or exhaust oxygen sensors. *(See Brettschneider).*

Automotive Master Technician

Wideband oxygen sensors

To improve the efficiency of the emission control system, some manufacturers use wideband oxygen sensors. This design of lambda sensor does not rely solely on the change of resistance in a zirconia ceramic. Instead, two chambers are created: one containing exhaust gas and the other open to air for reference.

A component called an oxygen pump is embedded in the wideband sensor, and through an electrochemical process, it tries to maintain a stable oxygen quantity in one chamber. The quantity of oxygen in this chamber is measured by the zirconia ceramic. The amount of current supplied to the oxygen pump to maintain the correct oxygen content is proportional to the amount of oxygen in the exhaust gas. From this information, the engine management ECU is able to maintain a very stable air/fuel ratio.

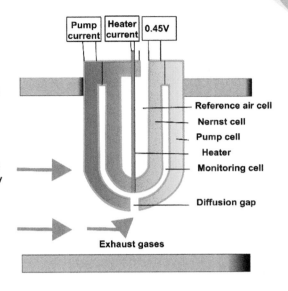

Figure 2.36 Wideband oxygen sensor

Fault diagnosis on modern engines

In order to conduct effective diagnosis on advanced engine systems, it will be necessary to use specialist equipment (*see Chapter 1*).

E-OBD

European legislation states that any faults with an engine management system which might lead to excessive exhaust pollutants being released to atmosphere must be stored as a diagnostic trouble code. A standardised list of codes and a diagnostic connector were produced to be used by manufacturers selling cars in Europe. In this way, information was made available to all service and maintenance repair facilities. This system has become known as E-OBD and is based on the American OBD - II **protocol**. Early OBD systems were limited to sensor values of voltage too high and too low, which gave an indication of circuit faults such as high resistance, short circuit and open circuit. They didn't monitor the operation of the sensor or the actuator. As technology has improved, diagnostic trouble codes are now available which indicate issues of 'range'. This often shows that a sensor or actuator may be functioning, but is not operating as expected by the engine management system.

Protocol - a system of rules that define how something is to be done. In computer terminology, a protocol is usually an agreed-upon or standardised method for transmitting data and/or establishing communications between different devices.

E-OBD code identification

All E-OBD codes are generic, meaning they are the same number for each individual fault, no matter which vehicle make or model is being tested.
An E-OBD code can be easily identified as they will all start with the letter and number 'P0'. The letter 'P' indicates that the fault is connected with the powertrain (i.e. engine and transmission) and the '0' indicates the code is generic (i.e. the same for every manufacturer make and model).
The 'P0' will then be followed by three numbers or combination of letters.

Advanced Light Vehicle Technology

The first of the three numbers indicates the area within the engine and transmission:

1- Fuel and air metering
2- Fuel and air metering
3- Ignition system or misfire
4- Auxiliary emission controls
5- Vehicle speed and idle control
6- Computer outputs
7- Transmission
8- Transmission

The final two numbers or letters of the diagnostic trouble code relate to the specific code designation.

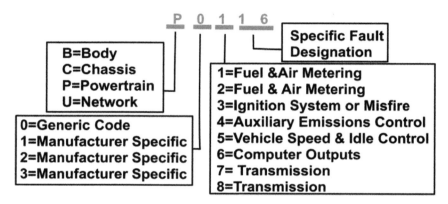

Figure 2.37 E-OBD code identification

It is only the generic E-OBD codes that start 'P0'. If the DTC starts 'P1', 'P2', or 'P3' it is still related to the vehicles powertrain (engine or transmission), but it is now manufacturer specific and the code has to be correctly matched to the one given for that particular make and model.

Always read the full DTC description: voltage high/low or range.

Voltage high or low often indicates a circuit fault.

Range often indicates a component fault.

However, a full assessment of the circuit or component should always be conducted to confirm your diagnosis.

Automotive Master Technician

Malfunction indicator lamps MIL

As part of the E-OBD system, an engine management malfunction indicator lamp MIL is included to warn the driver of a detected problem which may cause the engine to run badly and produce higher than expected exhaust emissions. If there are no detected faults, the light illuminates when the ignition is switched on, as a test to see if the bulb is working, and then extinguishes when the engine is started. If faults are detected, the warning light remains illuminated once the engine has started or while it is running.

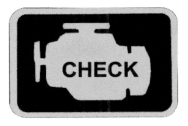

Figure 2.38 Malfunction indicator lamp MIL

- MIL remains illuminated when the engine is running - The ECU has detected a fault which may allow high emissions to be produced by the vehicle and can be driven to a repair workshop.

- MIL constantly flashes whilst the engine is running - The ECU has detected an engine misfire which has the potential to damage the catalytic converter. The vehicle should be taken immediately to a vehicle workshop for repair avoiding hard driving conditions.

Two trip detection logic

It is possible that an intermittent fault (glitch) may cause the engine management light to illuminate and store a trouble code. Once the engine has been switched off and re-started, if the fault is no longer present, the warning light will go out, but the stored code will remain as a pending fault. This is known as two trip detection logic. It ensures that transient faults are recognised but do not cause the vehicle to have unnecessary diagnosis.
If on the second **drive cycle**, the fault is no longer present, the code will be deleted.
If the fault remains, the MIL will remain illuminated and the diagnostic trouble code will become permanent.

Pending trouble codes are very useful for evaluating repairs to engine management systems. Following any repairs, the vehicle should be road tested over a full drive cycle, where possible, and re-scanned. Pending codes should be then double checked to ensure that no issues have been recorded that may not have put the warning light on.

Drive cycle - a set of full driving conditions that must be met, in order for an ECU to conduct a full self-diagnosis.

A drive cycle is designed by the manufacturer to ensure that the vehicle has met all of the conditions needed in order for an ECU to conduct its own self-diagnosis. It will require that the vehicle is driven in a particular way, and for some considerable distance. A drive cycle is often unique to a vehicle make and model, and the conditions required will need to be sourced from OEM data.
Regardless of conducting a full drive cycle, readiness monitors should always be checked as an initial indication that the vehicle systems are ready to be scanned.

Advanced Light Vehicle Technology

As E-OBD is only related to faults that are connected with emission related problems; it is not always the case that the engine malfunction indicator lamp MIL will illuminate. An effective method for using a diagnostic scan tool is listed below:

1. Question the driver to establish when/how/where the MIL became illuminated (also whether it is on continuously or flashing).
2. Locate the data link connector DLC (which should be inside the passenger compartment within reach of the driver, usually between the centre line of the vehicle and the driver's seat).
3. Plug in the scan tool and were possible set up using the OEM vehicle data (If this is not possible then the generic EOBD setting can be used, but this may not give a full list of diagnostic trouble codes).
4. Check the readiness monitors (this will give you an indication that all the required systems have met the prerequisite conditions for testing. If not the vehicle should be road tested until the required conditions have been met).
5. Read and record all diagnostic trouble codes (permanent and pending).
6. Read and record all freeze-frame information (this will give an indication of the vehicle running conditions at the time the last fault code was stored and help you recreate the situation during a road test).
7. Clear all diagnostic trouble codes.
8. Fully road test the vehicle, as long as it is safe to do so (try to recreate the conditions indicated by the freeze-frame data).
9. Re-scan for trouble codes and concentrate your diagnosis on any codes that have returned (remember to recheck the freeze-frame data).
10. Use further diagnostic testing to check the system and component indicated by the diagnostic trouble code (do not guess or conduct diagnosis by substitution).
11. Find the root cause of the fault and conduct repairs that ensure, as far as is reasonably practicable, the fault will not occur again.
12. Clear any trouble codes that may have been generated during the repairs, and reset ECU adaptions.
13. Fully road test the vehicle, over a complete drive cycle where possible, and re-scan to ensure that no trouble codes have returned (permanent or pending).

Offering to rescan a car 'free of charge' following a repair is not only good customer service, but also helps to reduce 'come-backs'.

Live data

Most diagnostic scan tools have the ability to provide live serial data from sensors or actuators. This can be in the form of a PID (parameter identifier) list or as a graph. This live data can then be used to help identify the correct function and operation of testable components.

Use live data (parameter identifiers PID) to check mechanical operation of components, (force a change by disconnecting or adjusting a component). Remember to only conduct this test if safe to do so and any diagnostic trouble codes or adaptions caused by this action will require re-setting following your repairs.

Automotive Master Technician

The graphing information shown on a scan tool is an interpretation of serial data produced by the ECU, and although helpful for quick analysis, is not as accurate as the waveforms produced when using an oscilloscope.

Functional testing

Some scan tools have the capacity to conduct component functional tests. When the correct menu is chosen, a list of available components will be shown which can be operated and their physical function may then be checked.

During diagnosis, the information provided by the vehicles short term fuel trim STFT and long term fuel trim LTFT, from freeze-frame or live data may sometimes give an indication of whether the fault is related to either a mechanical or engine management issue.

• Large out of value readings from short term fuel trim will often indicate an engine management fault
• Large out of value readings from long term fuel trim will often indicate a mechanical fault

NB: this information should only be used as a rough guide to assist with your correct diagnosis of the fault from associated symptoms.

Tooling

No matter what diagnostic task you are performing, you will need to use some form of tooling. As a Master Technician you may discover innovative methods when using certain tools for diagnosing faults for which there is no prescribed method. If used appropriately, these tools can be utilised to reduce overall diagnostic time and help locate the root cause of various symptoms, leading to a first time fix.

Always use the correct tools and equipment and never misuse tooling so that it is likely to cause personal injury or damage.

The following Table shows a suggested list of diagnostic tooling that could be used when testing and evaluating advanced internal combustion engine systems. Due to the complex nature of many modern system faults, you will experience different requirements during your diagnostic and repair routines and so you will need to adapt the list shown for your particular situation.

Advanced Light Vehicle Technology

Table 2.9 Diagnostic tooling

Tool	Possible use
Oscilloscope 	Used for retrieving engine management sensor waveforms, confirming correct function and signal production.
Specialist tools 	Conducting diagnosis of engine systems, mechanical, hydraulic, pneumatic and electrical with the minimum of stripping down.
Code reader/scan tool 	Scanning for diagnostic trouble codes to help with your diagnosis and also for checking the effectiveness of any repairs conducted.
Hand tools 	Conducting effective repairs that reduce the risks to health and safety and prevent damage to the vehicle or components.
Multimeter 	Conducting a volt drop test to check for open circuits or high resistances (*see Chapter 1*).

All tools and equipment used when working on advanced internal combustion engines should be accurately calibrated before each use and fit for purpose. For a description of the setup and use of these diagnostic tools, *see Chapter 1*.

Automotive Master Technician

Chapter 3 Advanced Vehicle Driveline and Chassis Technology

This chapter will help you develop an understanding of the construction and operation of advanced vehicle driveline and chassis technology. It will cover the relationship between engine and transmission and the effects that it has on the delivery of power and torque. It also gives an overview of the construction and operation of chassis systems, such as steering, suspension and brakes, showing how the inclusion of electronic control has improved vehicle safety, comfort and control. It will also assist you in developing a systematic approach to the diagnosis of complex system faults for which there is no prescribed method. This is achieved with hints and tips which will help you undertake a logical assessment of symptoms and then uses reasoning to reduce the possible number of options, before following a systematic approach to finding and fixing the root cause.

Contents

Information sources

Many of the diagnostic routines that are undertaken by a Master Technician are complex and will often have no prescribed method from which to work. This approach will involve you developing your own diagnostic strategies and requires you to have a good source of technical information and data (*see knowledge is power*). In order to conduct diagnosis on advanced chassis systems, you need to gather as much information as possible before you start and take it to the vehicle with you if possible.

Sources of information may include:

Table 3.1 Possible information sources

Verbal information from the driver (*see the interrogation room*)	Vehicle identification numbers
Service and repair history (*see the interrogation room*)	Warranty information
Vehicle handbook	Technical data manuals
Workshop manuals/Wiring diagrams	Safety recall sheets
Manufacturer specific information	Information bulletins
Technical helplines	Advice from other technicians/colleagues
Internet	Parts suppliers/catalogues
Jobcards	Diagnostic trouble codes
Oscilloscope waveforms	On vehicle warning labels/stickers
On vehicle displays	Temperature readings

Always compare the results of any inspection or testing to suitable sources of data. Remember that no matter which information or data source you use, it is important to evaluate how useful and reliable it will be to your diagnostic and repair routine.

Advanced Light Vehicle Technology

Electronic and electrical safety

Working with any electrical system has its hazards. The main hazard is the possible risk of electric shock, but remember that the incorrect use of electrical diagnostic testing procedures can also cause irreparable damage to vehicle electronic systems. Where possible, isolate electrical systems before conducting the repair or replacement of components.

Always use the correct tools and equipment. Damage to components, tools or personal injury could occur if the wrong tool is used or misused. Check tools and equipment before each use.
If you are using electrical measuring equipment, always check that it is accurate and calibrated before you take any readings.

If you need to replace any electrical or electronic components, always check that the quality meets the original equipment manufacturer (OEM) specifications. (If the vehicle is under warranty, inferior parts or deliberate modification might make the warranty invalid. Also, if parts of an inferior quality are fitted, they might affect vehicle performance and safety). You should only carry out the replacement of electrical components if the parts comply with the legal requirements for road use.

Table 3.2 The operation of electrical and electronic systems

Electrical/Electronic system component	Purpose
ECU	The electronic control unit is designed to monitor the operation of vehicle systems. It processes the information received and operates actuators that control system functions. An ECU may also be known as an ECM (electronic control module).
Sensors	The sensors are mounted on various system components and they monitor the operation against set parameters. As the vehicle is driven, dynamic operation creates signals in the form of resistance changes or voltage which are sent to the ECU for processing.
Actuators	When actioned by the ECU, motors, solenoids, valves, etc. help to control the function of vehicle for correct operation.
Digital principles	Many vehicle sensors create analogue signals (a rising or falling voltage). The ECU is a computer and needs to have these signals converted into a digital format (on and off) before they can be processed. This can be done using a component called a pulse shaper or Schmitt trigger.

Automotive Master Technician

Table 3.2 The operation of electrical and electronic systems

Duty cycle and PWM 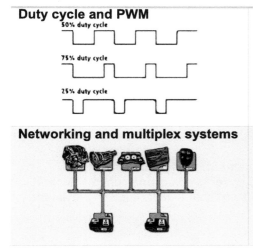	Lots of electrical equipment and electronic actuators can be controlled by duty cycle or pulse width modulation (PWM). These work by switching components on and off very quickly so that they only receive part of the current/voltage available. Depending on the reaction time of the component being switched and how long power is supplied, variable control is achieved. This is more efficient than using resistors to control the current/voltage in a circuit. Resistors waste electrical energy as heat, whereas duty cycle and PWM operate with almost no loss of power.
Networking and multiplex systems	Many modern vehicle systems are controlled using computer networking. In these systems a number of ECU's are linked together and communicate to share information in a standardised format. The most common network system is the Controller Area Network (CAN Bus). *See Chapter 1.*

Electronically controlled transmission systems

No matter whether an engine is petrol or Diesel, it only delivers its peak torque at a certain rev range. This will compromise overall performance, meaning that the power produced and the economy achieved will only be maximised if the engine is run at a certain speed. Some of these issues can be overcome if the engine is coupled with an appropriate gearbox. Mechanical gearing is able to multiply turning effort, and therefore, depending on the ratio of gears used, the peak area of torque delivery can be reduced to match road speeds.

For many years, it was normal for the manual selection of gearing to be controlled by the driver, using his or her skill to match the appropriate gear ratios to driving conditions. Since the introduction of electronics onto many vehicle systems, it is natural for transmissions to be included in the utilisation of this advanced technology. Not only can assistance be provided to the operator in order to select the most appropriate gear ratio for any driving condition, it can also improve comfort and safety.

Multiplexing technology allows sensors from other vehicle systems to be used to help control the delivery of power from the engine to the road wheels, accurately matching the demands of dynamic driving situations.

Electronic control system

As with all electronic control systems, the management of transmission clutch or gear selection involves three main processes:

- **Input:** This comes from various sensors, normally related to load and speed.
- **Processing:** The ECU takes the sensor information and calculates the best gear ratios for a given driving situation, pre-programmed into its memory.
- **Output:** Signals are sent from the ECU to various actuators, **servos** and solenoids which control the operation of clutches, selectors or brake bands and allow gear change to take place.

Advanced Light Vehicle Technology

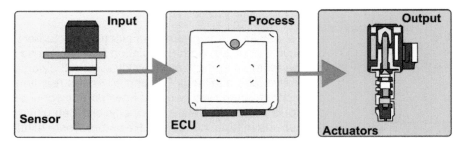

Figure 3.1 Electronic control system

The transmission control ECU will have a number of different shift patterns programmed into its memory. This allows the driver to select an economy or sport mode, for example. Some transmission ECU's are able to learn a driving style and adapt the shift pattern to suit the driver.

Servo – a mechanism that converts a small mechanical motion into a larger movement with greater force.

Clutches

The purpose of a clutch is to:

- Link up the power from the engine to the rest of the transmission system.
- Separate the power from the engine to the rest of the transmission. In this way, the clutch allows the drive of the vehicle to be stopped and started.

A clutch provides three main functions:

1. It provides a smooth take-up of drive (going from stationary to moving).
2. It provides a 'temporary position of neutral' (allows the car to come to a stop without taking it out of gear or stalling the engine).
3. It allows the engine to be disconnected from the gearbox so that gear change can take place.

Clutches make use of the principles of friction. **Friction** is a force resisting the relative motion of solid surfaces, fluid layers, and elements sliding against each other. There are several types of friction:
Dry friction resists relative movement between of two solid surfaces in contact. Dry friction can be divided into **static** friction "**stiction**" between non-moving surfaces, and **kinetic** friction between moving surfaces. Dry friction is common in the use of light vehicle manual clutches, however dual clutch transmission systems DCS will often use a lubricated wet clutch set-up.

Fluid friction describes the friction between layers within a viscous fluid that are moving relative to each other. Fluid friction is common in the use of automatic gearbox torque converters.

Lubricated friction is where a fluid creates friction between two separate solid surfaces. Lubricated friction is common in the use of automatic transmission clutch packs which operate the epicyclic gearing.

Automotive Master Technician

Friction - the grip produced between two surfaces in contact.

Static - stationary, not moving.

Stiction - static friction.

Kinetic - movement energy.

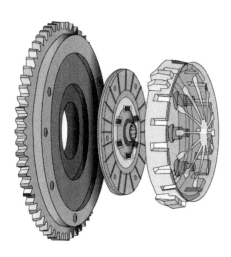

Figure 3.2 Standard manual transmission friction clutch

A standard manual transmission friction clutch is made up from a number of components. An overview of the main parts are shown in Table 3.3.

Table 3.3 Friction clutch components

Friction clutch component	Purpose
Drive surface	This is often the flywheel of an engine which rotates with the crankshaft and forms a flat surface to drive the clutch. In a multi-plate or dual clutch system, the drive surface may be a clutch basket or housing which is driven directly or indirectly by the engine. If a multi-plate clutch is used, a number of **plain plates** are inserted between the friction plates to increase the surface area and number of surfaces in contact.
Clutch cover	The clutch cover houses the components of a **dry clutch** system. It is bolted to the flywheel and rotates at crankshaft speed, transferring engine rotation to the pressure plate.

Advanced Light Vehicle Technology

Table 3.3 Friction clutch components

Component	Description
Pressure plate 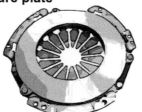	The pressure plate provides the clamping surface (operated by springs or hydraulic force) to drive the friction plate. In a multi-plate clutch, the pressure plate compresses all of the major drive components (friction and plain plates).
Friction plate 	Clamped between the pressure plate and the flywheel, the friction plate transfers drive to the input shaft of the gearbox. In multi-plate and dual clutch systems, a number of friction plates are used to keep the components compact, while still transmitting large amounts of torque. Some friction plates are designed to operate in oil (lubricated friction) to help control friction and heat. These are known as **wet clutches**.
Release bearing 	The release bearing operates against the clutch springs while the engine is turning to remove the clamping force on the pressure plate and disengage the clutch. A clutch release bearing is not always needed, as some systems use **hydraulic** forces to engage and disengage the clutch plates.
Release fork/servos and valves 	The release fork operates against the release bearing when the pedal is pushed to actuate the clutch mechanism. If hydraulic pressure is used to engage and disengage the clutch, valves and servos may be used to control this action.

Plain plates – a set of smooth metal discs used in a multi-plate clutch to increase surface area.

Dry clutch – a clutch where friction surfaces are operated dry (with no lubrication).

Wet clutch – a clutch where the friction surfaces operate in oil to control heat and grip.

Hydraulic - the science of movement of liquids or fluids.

Automotive Master Technician

Coil and diaphragm spring clutches

A number of different construction designs are used in the operating mechanisms of clutches which clamp the pressure plates against other friction components. The design will normally involve the use of coil or diaphragm springs.

Diaphragm spring

Most dry clutch systems use a diaphragm spring. This is a single metal plate, made into a series of sprung steel fingers. It is slightly dished in shape. When one end of the fingers is pressed by the clutch release bearing, the fingers pivot about a **fulcrum**. This moves the opposite end of the diaphragm fingers in the other direction. When this happens, the pressure plate is moved away from the friction plate and disengages the clutch.

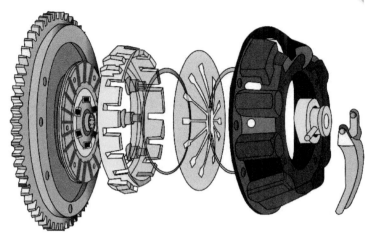

Figure 3.3 Diaphragm spring clutch

Fulcrum – a pivot point, like the one on a seesaw.

Preload - an initial setting or tension applied to a component before it is operated.

When the driver lifts their foot off the clutch pedal, the ends of the fingers on the diaphragm spring are released. Because the steel fingers are sprung, they return to their original position and reapply pressure to the friction plate. This then reconnects the drive. Because the spring diaphragm fingers are made from a single piece of metal, an even clamping force can be produced.

Self-adjusting clutches

Many manufacturers are now producing self-adjusting diaphragm spring clutches. In this design the main diaphragm spring is not permanently attached to the clutch cover. Instead it rests against another diaphragm spring which keeps it in tension against the pressure plate. This way, as the clutch friction plate wears, any excessive free play is taken up by the **preload** of the second diaphragm spring. This will give a consistent feel to the clutch operation during its normal lifespan.

It is often necessary to use a specialist tool to align and provide initial tension on the self-adjusting mechanism if clutch units are replaced. If this is not done, judder, premature wear and failure may occur. Always follow manufacturer's instructions.

Figure 3.4 Clutch alignment/tension tool

Advanced Light Vehicle Technology

Coil spring

A number of early dry clutches used coil springs to provide the clamping effort. Because of wear caused by normal operation, coil spring tension and length will change over a period of time. This can result in an uneven clamping force being produced on the friction plate. This uneven clamping force can lead to clutch **drag**, **slip** or vibrations (often referred to as clutch **judder**), caused as take-up of drive is required. It is now more common for coil springs to be used as the release mechanism in a multi-plate clutch where hydraulic pressure has been used to provide the clamping force. When the hydraulic pressure is released, the coil springs react to release and separate the friction plates, allowing the disengagement of the clutch.

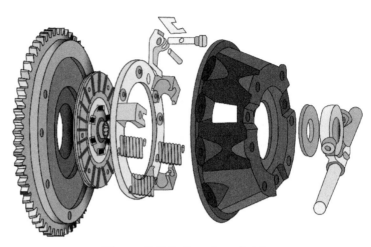

Figure 3.5 Coil spring clutch

Drag – when the clutch is not fully disengaged and the friction surfaces rub against each other.

Slip – when the clutch is not fully engaged and the friction surfaces slide over each other.

Judder – a vibration felt during the take-up of drive.

Clutch by wire (CBW) and semi-automatic gearbox clutch operation

The operation of some standard friction clutches have been modified in order to incorporate electronic control. Two main systems are used.

In a clutch by wire system, the physical connection between the pedal and the clutch components has been removed. The pedal is attached to a sensor mechanism which simulates the feel of a standard clutch. As the driver moves the pedal to engage and disengage the clutch, an ECU calculates the best possible operation of the clutch mechanism depending on engine/vehicle load and speed.

This still allows the driver to control when the clutch is engaged and disengaged, but can overcome many of the problems caused by poor manual clutch control. Because of this, vibrations created when pulling away are eased, strain and component wear are reduced. The integration of other system information, such as engine management data, ABS and traction control, work in

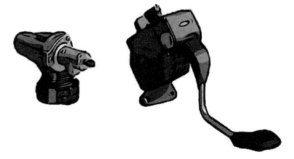

Figure 3.6 Clutch by wire CBW

conjunction with movement sensors on the clutch to provide the best possible delivery of power through the clutch.

Automotive Master Technician

Some systems have removed the clutch pedal completely. Various engine and transmission sensors work together and control hydraulic fluid pressure to the clutch slave cylinder. The fluid pressure is created by a pump and stored in an accumulator unit for use at a moment's notice. Depending on the actions of the driver and the requirements of the vehicle, the clutch can be operated in a fully electronic automatic manner. As the driver applies or releases the brakes and moves the gear shift, sensors decide if the clutch should be engaged or disengaged. The main actions of the clutch (providing a smooth take-up of drive, providing a temporary position of neutral and disengaging the engine from the gearbox to allow for gear change) are all controlled by the ECU, giving better vehicle control and driver comfort.

Notes

If the car is run in semi-automatic mode, a system is normally incorporated to inform the driver if they have selected the wrong gear. Many systems have a gear number indicator on the dashboard or may cut fuel injection in and out, so the engine hesitates and prompts the driver to change to a more appropriate gear ratio.

Dual mass flywheels (DMF)

As engine technology and driver comfort have advanced, many manufacturers now use dual mass flywheels (DMF) in the design of their transmission systems to help reduce driveline vibrations, particularly in Diesel or direct injection petrol engines.

The flywheel forms the driving surface for most clutch systems, although the main purpose of a flywheel is to store kinetic energy and keep the crankshaft turning during the engine's non-power strokes.

The delivery of power strokes to the crankshaft is not smooth and can create **pulsations** in the transmission system. These pulsations can normally be seen as vibrations at the gearstick. Over a period of time, the shaking created through the transmission will lead to premature wear and damage of gearbox components.

To help reduce judder during the take-up of drive, many clutch friction plates have **torsional** damping springs incorporated in their design. A DMF works on a similar principle. It is essentially two separate flywheels connected by a series of torsion springs and dampers. A friction ring between the two main flywheel sections allows them to slip across each other in a **radial direction**. Both flywheel sections are supported in the centre by a bearing which carries most of the load. During normal operation, pulsations from the crankshaft are smoothed out, giving a more comfortable ride and increasing the overall lifespan of transmission components. The main benefits of a DMF include:

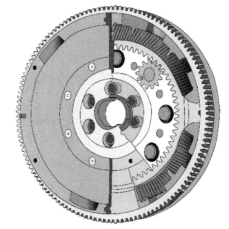

Figure 3.7 Dual mass flywheel DMF

- Elimination of gear rattle.
- Reduced drivetrain noise.
- Less need to change up and down gear so often.
- Less synchroniser wear.
- Lower engine operating speeds can be used, saving fuel and reducing emissions.
- Less drivetrain torque fluctuation.

A disadvantage of DMFs is that, if the damping springs and connecting components between the two flywheels wear out, pulsations from the crankshaft are exaggerated. This will create rapid wear in other transmission components. This means that when you a changing a clutch, you need to examine DMFs carefully and replace them if necessary.

Terms

Pulsation – a rhythmic vibration or oscillation.

Torsional – using a twisting action.

Radial direction- moving in a circular motion.

Notes

Dual mass flywheels are not normally backwardly compatible, meaning that if they are damaged, they cannot simply be exchanged for a standard flywheel and clutch. This is because the engine and its operation have been specifically designed around the use of a DMF and will not function correctly if a solid flywheel is used.

Tip

A worn dual mass flywheel (DMF) can often cause vibrations to be seen at the gear stick.

Launch control

Some manufacturers are now producing road cars with a system of launch control. This is a system that has its origins in racing and motor sports. To achieve the quickest possible start and acceleration from rest, the engine and transmission ECU's work together to control the throttle and clutch. Once the driver has set a switch inside the vehicle, the throttle can be held wide open. When the driver releases the brake, the car will pull away in the most efficient manner, avoiding spinning the wheels or over-revving the engine and damaging the clutch or gearbox.

Seamless shift dual clutch systems (DCS)

Some manufactures now use a paddle shift gear change mechanism. Instead of the traditional gear lever and clutch pedal, a pair of flat levers or 'paddles' are mounted behind the steering wheel. When the driver wants to select a different drive ratio, they simply operate one of the paddles to change up or down. The paddles send a signal to a transmission control ECU, and electronics, actuators and hydraulics then perform the actual gear change in the gearbox.

To improve the speed and smoothness of the gear change in this type of transmission, a dual clutch, seamless shift design is used. The gearbox has two different shafts to support the drive gears. One shaft is hollow with the other shaft running through the middle, and they can rotate independently of each other, although from the outside it may look like a single shaft. The odd-numbered gears are mounted on one shaft and the even-numbered gears are mounted on the other. Each shaft has its own multi-plate clutch pack, leading to the term 'dual clutch'.

- When the driver selects first gear, actuators in the gearbox move the selector hub, which locks first to the drive of the gearbox. The multi-plate clutch connected to the odd-numbered shaft is engaged by hydraulics and drive is taken up.
- When the driver wants to change up to second gear, they operate a paddle behind the steering wheel. Actuators in the gearbox move the hub to select second gear. Once second gear is selected, the multi-plate clutch pack on the odd-numbered shaft is disengaged and the multi-plate clutch pack on the even-numbered shaft is engaged hydraulically.

The smooth change over between the clutch packs and gear shafts gives an almost seamless gear change. This process continues for all other gears in a sequential pattern, up and down the gearbox.

Automotive Master Technician

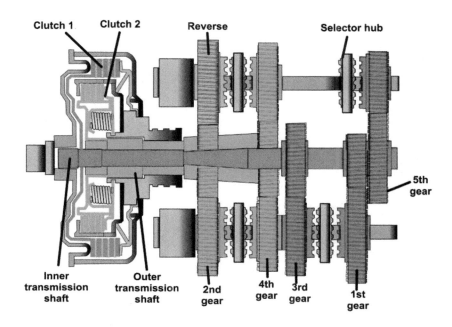

Figure 3.8 Seamless shift dual clutch system

In a standard transmission system, if two different gear ratios are selected at the same time, because they would try and rotate a different speeds, the gearbox would lock solid. In a dual clutch seamless shift transmission, because the gears are mounted on separate shafts, the next gear up or down (depending on which paddle has been operated) can be pre-selected/engaged and will only come into operation when the DCS system swaps over input drive.

Fluid couplings

An automatic transmission system also needs a method of transferring drive from the engine to the input shaft of the gearbox, but a friction clutch is not always appropriate. A fluid coupling is a component that uses hydraulic forces to create drive. A fluid coupling is still able to provide a temporary position of neutral so that the vehicle can be held stationary while in gear without stalling the engine. One of the most common fluid couplings is the torque converter.

A torque converter is mounted at one end of the engine crankshaft, in a similar position to a standard friction clutch. It consists of three main components (as shown in Figure 3.9):

- Impeller
- Turbine
- Stator

Advanced Light Vehicle Technology

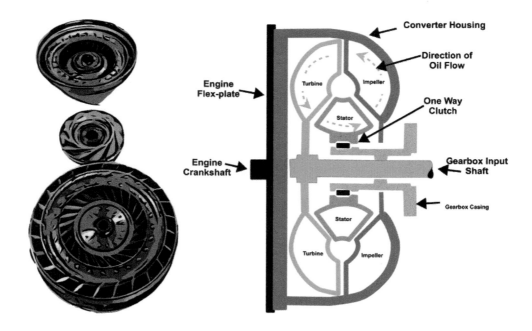

Figure 3.9 Torque converter

These components are sealed inside the torque converter casing. The casing is pressurised with automatic transmission fluid (ATF) from a crankshaft-driven oil pump. When the torque converter is spun by the engine crankshaft, fluid is taken into the impeller blades and thrown outwards by **centrifugal force**.

Centrifugal force – a force that makes rotating objects move outwards.

Torque converter operation

Fluid exiting the impeller at the outer edge strikes the blades of the turbine, making it spin. The centre of the spinning turbine is connected to the input shaft of the gearbox which now also turns.

The hydraulic fluid (ATF) now leaves the turbine and strikes the blades of the stator, which direct it back into the impeller at high speed. This force helps to multiply the torque provided by the crankshaft and leads to the name torque converter.

The largest amount of torque multiplication happens when there is the greatest difference in speed between the impeller and the turbine. This is usually when the vehicle is starting to pull away and is sometimes known as **stall**. The **torque multiplication** at this time can be around 2.2:1.

As the speed of the impeller and the turbine begin to synchronise, torque multiplication falls to zero. This is called **coupling point**. At coupling point, fluid is leaving the centre of the turbine blades with such speed that it would create drag as it strikes the blades of the stator. The stator is mounted on a one-way clutch, and as coupling point is reached, the action of the hydraulic fluid striking the blades makes the stator **freewheel** and prevents drag.

Slip between the turbine and impeller can reduce performance, so as the torque converter reaches coupling point, a hydraulically operated lock-up clutch can be used. This holds all internal components together and prevents slip.

Automotive Master Technician

This helps improve fuel economy, reduce emissions and maintain engine performance. Many modern systems use electronic control to activate the lock-up clutch inside a torque converter for most efficient operation.

If the car is held stationary, using the brakes, the turbine is also held still while the impeller spins. As the brakes are released, the hydraulic action of the transmission fluid striking the turbine blades makes them turn and provides a smooth take-up of drive.

Torque multiplication – an increase in engine turning effort.

Stall – the point of greatest torque multiplication, when the impeller and turbine are moving at different speeds (usually moving away from rest).

Coupling point – when the impeller and turbine are turning at the same speed and torque multiplication falls to zero.

Freewheel – to spin freely with no drive connection.

Creep – movement of the vehicle as the brakes are released, caused by drag inside the torque converter.

Early fluid flywheels did not contain a stator to help redirect fluid flow inside. As a result, the drag created when pulling away made them very inefficient.

A feature of the torque converter is that when the brake pedal is released and fluid from the impeller begins to react against the turbine and drive to the gearbox begins. This means that the vehicle will start to move, unless it is held stationary by the brake. This movement is often referred to as **creep**.

If an automatic vehicle is to be left stationary for any period of time with the engine running, you must take it out of gear and select the neutral or park position. The neutral position will simply remove the connection to the gearing inside the box, but the park position has the added advantage of locking a lever mechanism, known as a 'pawl', into the gearing to physically prevent any further movement.

Advanced Light Vehicle Technology

Table 3.4 lists some of the advantages and disadvantages of torque converters.

Table 3.4 Advantages and disadvantages of torque converters

Advantages	Disadvantages
Multiplication of engine turning effort which is not accomplished using other forms of clutch.	Drag and slip created inside a torque converter reduce overall performance, making it inefficient.
Comfort and convenience, as no driver interaction is required to operate the smooth take-up of drive.	Creep created by the torque converter when the vehicle is placed in gear can allow the car to move unexpectedly and cause an accident.
Normally very long lasting, as little wear takes place in the main torque converter components.	If the torque converter goes wrong, it can be a very expensive component to replace.

Manual and automatic gearboxes

Depending on the design of an engine, it delivers **torque** in a very narrow rev band. This means that the greatest effort being produced by the engine is only available when the engine is running at certain speeds. When moving from rest, climbing a hill or transporting heavy weights, the vehicle will be under load and require large amounts of torque. To overcome this, the gear ratios need to be raised to a point where torque is increased and speed is decreased. When travelling at speed under light loads, very little torque is required to maintain **momentum**. To ensure good fuel economy and low exhaust emissions, the gear ratios need to be reduced to a point where speed is increased and torque is decreased. This is known as **overdrive**. In reality, a combination of these conditions exist during normal driving. A gearbox is needed in order to make the vehicle drivable in all situations.

Figure 3.10 The need for a gearbox

Torque - turning effort or force.

Momentum – movement created by the speed and weight of the vehicle.

Overdrive – when the output speed of a gearbox is higher than the input speed.

Gearbox requirements for different engine types

A Diesel engine produces torque low down in its rev range. This means that the crankshaft is turning more slowly when the greatest amount of torque is created. As a result, the gearbox has to be designed so that this torque can be transmitted to the road with sufficient speed for general use.

A petrol engine produces torque higher in its rev range. This means that the crankshaft is turning fast when the greatest amount of torque is created. As a result, the gearbox has to be designed so that speed is reduced and torque is multiplied. It is then transmitted to the road, depending on load conditions, so that it is suitable for general use.

A hybrid vehicle using a combination of a petrol engine and an electric motor produces a combination of torque. An electric motor gives its greatest amount of torque when starting from rest. A petrol engine gives its greatest amount of torque at speed. Many manufacturers use a continuously variable transmission (CVT) with hybrid vehicles which will deliver the most efficient amount of torque and speed to the road wheels no matter which motor is driving at the time.

A series hybrid car doesn't need a gearbox to achieve movement. The internal combustion engine acts as a generator to charge batteries or create electricity which operate the electric traction motors to drive the wheels. As there is no direct connection between the engine and wheels, it doesn't need to have the torque multiplied as this is supplied directly by the electric drive motors. Reverse will be achieved by allowing the motor drive to operate in reverse or by transmitting movement through a counter gear. (*See Chapter 5*)

The construction and operation of manual gearboxes

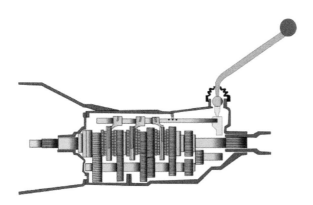

Figure 3.11 Manual gearbox

A transmission casing mounted between the engine and final drive unit contains gears of varying sizes. When engaged with each other, these gears multiply torque through the principle of leverage.

- When a small drive gear is connected to a large driven gear, torque is multiplied and speed is reduced.
- When a large drive gear is connected to a small driven gear, torque is reduced and speed is increased (overdrive).

Figure 3.22 Torque multiplication using gears

Advanced Light Vehicle Technology

Gear selection

Because gears of different sizes are used in the construction of a standard manual gearbox (i.e. non-dual clutch DCS) no two different gear ratios can be engaged at the same time as this would cause the gearbox to lock solid. Traditionally a series of rods and levers were used to operate selector forks, which moved the synchro hubs connecting the gears to the output shafts, in and out of engagement. This way, only one drive gear is used at any one time.

In modern electronically controlled manual transmission systems, when the driver operates the selector inside the car, instead of manually pushing rods and levers, the selector forks are moved by solenoids or hydraulic pressure. This gives quick, efficient and accurate gear selection. Because the function of the solenoids or hydraulics is electronically controlled, dynamic vehicle operation can be analysed by the management system and gear change timed or altered as required. Some systems will not allow gear change to take place until certain driving conditions have been met, with regard to speed or load. Other systems are able to override the driver if required, i.e. changing into first gear for example when the vehicle becomes stationary, ready to pull away.

Badly fitted radio transmission equipment such as telephones can affect gear selection on electronically controlled transmission systems.

Sequential gearbox operation

A **sequential** gearbox is one where the gears must be selected in order, one after another. To move up and down the gears, the driver operates a gear lever, but instead of acting on a series of selector rods, a ratchet mechanism rotates a selector drum. The selector drum is a cylindrical component with angled grooves machined in the outer surface. A selector fork is located by a peg in each of the machined grooves.

As the drum is rotated, the selector forks are forced to move in the direction of the machined grooves, causing the synchromesh selector hub to engage a single gear. The grooves on the surface of the selector drum are designed so that as one gear is disengaged, another is engaged. This removes the need for an **interlock** device.

An overshift limiter mechanism is often incorporated in the design of a sequential gearbox. It only allows the driver to change up and down the box one gear at a time in sequence, so gears cannot be skipped.

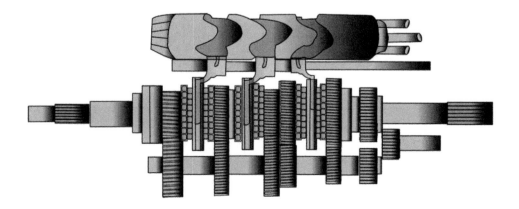

Figure 3.23 Sequential gearbox

Automotive Master Technician

> **Sequential** – in sequence, one after another, following a set pattern.
>
> **Interlock** - a mechanism that prevents two gears being accidentally selected at the same time.

Automatic transmission

Unlike a manual gearbox, where the driver selects the gear ratio for the driving situation, in an automatic gearbox, engine speed and load are detected by the transmission and the system itself chooses the most appropriate gear.

Many automatic transmission systems use a different type of gearing method from that of manual transmission systems. Automatic transmissions still need to achieve the appropriate ratios required for the multiplication of torque. In a standard automatic transmission, instead of **spur** or **helical gears** which are engaged and disengaged, a system called an epicyclic gear train is used.

Figure3.24 An automatic transmission

Epicyclic gear train

An epicyclic gear train uses gears that are constantly in mesh and consist of:

- A large outer ring gear, often called the 'annulus'.

- A central gear, often called the 'sun gear'.

- A series of intermediate gears (that sit between the sun gear and the annulus), called the 'planet gears'. These are supported on spindles attached to a planet carrier.

Annulus or ring gear

Planet gear

Planet carrier

Sun gear

Figure 3.25 Epicyclic gearing

Advanced Light Vehicle Technology

Spur gears - gear wheels with straight cut teeth.

Helical gears - gear wheels with teeth cut on an angle or spiral (helix).

To select an appropriate gear ratio, one section of the epicyclic gearing will be locked to another part of the transmission. This section of gearing then becomes an idler gear with no direct effect on the gear ratio, meaning the remaining two gears become input and output. The differing numbers of teeth on the input and output are now able to provide various gear ratios, including reverse. The gear ratios show how much the torque is increased (multiplied).

A single epicyclic gear mechanism is able to provide three forward gear ratios and one reverse, as shown in Table 3.5.

Table 3.5 example of epicyclic gear selection, torque increase and direction of travel

Stationary	Input	Output	Ratio and direction
Annulus	Sun gear	Planet carrier	3.4:1 (forward)
Sun gear	Planet carrier	Ring gear	0.71:1 (forward)
Planet carrier	Sun gear	Ring gear	2.4:1 (reverse)
When annulus, planet carrier and sun gear are locked together			1:1 (forward)

These gear ratios are not suitable for all driving conditions and, as a result, many systems use at least two epicyclic gear sets joined together to form a compound gear train. Two main types of compound gear set are common:

- **Simpson gear set**: A single long sun gear is used between the two epicycles to join the gear sets together.
- **Ravigneaux gear set:** In this design, it is the planet carriers that are connected to join the gear sets together.

Most automatic gearboxes use hydraulic fluid pressure to control the operation and selection of gears. Automatic transmission fluid ATF under pressure is directed through a series of channels and galleries by valves. The hydraulic fluid will then operate brake bands or multi-plate clutches and provide engagement of a particular gear ratio. Early systems relied on fluid pressure created by an engine driven oil pump to sense load and speed and initiate gear change. Modern systems now use engine management and transmission sensor data to control the operation of solenoid valves in a valve block. The sensor information is processed by a transmission ECU, which actuates the solenoid valves controlling the hydraulic system of the automatic gearbox. In this way, gear ratio and shift timing can be accurately matched to all road situations and driver demands.

Figure 3.26 Automatic transmission valve block

Three main methods are used to control the selection of gearing inside an automatic transmission system:

- Brake bands.
- Multi-plate clutch packs.
- Unidirectional clutches.

Brake band

Brake bands are an actuator system that hold a section of the epicyclic gearing stationary by anchoring it against the transmission casing. A brake band is similar to a belt that is wrapped around the outside of one of the gearing components. One end of the brake band is fixed to the transmission casing by an adjustable mounting. The other end of the brake band is connected to a hydraulic servo, which when operated, will try to squeeze the ends of brake band together. The inside of the brake band usually has a friction material attached so that when it is pinched together, it provides grip and stops part of the gear set rotating (*see Figure 3.27*).

Over a period of time the friction material will wear, and grip on the gear set will be reduced. You can usually access the fixed end of the brake band from the outside of the gearbox casing to adjust it during maintenance procedures.

Figure 3.27 Brake band

Multi-plate clutch packs

To connect two rotating gear sets in an epicyclic gear mechanism, a hydraulically operated multi-plate clutch pack can be used. A number of friction plates, interspaced with plain plates, are housed in a clutch basket. A pressure plate is mounted on the outside end of the clutch basket. Hydraulic pressure acts on the pressure plate, squashing the component parts together to create friction. The friction provides the grip that locks parts of the rotating gear set together.

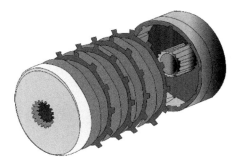

Figure 3.28 Multi-plate clutches

Unidirectional clutches

To prevent parts of an epicyclic gear set rotating in the wrong direction, a unidirectional (one-way) clutch can be used. This is similar in construction to ball bearing race, but instead of balls, a set of sprags are used. A sprag is a small 'S'-shaped wedge held between an inner and outer bearing race. When rotated in one direction, the 'S'-shaped sprag tilts to one side, allowing the two bearing races to turn freely. When rotated in the opposite direction, the 'S'-shaped sprag wedges itself tightly between the two bearing surfaces and stops it from turning.

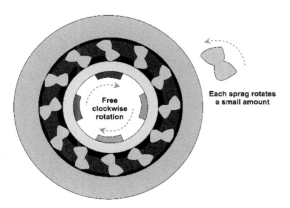

Figure 3.29 Unidirectional clutch

Clutch and brake band operation for automatic transmissions

Table 3.6 gives an example of how brake bands, multi-plate clutches and unidirectional clutches are combined to connect epicyclic sets for gear selection.

Table 3.6 Clutch and brake band operation for automatic transmissions

Gear	S1	S2	S3	C0	U0	C1	C2	B0	B1	B2	U1	B3	U2
Park	Active			Active									
Reverse	Active			Active			Active					Active	
Neutral	Active			Active									
First	Active			Active	Active	Active							Active
Second	Active	Active		Active	Active	Active				Active	Active		
Third		Active		Active	Active	Active	Active			Active			
Overdrive			Active			Active	Active	Active		Active			
Low Range Second	Active	Active		Active	Active	Active			Active	Active	Active		
Low Range First	Active			Active	Active	Active						Active	Active

Key:

S = Electronic control solenoid	C = Multi-plate clutch pack	B = Brake band	U = Unidirectional (one-way) clutch

To help with the diagnosis of automatic transmission system faults, the vehicle can be road tested through all of the gears and under different driving conditions to see how it reacts. Once this has been completed, a chart of operation such as the one shown in Table 3.6 can be used to help locate the approximate position of the fault before any stripping down is started (*see crime scene investigation CSI*).

Automotive Master Technician

Tip

If automatic transmission shift solenoids are disconnected it can often force the transmission into fail safe for the checking of the gearbox mechanical operation. Remember to only conduct this test if safe to do so and any diagnostic trouble codes or adaptions caused by this action will require re-setting following your repairs.

Continuously variable transmission (CVT)

A continuously variable transmission (CVT) gearbox is a form of automatic transmission. Although the design has been around for over 100 years, many manufacturers are now offering this type of gearbox as an option because of its efficient delivery of torque and power. Instead of using a fixed set of five or six gear ratios, this type of transmission is able to offer an infinitely step-less ratio between an upper and lower limit. This means that when coupled to an engine, it is always able to run within its optimum range. Two main types of CVT are used:

- Variable diameter pulley (VDP)
- Toroidal CVT

Variable diameter pulley (VDP)

Instead of using mechanical gear sets, as in an epicycle, VDP uses a drive belt held between two pulleys similar to the chain and sprockets on a bicycle. Originally this belt was made of rubber but, as technology has developed, a steel drive belt has replaced the original design.

A bicycle is able to vary its gearing by changing the size of the sprocket on which the drive chain runs. VDP operates by changing the size of the drive pulleys, which allows different gear ratios to be created. To do this the drive pulleys expand and contract. In this way, the drive belt is able to ride up and down within the pulleys, varying their size and therefore the gear ratio. As one pulley expands the other will contract equally.

The steel drive belt is made up of many small links held together on a metal band. As the drive pulleys rotate, the metal links are forced into compression, causing the belt to push rather than pull.

Because these pulleys do not rely on fixed gear sizes, a step-less gear ratio can be achieved, which maintains optimum efficiency for any engine speed or load. The output from the drive pulleys is normally transmitted through a further reduction gear, which can be of epicyclic design. This allows a reverse gear to be included.

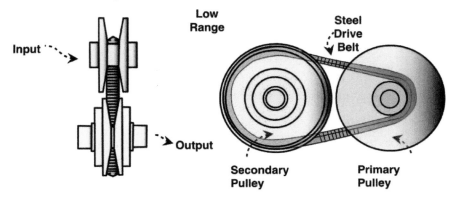

Figure 3.30 Variable diameter pulley low range

Advanced Light Vehicle Technology

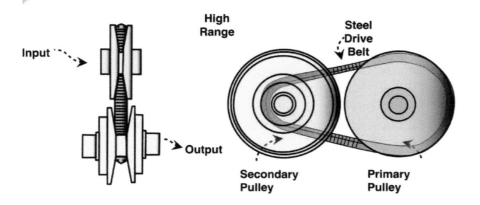

Figure 3.31 Variable diameter pulley high range

Toroidal CVT

A toroidal CVT has a tapered input disc and output disc which are placed face-to-face to form a **concave** driving surface. The input and output discs are able to turn independently of one another and are connected using **torus**-shaped rollers. The rollers are able to ride up and down against the concave surface of the input and output drive discs and transfer turning effort between the two.

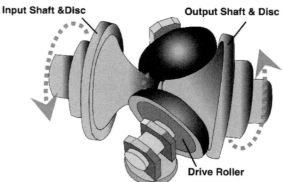

- When the roller is touching a low point on the input disc curve and a high point on the output disc curve, a low gear ratio is achieved.

- When the roller is touching a high point on the input disc curve and a low point on the output disc curve, a high gear ratio is achieved.

By moving the rollers across the surfaces of the input and output discs, a continuously variable transmission (CVT) ratio can be achieved. The output from the drive disc is normally transmitted through a further reduction gear, which can be of epicyclic design. This allows a reverse gear also to be included.

Figure 3.32 Toroidal CVT

Concave – curved inwards.

Torus – ring-shaped like a doughnut.

Because a continuously variable transmission is able to transmit turning effort to the road wheels in a very efficient manner while keeping the engine at its optimum speed, this design has become very popular with manufacturers of hybrid vehicles (*see Chapter 5*).

Automotive Master Technician

Drive connection

To join the engine to a continuously variable transmission system and allow a smooth take-up of drive and a temporary position of neutral, some form of clutch or torque converter is required. Three main methods are used to achieve this:

- **Torque converter**: If a CVT is connected to the engine via a standard torque converter some efficiency can be lost due to the drag created by the automatic transmission fluid and its overall weight.

- **Centrifugal clutch**: Early systems used a centrifugal clutch, where a set of clutch shoes, similar in construction to brake shoes, were rotated by the engine. Centrifugal force acting on the shoes moved them outwards until they contacted a drive drum. This drive was then transmitted to the CVT system. This was also a very inefficient drive connection between the engine and gearbox.

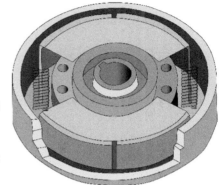

Figure 3.33 Centrifugal clutch

- **Electromagnetic clutch:** In this system, a housing of similar size and shape to a standard clutch is mounted on the end of the engine crankshaft. The housing contains a metallic powder, which when energised by an electromagnet, bonds together to provide drive to the CVT. The electromagnet is managed by a transmission ECU, which is able to vary the strength of the magnetic field and therefore control the take-up of drive.

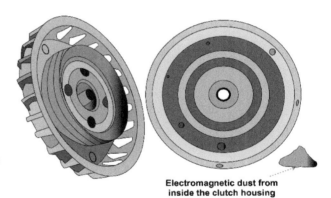

Electromagnetic dust from inside the clutch housing

Figure 3.34 Electromagnetic clutch

Differentials

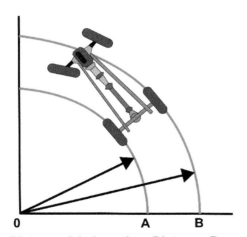

Distance A is less than Distance B

RPM of the inside wheel is slower than the RPM of the outside wheel

When a car is travelling along a straight piece of road, both of the driven wheels cover the same amount of ground at the same speed. When it comes to take a bend, the inner driven wheel doesn't have to travel as far as the outer one, and so needs to travel more slowly. If both wheels rotated at the same speed when trying to turn the corner, the inner wheel would be forced into a skid.

On a bend, drive must be transmitted at different speeds, and this is done by using a **differential** unit housed inside the final drive casing. When needed, some of the driving force from the inner wheel is transferred to the outer wheel. This speeds up the outer wheel and slows down the inner wheel.

Figure 3.35 The need for a differential

Advanced Light Vehicle Technology

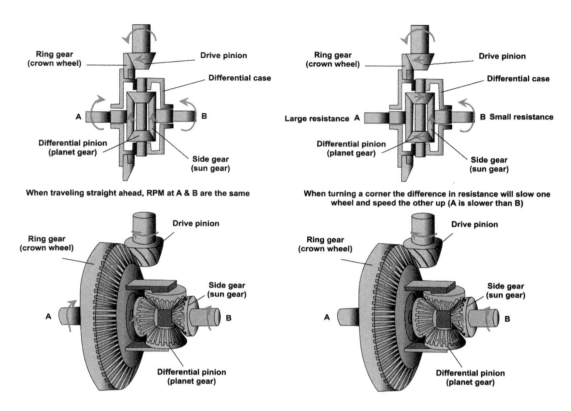

Figure 3.36 Differential operation

A differential unit allows this because of its internal gearing system. The turning effort taken from the crown wheel is transmitted to the differential casing, where a metal drive pin is fixed. As the differential casing turns, the drive pin moves end over end. On the drive pin are mounted two small gears often called 'planet gears', which are in constant mesh with two side gears, often called 'sun gears' (*see Figure 3.36*).

Differential – a mechanism that allows one driven wheel to travel faster than the other when the car goes around a bend.

Automotive Master Technician

Driving in a straight line

When the car is travelling in a straight ahead direction, the drive pin turns end over end, locking the planet gears directly to the side gears and driving them all at the same speed, as shown in the first flow chart. *(Figure 3.37)*

Turning a corner

As the vehicle turns a corner, the extra load tries to slow down one wheel and reduces the speed at one sun gear. The drive pin still turns end over end, providing torque or turning effort to the sun gears, but the planet gears now rotate on the pin, allowing more drive to be transmitted to one wheel than the other. This allows one wheel to travel faster than the other but still transmits drive with the same torque.

Turning a left hand bend

The second flow chart shows the process involved when turning a left hand bend. *(Figure 3.38)*

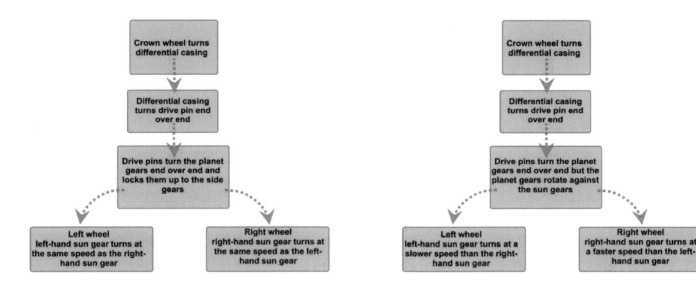

Figure 3.37 Driving straight ahead   **Figure 3.38** Turning a left-hand bend

Limited slip differentials (LSD)

A disadvantage of a standard differential is that if one wheel loses traction, nearly all of the turning effort will be directed away from the wheel with grip. This means that one wheel will spin uncontrollably and no drive will be transmitted.

A limited slip differential (LSD) is designed to transmit an equal torque to both driving wheels when the car is travelling in a straight ahead direction, but still allows standard differential action when going round a corner. Because of this, if one wheel loses traction, the other will still have some drive transmitted and this will give greater vehicle control. Three main methods of limited slip differential are in common use:

- Clutch type LSD
- Viscous coupling type LSD
- Torsion wheel type LSD

Advanced Light Vehicle Technology

Clutch operation

A series of multi-plate clutches are included in the design of the clutch type limited differential unit. Four planet gears are mounted on two drive pins. The ends of the drive pins are **tapered** and sit in corresponding tapered grooves in the differential casing.

When travelling in a straight ahead direction, forces acting on the planet gears, cause the drive pins to move up and outwards in the tapered grooves of the casing. This movement creates pressure on the clutches, clamping them together to create friction and transmit drive to both road wheels.

As the vehicle turns a bend, the load on the clutches is reduced as the drive pins move back down into the grooves. Pressure is reduced on the clutches, allowing them to slip; this lets normal differential action take place.

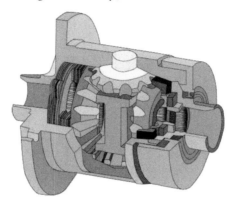

Figure 3.39 Clutch type limited slip differential LSD

Tapered – angled inwards slightly (getting narrower).

Viscous coupling operation

A chamber inside the differential is constructed that contains a **viscous** (thick) liquid and a series of rotor blades. When travelling in a straight ahead direction, the viscous liquid creates drag and reduces slip in the differential gears and attempts to transmit an even torque to both drive shafts. As the car turns a bend, normal differential action can take place because the rotor blades are able to **shear** through the viscous liquid, creating an acceptable amount of slip.

Automotive Master Technician

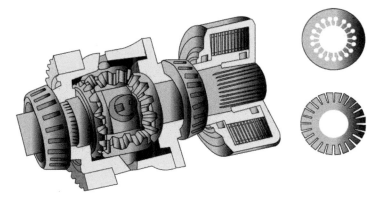

Figure 3.40 Viscous coupling type limited slip differential LSD

Viscous – describes a liquid with a high resistance to flow (a thick liquid).

Shear – to slice through something with a similar action to scissors.

Worm gear – a gear consisting of a spiral threaded shaft and a wheel with teeth that mesh into it.

Torsion wheel operation

A series of **worm gears** are included in the design of this differential unit. In a worm and wheel gear set up, the worm can easily drive a gear, but the gear is unable to drive the worm (this allows drive in one direction only). When travelling in a straight ahead direction, gear teeth lock against the worm drive, providing equal torque to both wheels at the same time. As the car turns a bend, one wheel slows down while the other speeds up. When this happens, the worm turns the gear, allowing one wheel to travel faster than the other and provide normal differential action.

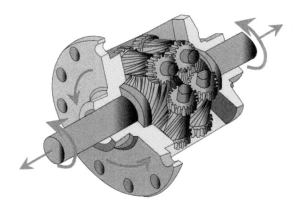

Figure 3.41 Torsion wheel limited slip differential LSD

Advanced Light Vehicle Technology

It is advisable not to test the brakes of a vehicle with a limited slip differential in a brake rolling road. If you are operating or testing wheels individually, you may obtain inaccurate readings or damage could be created in the LSD.

Automatic brake differential (ABD)

An automatic brake differential (ABD) is a method of creating the operating characteristics of a limited slip differential, but with the ability to add electronic control. A series of multi-plate clutches are mounted in a similar position to those found in a clutch type limited slip differential. When engaged, these clutches are able to transmit driving torque to the wheel which has the greatest amount of grip.

Pressure to activate the clutches in an ABD system is normally created by an electromagnet. The current flowing in the electromagnet can be varied by an ECU to manage the amount of slip needed at the clutches and provide the best delivery of torque to the most appropriate wheel.

An advantage of this type of system, when compared to a standard limited slip differential, is that it can work in harmony with other active safety systems to provide dynamic vehicle control.

Other systems that ABD is able to interact with include:

- Anti-lock braking systems (ABS).
- Traction control systems (TCS).
- Electronic stability programs (ESP).
- Electronic brake force distribution (EBD).
- Active yaw control (AYC).

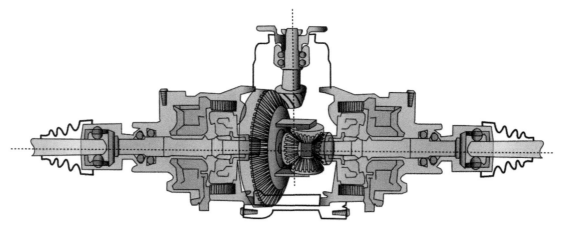

Figure 3.42 Automatic brake differential ABD

It is sometimes possible to determine if a car has a limited slip differential by raising the driven wheels off the ground and placing the transmission in gear. If one of the driven wheels is turned in a particular direction and the opposite wheel rotates in the reverse detection, a standard differential has been used. If both wheels rotate in the same direction, a limited slip differential LSD has been used.

Automotive Master Technician

Four wheel drive

In a four-wheel drive vehicle, as drive exits the gearbox, an extra unit called a transfer box is often employed, which splits the drive so that it can be used by the front and rear axles. A four-wheel drive vehicle has at least two differential assemblies, one for the front wheels and one for the back. In addition to this, vehicles with permanent four-wheel drive often have a central differential that splits the drive from front to rear.

When a vehicle is travelling with forward motion, the front axle reaches a bend first. Because of this, the front axle needs to be travelling at a different speed to the rear axle. A central differential is sometimes used to allow for the difference in speed between the two axles.
The switching between two wheel drive and four wheel drive may be manually selected, or electronically controlled using vehicle dynamic sensor information.

Haldex coupling

Some manufacturers do not use a central differential in the design of their four-wheel drive vehicles. A method is still needed to control the distribution of torque between the front and rear axles. This can be achieved using a Haldex coupling.

Normally mounted on or near the rear axle, the Haldex coupling is a multi-plate clutch unit which is able to control the amount of torque delivered to the rear axle by managing slip or drag created at the clutch plates. Pressure to engage the clutches relies on hydraulic forces created by the Haldex coupling's own internal oil pump. When the vehicle is operating, fluid pressure is controlled by a series of electronic valves activated by an ECU which is processing dynamic information from various chassis and drive sensors. This means that the vehicle can instantly respond to changes in road surface grip, giving the best amount of traction for different driving situations.

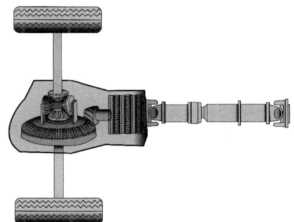

Figure 3.43 Haldex coupling

Notes

If a vehicle is to be used off road, some manufacturers include a mechanical locking mechanism in the design of their differential. This is a means of eliminating differential action on loose and slippery surfaces and improves overall traction.

Safety

A differential lock should not be used when driving on a normal road surface. Grip created at the road wheels can cause transmission 'wind-up' as the vehicle goes around a bend. A reverse torque is created in the transmission parts which places undesirable stress on shafts and gears and may cause components to break.

Advanced Light Vehicle Technology

Electronically controlled chassis systems

Four wheeled steering systems

In order for a vehicle to manoeuvre and handle correctly, all four road wheels must work together to provide true rolling motion. If this doesn't happen, safety may be compromised and rapid tyre wear could occur. Cars are primarily steered from the front axle, as this way the rear of the vehicle will follow its direction in a predictable manner. If the car were to steer solely from the rear, this would provide excellent manoeuvrability at slow speeds, but every time the steering was turned, the rear end will move sideways. The swinging outwards of the rear end will prevent a vehicle pulling away in a forwards direction when parked parallel to an object. Rear wheel steer will also be unstable during high speed driving.

Figure 3.44 Compliance bush

To improve the manoeuvrability and handling of cars, some manufacturers provide a system that allows all four wheels to be steered, combining both front and rear wheel steering.

Many cars use a passive system called **compliance** to provide small amounts of four-wheel steering. Rubber suspension bushes that support the non-steered axle are manufactured with holes known as **voids**. These voids allow the axle to flex slightly during cornering, giving a very basic form of four-wheel steering. These bushes are carefully aligned during production so that the voids are positioned correctly and will often be marked with the word 'top' so that if they are replaced during maintenance, they are correctly fitted and do not compromise handling.

Tip

Download a free app and use your mobile phone as a spirit level or 'clinometer 'to gain a quick idea if steering geometry is different side to side, by holding it against various steering and suspension components.

Terms

Compliance – the willingness to follow instructions or commands (such as steering movements).

Voids – holes manufactured in rubber bushes to allow them to flex in a certain direction.

Some manufacturers produce active four-wheel steering systems where a steering rack is provided for both the front and the rear axles.

The front and rear steering racks are connected mechanically or more commonly driven electronically by motors. The rear steering rack does not normally pivot the wheels to the same degree as the front rack, but the **Ackerman principle** and toe-out on turns are still used. When the driver rotates the steering, the front rack is controlled in the normal manner. Depending on vehicle speed and steering angle, the rear rack is operated to provide a degree of rear-wheel steering.

Automotive Master Technician

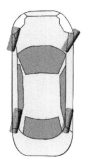

Wheels turn in the same direction

The steering racks are phased, so that when that front wheels are turned, the rear wheels initially pivot in the same direction by a small amount. In this way, when changing direction at speed, the car will have a tendency to crab sideways across the road, giving good directional stability. As the front wheels are turned at a greater angle, the rear wheels straighten up and then begin to turn in the opposite direction to the front wheels. This means that very tight turning circles can be produced for slow speed manoeuvres.

If a fault occurs in a four-wheel steering mechanism, the system is designed as failsafe. The rear axle will return to a neutral position and the steering will act only on the front wheels, as with a standard car.

When diagnosing four-wheel steering system faults, you should apply the same geometry and alignment principles to the rear steered axle as to the front steered axle.

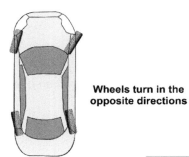

Wheels turn in the opposite directions

Figure 3.45 Four-wheel steer phasing

Terms

Ackerman principle - a series of steering geometry angles that allow the inner wheel to turn through a tighter arc when cornering. The Ackerman steering angle produces toe-out-on-turns (TOOT) and allows for true rolling motion with minimal tyre scrub.

Safety

Take care when manoeuvring a four-wheel steer car in the workshop, as the rear axle can make the back end of the car move sideways, which could catch people or equipment and cause injury or damage.

Figure 3.46 Electronic power assisted steering EPAS

Electronic power assisted steering systems

As the design of vehicles has improved, the introduction of wider, low profile tyres has placed strain on the vehicle steering system and on the driver. To help improve driver comfort and control, many modern vehicles incorporate a power-assisted steering (PAS) mechanism. These mechanisms are designed to reduce driver effort during slow-moving manoeuvres and to help maintain control of the vehicle at speed.

Early systems used hydraulics, fed from an engine-powered pump to drive pistons in the steering mechanism, which helped the driver undertake manoeuvres. More modern systems use either electro/hydraulic or fully electric power assistance. When this is linked with active safety systems, such as electronic stability programmes (ESP), it improves handling and vehicle control.

Advanced Light Vehicle Technology

With fully electronic power assisted steering, motors are used instead of hydraulics. As the driver applies effort at the steering wheel, movement and turning effort are registered by a torque sensor. An ECU then uses this information to operate a DC electric motor attached to the steering column or rack. By altering the supply voltage applied to the motor using duty cycle, the amount of assistance provided by this system can be varied to suit road and driving conditions.

The advantages of this system are:

- The motor is only operated when the steering is turned – this reduces loads, improves fuel economy and reduces engine emissions.
- The motor and control system is very compact and can be used unobtrusively, even on small cars.
- Assistance can be easily varied to provide greater help when parking.
- Less maintenance is needed as there is no fluid system or leaks.
- If combined with a vehicle radar system, it can be used to provide a self-parking function or active assistance in a lane departure safety control.

To diagnose faults with electronic power-assisted steering systems, you can often use a scan tool to retrieve diagnostic trouble codes (DTC), which you can then use to direct your fault finding routines. You can then check the input and output signals from the ECU using an oscilloscope (*see Chapter 1*).

Self-parking

Parallel parking is a slow speed manoeuvre that many drivers find difficult. As a response to consumer demand, car makers are starting to design and sell self-parking cars.
Advantages of self-parking cars include:

- Choosing a parking space is not restricted by the driver's skill at parallel parking.
- A self-parking car can often fit into smaller spaces than most drivers can manage on their own, which allows the same number of cars to take up fewer spaces.
- Parking takes less time, which helps to keep traffic moving.
- Minor damage created by parking is reduced.

Many systems operate with the driver controlling vehicle speed and direction using the normal driving controls. They have sensors distributed around the front and rear bumpers of the car, which act as both transmitters and receivers. These sensors send out signals, which bounce off objects around the car and reflect back to them. The car's ECU then uses the amount of time that it takes those signals to return to calculate the location of the objects. The electronic power-assisted steering then manoeuvres the car into the parking space.
Manufacturers are now designing vehicles that are completely autonomous and will control the drive as well as the steering. This means that the driver simply has to select an appropriate parking spot and position the vehicle close to the space. Having pressed a button, the car can then park completely by itself.

Operation of electronic ABS and EBD braking systems

To slow down and eventually stop the movement of a car, a braking system is used. Friction between brake pads and discs or brake shoes and drums is used to convert kinetic movement energy into heat.
The braking system will normally consist of:

- A brake pedal.
- A brake servo or brake booster.
- A brake master cylinder.
- Brake pipes and hoses.
- Brake callipers, discs and pads.
- Wheel cylinders, brake shoes and drums.

Automotive Master Technician

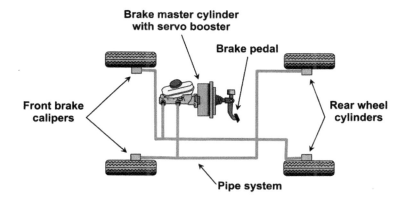

Figure 3.47 Standard brake system layout

When the brake pedal is pressed, the master cylinder, assisted by the brake servo, forces brake fluid through the brake pipes and hoses to the wheel assemblies. At the wheel assemblies, brake pads are clamped against brake discs or brake shoes are forced against rotating brake drums. This creates friction and converts movement into heat.

The slowing down of the vehicle however, can only be as good as the grip between the tyre and the road surface. If the tyre skids, vehicle control is lost and steering, braking and acceleration are no longer possible. This means that the efficiency of the braking system is compromised.

Anti-lock braking system (ABS)

If the hydraulic pressure in the brake lines can be regulated so that the tyres are prevented from skidding, then vehicle control can be maintained. This is the job of an anti-lock braking system.

Anti-lock braking uses electrics and electronics to regulate the pressure in the hydraulic system which provides an artificial form of **cadence braking**. This is an active safety system which will help the driver maintain control in emergency situations. A modern anti-lock braking system is so efficient that it is able to control the braking of the car within a 10–30 % slip tolerance. This means that the tyres are kept on the point of skidding, but the wheel should never fully lock up.

Cadence braking – a form of braking in which the pedal is rapidly pressed and released by the driver.

Electronically controlled ABS has a considerable advantage over cadence braking. Cadence braking is only able to operate in an on or off state, and acts on all four wheels at once; whereas ABS can create a regulated braking action to individual wheels many times a second.

A vehicle fitted with anti-lock brakes uses all of the parts normally associated with a hydraulic disc brake system. Three extra components are needed so that anti-lock braking can be achieved:

1 wheel speed sensors   2 an electronic control unit (ECU)   3 an ABS modulator valve block

Advanced Light Vehicle Technology

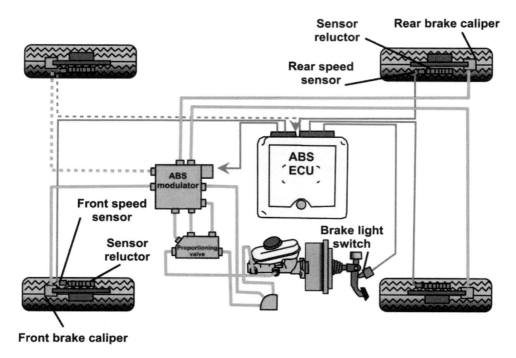

Figure 3.48 Anti-lock braking system layout ABS

These components work in partnership with the rest of the brakes so that an anti-lock system is achieved. Although sometimes mounted **inboard** on a driveshaft, most wheel speed sensors are located at the hub assembly of each road wheel. They consist of two main components:

- A toothed wheel called a reluctor or a magnetic pulse ring
- A wheel speed sensor unit.

The toothed reluctor or pulse ring is mounted on the hub so that it rotates with the wheel.

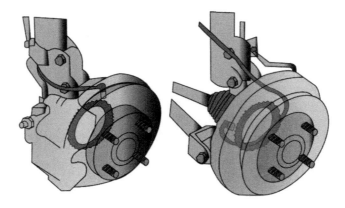

Figure 3.49 Wheel speed sensor location

Automotive Master Technician

Inboard – towards the middle of the car.

Reluctor – a toothed wheel used with a sensor to detect speed and position.

Three man types of wheel speed sensor are in common use:

- Inductive sensors (normally having two wire connections)
- Magnetic resistance element MRE sensors (normally having two wire connections)
- Hall effect sensors (normally having three wire connections)

Inductive sensors

A permanent magnet inductive sensor is mounted close to the reluctor. It consists of a thin coil of copper wire wrapped around a permanent magnet. While the road wheel rotates, the small magnetic field produced by the inductive sensor is disrupted. As a reluctor tooth comes towards the sensor, the movement of the magnetic field produces a small electric current in the copper winding which is generated in one direction. As a reluctor tooth moves away from the sensor a small electric current in the copper winding is generated in the opposite direction. If an oscilloscope is connected to a wheel speed sensor and the wheel is rotated, an AC waveform (sinewave) will be seen on the screen. When the wheel speed increases, the amplitude (voltage) or the height of the sinewave will increase and the frequency (time signal) will also increase, making the waves appear closer together.

Sinewave – a smooth oscillating (normally up and down) repeating waveform often seen on an oscilloscope.

Frequency – how often something happens; the distance/time between the peaks of a waveform.

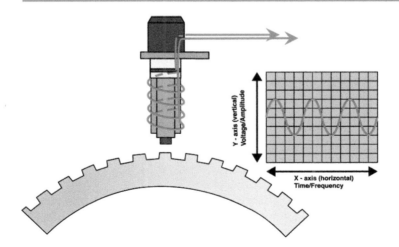

Figure 3.50 Inductive wheel speed sensor

Once the signal from the wheel speed sensor has been created, the analogue sinewave is converted into a digital signal so that it can be processed by the electronic control unit. When this is complete, the ECU will detect the frequency and convert it into a reading of speed.

Advanced Light Vehicle Technology

Magnetic resistance element MRE sensors

At first glance, an MRE sensor looks similar to an inductive wheel sensor, but unlike the inductive sensor uses a magnetic pulse ring instead of a toothed reluctor. The pulse ring is a circular metal plate with a number of segments that have been magnetised with alternating north and south poles. The sensor contains a magnetic sensitive resistor which is affected by the change in magnetic flux as the pulse ring is rotated past it. Unlike the inductive sensor, the MRE doesn't generate its own electric current, but instead relies on a feed from the ECU. As the wheel rotates, the change in resistance causes the voltage feed to be alternated creating a sinewave which is similar to that produced by an inductive sensor, however its amplitude (voltage) will not change with wheel speed. Once the signal from the wheel speed sensor has been created, the analogue sinewave is converted into a digital signal so that it can be processed by the electronic control unit. When this is complete, the ECU will detect the frequency and convert it into a reading of speed.

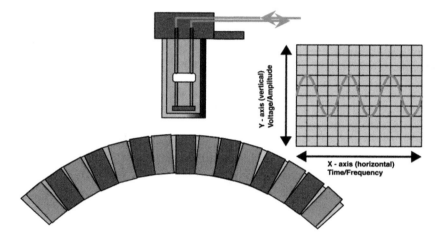

Figure 3.51 MRE wheel speed sensor

To check the magnetic pulse ring of an MRE ABS system it can be removed and laid flat. A piece of paper can then be placed over the top and metal filings scatter across the surface of the paper. The filings can now be used to identify any faulty magnetic sections.

Hall effect wheel speed sensors

Some manufacturers use Hall effect sensors to create the speed signal from the wheel. A rotating drum attached to the hub interrupts a magnetic field from a small integrated circuit, which creates a digital or square waveform output. The advantage of this system is that the signal does not have to be converted before it can be processed by the ECU. This gives the system a far greater accuracy because it only has to measure frequency rather than amplitude.

Automotive Master Technician

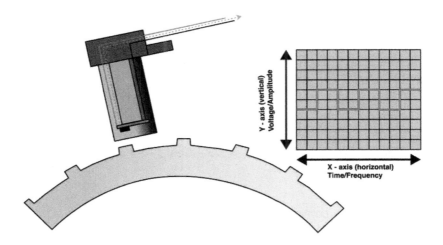

Figure 3.52 Hall effect wheel speed sensor

Analogue – a signal with a variable rising and falling voltage.

Digital – a signal with a voltage that is switched on and off.

Square wave – an angular/square oscillating (normally up and down) repeating waveform often seen on an oscilloscope.

An advantage of MRE and Hall effect wheel speed sensors when compared to an inductive type is that they are able to detect wheel movement at very slow rotational speeds. This makes them more accurate and allows greater utilisation of their sensing capabilities for vehicle stability and safety control.

ABS modulator unit

The ABS modulator unit contains a series of electrically controlled valves which regulate the hydraulic pressure in each part of the car's braking system. The valves are of a **solenoid** type – they contain a small coil of copper wire which, when energised with electricity, creates an invisible magnetic field. This magnetic field moves an **armature**, which operates a valve that restricts the flow of brake fluid to an individual brake. When the solenoid valve is switched off, a return spring moves the valve back to the open position. This means that the system is **failsafe** in the event of electrical malfunction. If the solenoids are not energised they will be held in the open position and standard braking is achieved.

Many modulators also contain an electric motor and pump which is used to return fluid to the master cylinder via an **accumulator**.

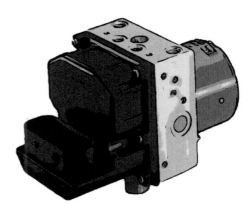

Figure 3.53 ABS modulator unit

Advanced Light Vehicle Technology

Solenoid – a linear motor (one that moves in a line rather than rotating).

Armature – the central shaft of a motor.

Failsafe – a system that will still have basic operation if it malfunctions.

Accumulator – a storage area or reservoir.

Electronic control unit (ECU)

The electronic control unit (ECU) is the computer that is used to monitor the speeds of the individual wheels and to make calculations which will decide if one of the wheels is about to lock up or skid. It then acts as a switch to turn the solenoid valves in the ABS modulator unit on and off. By doing this, it enables the ABS system to provide an electronic version of cadence braking but to each individual wheel instead of all four. The ABS ECU is also able to put wheel speed information onto the vehicles network as a data packet, allowing other systems to make use data.

ABS operation

During normal driving, the ECU monitors the speed from each wheel speed sensor but remains passive because the brake pedal has not been pressed. The ABS ECU is able to monitor the operation of the brake pedal via the brake light switch or the brake light circuit using network communication. As soon as the driver presses the brake, the ABS system becomes active. Any rapid deceleration from an individual wheel, when compared with the others, may cause that particular wheel to lock up or skid. If this happens the ABS will take the following three actions:

1 ABS pressure holding phase   2 ABS pressure reduction phase   3 ABS pressure increase phase

ABS pressure holding phase

In the event of rapid deceleration from an individual wheel, the ABS ECU will operate the solenoid valve that relates directly to the wheel that is rapidly decelerating. Using pulse width modulation (PWM) the ECU regulates the current flowing to the solenoid valve. This regulated current partially operates the valve into a position where it blocks the hydraulic input from the master cylinder. Pressure after the valve now remains constant, still allowing the wheel to slow down. No matter how hard the driver presses the brake pedal, hydraulic pressure in the calliper part of the system will not increase.

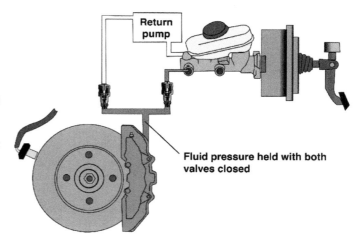

Figure 3.54 ABS pressure holding phase

Automotive Master Technician

ABS Pressure reduction phase

If the signal from the wheel speed sensor continues to indicate that the wheel is still decelerating too rapidly, the ECU will fully energise the solenoid valve. This will still block any extra pressure from the brake master cylinder, but will now open up a passageway, allowing the fluid in the brake calliper circuit to release and flow into an accumulator. As the accumulator fills up, a pump attached to the modulator unit returns the excess fluid to the master cylinder circuit. This will reduce pressure in the calliper circuit and allow the wheel to speed up again.

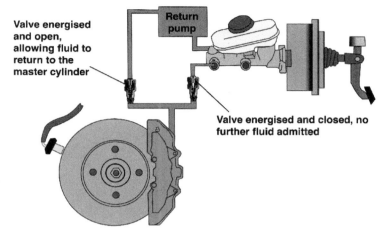

Figure 3.55 ABS pressure reduction phase

ABS pressure increase phase

As soon as the ECU receives a signal to say that the wheel is once again speeding up, it switches off the current to the modulator valve. A return spring resets the valve to the open position. This means that hydraulic pressure from the master cylinder circuit is allowed to increase and normal braking is resumed. If the wheel begins to slow rapidly again, the process is repeated. This will happen over and over many times a second for any individual brake that gives indications that it is about to skid.

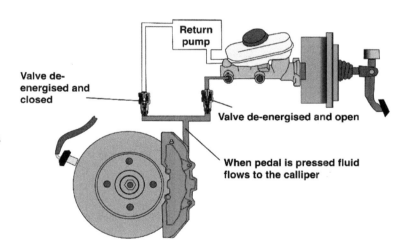

Figure 3.56 ABS pressure increase phase

When the ABS system operates, the driver can often feel a buzzing pulsation through the brake pedal. This is caused by the modulator pump returning brake fluid to the master cylinder circuit under pressure – this is a normal function of ABS operation.

Advanced Light Vehicle Technology

Brake-by-wire

Some manufacturers are now including brake-by-wire ABS systems in their vehicle design. Instead of the master cylinder applying hydraulic pressure directly to the brake calliper system, it operates as a pressure measurement sensor. This sensor simulates correct pedal feel so that brake operation feels normal to the driver. The signal from the brake pressure sensor is processed by an ECU, which operates a motor in a secondary master cylinder unit. In conjunction with the wheel speed sensors, the secondary master cylinder applies the maximum brake force required to slow the wheels without allowing them to skid. Because a brake-by-wire system does not use a modulator valve block assembly, the driver does not feel the characteristic pulsing and buzzing through the brake pedal when it is operating.

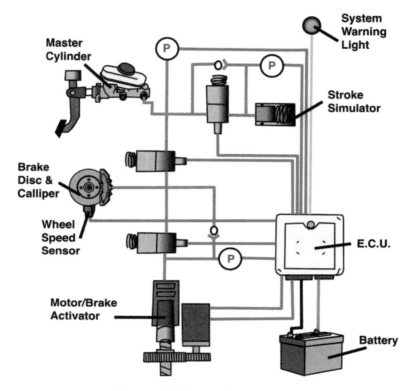

Figure 3.57 Brake by wire system

With brake-by-wire technology, the system must be failsafe. If a fault occurs in the system or power is lost, a safety valve will open, which allows standard braking with the master cylinder acting directly on the calliper circuit.

You must take precautions when bleeding the hydraulic system of an ABS equipped car, following the manufacturer's guidelines. It is possible that, unless a maintenance routine is selected using dedicated manufacturer equipment, brake fluid may not correctly pass through a hydraulic actuator unit.

Automotive Master Technician

Fault diagnosis

You should begin any fault diagnosis of an ABS system by inspecting the mechanical units. Because ABS is an electronic system, this stage is often overlooked. Table 3.7 gives an overview of what you need to inspect.

Table 3.7 Inspection of mechanical systems

Mechanical system	Possible fault
Wheels and tyres 	**Inspect wheels and tyres for:** Tread wear/pattern: Is it suitable for the road surface? Wheel and tyre size: Different size wheels and tyres on the same vehicle will give inaccurate wheel speed signals. Tyre pressures: These can cause issues with grip and overall wheel diameter. Correct wheel bearing adjustment: Loose wheel bearings can cause spacing issues with wheel speed sensors and reluctor rings.
Steering and suspension 	**Inspect steering and suspension joints:** Worn ball joints and suspension bushes: These will give unpredictable wheel movements that may reduce grip. Worn springs and dampers: These may allow the tyre to leave the road surface and lose grip. Worn steering gear and track rod ends: Play in steering systems will give unpredictable wheel movements that may reduce grip. Check wheel alignment and steering geometry: Incorrect wheel alignment and steering geometry can allow the tyre to scrub across the road surface and reduce grip.
Brakes 	**Inspect the brake system:** Check the pedal assembly: Check pedal height, free play, the operation of the pedal and ensure that the brake light switch operates correctly from the pedal mechanism. Remember that the ABS system relies on the brake light switch or circuit to show that the brakes are being operated. Check the hydraulic system: Ensure that there are no leaks and that the fluid level is correct. Braking efficiency can be lost if fluid contaminates brake friction surfaces. Ensure that there is no air in the hydraulic system causing the brakes to feel spongy. Check brake assemblies: Conduct a visual inspection of brake assemblies, ensuring that all friction material is in good condition. If the friction material is worn low it may not be able to dissipate heat correctly and as a result may overheat and cause **brake fade**.

Decide on a systematic diagnostic routine and stick to it. Examine the symptoms and take appropriate steps to locate the fault. Remember to fix the fault and not the symptom.

Figure 3.58 Roller brake testing

Advanced Light Vehicle Technology

Brake fade - a condition where the brake friction material has become so hot it can no longer transform kinetic (movement) energy into heat energy and braking efficiency is lost.

If the ABS system is not functioning correctly and the malfunction indicator light is not illuminated, check the following:

1 Power supply to the control unit
2 Earth connection of the control unit
3 Actuator/modulator power supply
4 Actuator/modulator earth connection
5 Actuator/modulator operation
6 Correct alignment and air gap of the wheel speed sensor

Check for brake light or brake light switch operation if ABS has faults or issues.

Wheel speed sensor alignment

In order for the wheel speed sensors to operate correctly and accurately, there must be no misalignment between the reluctor or pulse ring and the sensor pickup. Many manufacturers recommend a specific air gap that should be maintained between the sensor unit and the reluctor/pulse ring. This air gap is often adjustable, sometimes using elongated mountings or shims.

Electronic brake force distribution (EBD) and emergency brake assist

Once the vehicle has been equipped with an anti-lock braking system, it is a very small step to provide other electronic active safety devices.

It was once common practice to fit vehicles with a brake pressure proportioning valve. During heavy braking, weight transfer over the front wheels of the vehicle reduced the grip between the rear tyres and the road surface. This often led to the rear wheels entering a skid or creating oversteer. The purpose of the brake proportioning valve was to reduce hydraulic pressure in the rear brake circuit, which lessened the possibility of a rear wheel skid. In modern vehicles, electronic brake force distribution (EBD) is used instead. This makes use of the ABS system to direct hydraulic fluid pressure to the wheels with the most grip. In this way, stopping distances are reduced and vehicle handling and control are maintained.

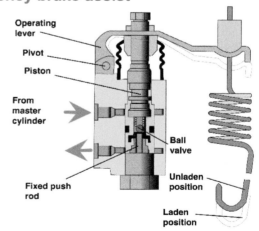

Figure 3.59 Brake pressure proportioning valve

Automotive Master Technician

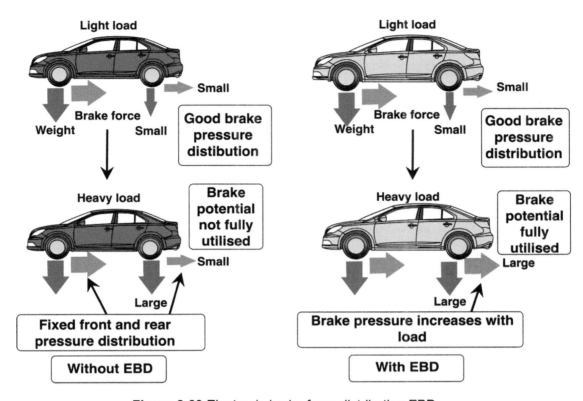

Figure 3.60 Electronic brake force distribution EBD

An emergency brake assistance system can increase pressure in the hydraulics using a solenoid attached to the brake servo unit, or pump actuated accumulator. The emergency brake assistance system senses the speed and rate at which the driver releases the throttle and presses the brake pedal in an emergency situation. It then responds by directing the extra hydraulic pressure to the wheels with the most grip in conjunction with the ABS system, helping to brake the car automatically.

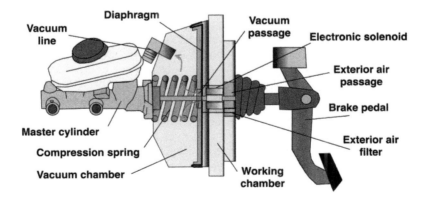

Figure 3.61 Emergency brake assistance servo

Advanced Light Vehicle Technology

Electronic parking brakes (EPB)

Many manufacturers are now including electronic parking brakes in the design of their cars. Two main types are in use:

- Cable actuated – this system uses a motor to draw on a standard cable-operated handbrake mechanism. Sensors detect how much force must be applied to correctly hold the brakes.
- Calliper type – this system has a mechanical actuator integrated into the brake calliper. When this is operated it applies pressure to the brake pads to hold the vehicle stationary.

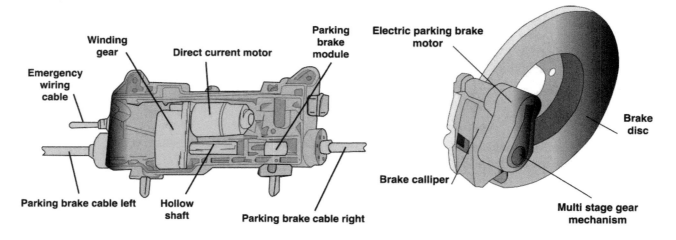

Figure 3.62 Cable type electronic parking brake   **Figure 3.63** Calliper type electronic parking brake

Both systems have sensors to measure the amount of force needed to correctly operate the parking brake, and an ECU to manage the brake mechanisms. They may also make use of the ABS wheel speed sensors to detect unwanted movement and reapply pressure to the parking brake mechanism.

To release the parking brake, the driver normally needs to switch on the ignition, apply the foot brake and press the release button. Some systems have an automatic drive away release. As the car is placed in gear and drive is taken up, sensors signal the ECU to release the parking brake and allow the car to drive away. An advantage of electronic parking brakes is that they can be integrated with other chassis management systems such as ABS and traction control to improve driver comfort, convenience and safety.

Some electronic parking brake systems include an emergency braking facility. At speeds greater than 4 mph, if the EPB button is pressed and held, the electronic stability programme applies the brakes to bring the car to a controlled stop. When the vehicle is stationary and the button is released, the parking brake is then applied.

Inspection, testing and repair of electronic parking brake systems may require special procedures and tooling. The efficiency of many designs can be tested in a standard brake rolling road. If the ECU detects that only the two wheels operated by the parking brake are rotating at low speed, it will enter a diagnostic inspection mode. You can then press the EPB button a number of times to gradually apply the brakes and check their operation.

Automotive Master Technician

Safety

The electronic parking brake mechanism remains active, even when the ignition is switched off. When you are servicing the brake system you must disable the EPB using a special service tool/procedure or scan tool. Once disabled, you can repair a cable operated system using standard procedures, but a calliper operated system will normally need a special service tool to wind back the piston. Following the replacement of brake pads, you may need to use a scan tool to adjust the electronic parking brake mechanism and remove any initial free play.
Ensure that that the EPB system is isolated before starting any work, as the brakes may operate unexpectedly and cause injury.

Traction control system (TCS)

Another active safety system that can make use of some of the ABS components is traction control (TCS). This system is designed to reduce the wheel spin during hard acceleration or manoeuvring. If the ABS wheel speed sensors detect a difference in rotational speed, particularly in the driven wheels of a car, it will interpret this information as wheel spin. It knows the difference between wheel spin that is caused by acceleration and slip that is caused by braking because it is also monitoring the operation of the brake light switch or circuit.
If wheel spin occurs, a number of actions can be taken by the traction control unit.

- Hydraulic brake pressure can be applied to the spinning wheel.
- Engine torque can be reduced by closing the throttle butterfly actuator.

- Ignition timing can be retarded.
- Fuel injection can be reduced.

Electronic stability control (ESC), electronic stability programmes (ESP) and active yaw control (AYC)

Other safety systems can make use of the function and operations of traction control and anti-lock braking systems. Electronic stability control (ECS) and electronic stability programmes (ESP) try to maintain the **attitude** of the vehicle suspension during cornering and manoeuvring procedures. Working in conjunction with active suspension, engine torque/drive and wheel spin/slip are monitored and adjusted where necessary to help improve overall vehicle handling.

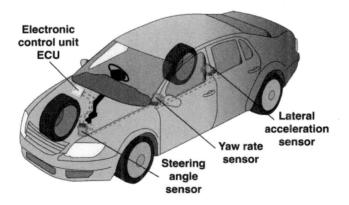

Figure 3.64 Electronic stability control

Advanced Light Vehicle Technology

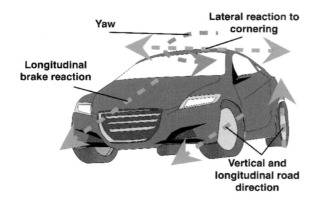

Active **yaw** control (AYC) is an electronic system that helps reduce the vehicle pivoting around a central point caused by driving, handling or dynamic road conditions. It uses traction control and ABS to achieve this.

Figure 3.65 Forces acting on suspension

Attitude – suspension height and angle which affect the position of the vehicle body.

Yaw – a twist or oscillation around a vertical axis.

Because the active safety ABS, EBD, ESP, AYC and TCS rely on the brake light system to monitor driver input to the braking system, it is always worthwhile checking the function and operation of the brake lights. A faulty brake light switch or blown brake light bulb can sometimes be enough to place these systems into limited operating strategy, causing drivability issues. These faults might not always show up as a diagnostic trouble code, so visually inspect the operation of the brake lights.

ABS, EBD, TCS, ESP and AYC are active safety systems. Active safety systems are designed to reduce the possibility of an accident occurring in the first place. These systems have also reduced the need for limited slip differentials in many vehicle designs.

Due to the advanced nature of modern braking systems, traditional methods of bleeding the hydraulics by pumping the brake pedal are no longer acceptable. Specialist tools and routines are often required in order to carry out maintenance and repair. To conduct brake bleeding you may require:
• Pressure bleeders
• Vacuum bleeders
• Diagnostic scan tools
Always follow manufacturer's recommended guidelines and maintenance routines.

Automotive Master Technician

Dynamic stability control systems

To help overcome undesirable vehicle body movements created during driving, some manufacturers are fitting active suspension systems which try to reduce the effect of these movements while maintaining comfort and handling. There are two main types of active suspension system:

- Self-levelling suspension.
- Ride-controlled suspension.

Self-levelling suspension

As forces are placed on suspension springs, usually due to the loading of weight inside the car, they are compressed and become shorter. This means that the effectiveness of the spring has been reduced, as its operating length has been compromised and it cannot move through its intended design distance as it travels over a bump. If the weight placed inside the car is distributed unevenly, this will also have an effect on vehicle attitude. The car may lean to one side, or be lower at the front or rear. As well as affecting the handling, this may have an effect on other systems, such as headlamp aim.

A self-levelling suspension system is designed to react against the loading of weight inside the car and return it to its original design ride height.

Self-energising suspension

A self-energising system is an older type of self-levelling device which uses a component that looks similar to a large telescopic suspension damper. It is mounted at the back of the vehicle, between the rear axle and the car body. If weight is placed in the rear of the car, the load will compress the rear suspension. This will lower the rear of the car and raise the front. When the vehicle is driven, and the car moves over bumps in the road, the self-energising unit acts like a pump, forcing gas between internal chambers and reacting against hydraulic fluid. This has the effect of raising the rear suspension to its original ride height without affecting normal suspension movement.

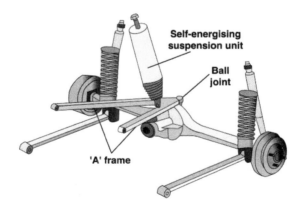

Figure 3.66 Self-energising system

A self-energising suspension unit is a sealed component and can only be visually assessed for damage or leaks. When diagnosing faults in this type of system, pay particular attentions to the mounting points of the self-energising unit. Excessive movement in bushes or connecting ball joints may cause noise during operation.

Advanced Light Vehicle Technology

Air assisted suspension

An air assisted self-levelling system uses airbags or chambers attached to the spring and damping units of a standard suspension. When weight or load is placed in the car, the compression of the suspension is detected by ride height sensors connected between the vehicle body and the suspension units. The ride height sensors are normally variable resistors or **potentiometers**, connected by a lever arm to the suspension. These sensors send a voltage signal to an ECU, which corresponds to how far the suspension has been compressed.

Potentiometer – a variable resistor.

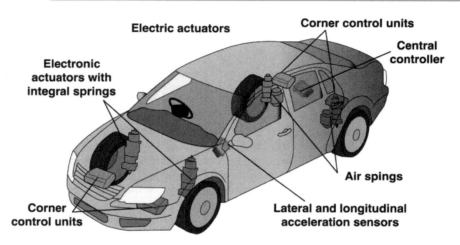

Figure 3.67 Air assisted suspension

The ECU monitors the suspension height and, if needed, operates a small on-board compressor to pump air and direct it to the appropriate self-levelling unit. This assists the springs and returns the car to its original ride height. If the load is removed and the suspension rises, the ride height sensors signal the ECU, and the air is expelled through solenoid exhaust valves, allowing the suspension to return to its original height.

So that the air assisted self-levelling system only reacts to load placed in the vehicle and not to normal suspension movement as the car is driven over bumps, the ECU is programmed with a short time delay before it operates the compressor or solenoid valves. This helps prevent unnecessary activation of the system during normal operation. Ride height sensors are normally adjustable so that the self-levelling system can be calibrated if any suspension components are replaced.

The compressor of an air assisted suspension system will normally contain a filter and dryer unit. The purpose of this is to clean the air entering the system and remove water moisture. If water entered the system, it could corrode the valves and as water is not compressible, it would stop the system operating correctly.

Automotive Master Technician

Testing an air assisted system for correct operation

1. Measure the ride height of the suspension with no load placed inside.
2. Place a heavy load in the car and recheck the ride height.
3. Start the engine and, after approximately two seconds delay, the compressor should begin to operate.
4. Check that the ride height has returned to its original position (this may take between 20 and 40 seconds to complete).
5. Remove the load from the car and the suspension should rise.
6. After approximately two seconds delay, the solenoid valves should operate and ride height should begin to fall.
7. Check that the ride height has returned to its original position (this may take between 20 and 40 seconds to complete).

Safety

Remember to calibrate any ride height sensors if repairs have been carried out on an air assisted system. This will ensure the safe and correct operation of the suspension system

Ride control suspension

Unlike self-levelling, a ride control suspension system is one that monitors dynamic car body attitude while the vehicle is in motion, and makes corrections in suspension stiffness to counteract undesirable vehicle body movement.

These systems can be manually switched by the driver, for different driving styles, or monitored by an ECU and automatically adjusted when required.

Variable rate damping

Dampers are fitted to a suspension system to reduce the bounce or **oscillation** of springs. They operate by forcing oil though valves inside the shock absorber. The resistance to flow of the oil through the valves contained in the shock absorber helps slow and control the movement of the suspension by turning the energy into heat. In a ride control system, variable rate damping is able to change the flow of the oil through the shock absorber valves, giving different damping characteristics under different driving conditions.

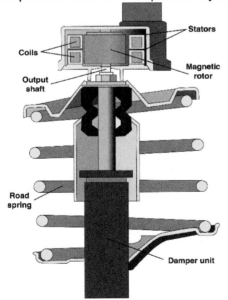

The damper valves are connected to the central piston rod of the shock absorber. This piston rod can be rotated by a small **stepper motor**, which changes the position of the valves and controls the oil flow through the damper.

• If oil flow is reduced, the suspension becomes stiffer and handling is improved.

• If oil flow is increased, the suspension becomes softer, giving a more comfortable ride.

Various suspension ride height, steering angle and acceleration sensors monitor the motion of the car and signal an ECU to adjust the stepper motors controlling the dampers.

Figure 3.68 Variable rate damping

Advanced Light Vehicle Technology

Oscillation – a continuous backwards and forwards movement.

Stepper motor – an electric motor that moves in small stages or 'steps'.

Active air suspension

Some manufacturers produce vehicles with active air suspension. The operation of this system is very similar to the air assisted self-levelling system, but in this design the air is used as the main suspension spring and there is no time delay between sensing and reacting.

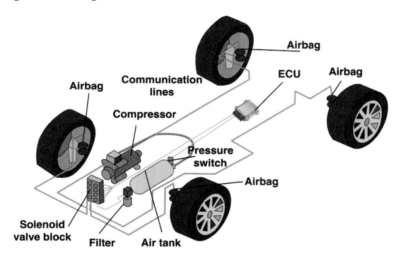

Figure 3.69 Active air suspension

Strong rubber airbags are connected between the car's suspension components and the vehicle chassis. Air supplied by an on-board compressor inflates the airbags and acts as the spring.
Suspension ride height, steering angle and vehicle acceleration forces are monitored by sensors as the vehicle is driven. If required, air can be added (from an air storage accumulator tank) or removed (by opening solenoid valves) so the correct vehicle attitude can be maintained under all driving conditions.

Automotive Master Technician

Active hydro-pneumatic systems

A hydro-pneumatic system uses a hydraulic liquid to transfer suspension movement to a sealed compressed gas chamber.

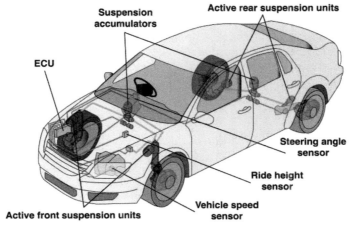

Figure 3.70 Active hydro-pneumatic suspension

Pneumatic – using gas to provide a source of mechanical force or control.

As the car travels over a bump, it operates a piston, which forces fluid through a series of pipes to a sealed compressed gas chamber.
The fluid pushes against a rubber diaphragm, which compresses the gas still further so that it reacts like a spring. The diaphragm and fluid are forced back in the opposite direction, causing the piston at the suspension unit to react against the bump and provide suspension action.

The gas used in the suspension of a hydro-pneumatic system is nitrogen. Nitrogen is an inert gas, which means that it is stable and non-reactive. It will not support combustion so is safe in the event of a fire.

Ride height and body movement can be controlled by adding or removing hydraulic fluid from the system.
An engine-driven pump draws fluid from a reservoir and uses it to charge up a pressurised accumulator tank. Suspension ride height, steering angle and vehicle acceleration forces are monitored by sensors as the vehicle is driven. When more fluid is required in one part of the suspension system to maintain vehicle attitude, a valve can be opened and fluid is added under pressure. If less fluid is required in one part of the suspension system to maintain vehicle attitude, another valve can be opened and excess fluid is returned to the reservoir.

Chapter 4 Advanced Vehicle Body Electrics

This chapter will help you develop an understanding of the construction and operation of advanced vehicle body electrics. It will cover the relationship between the electrics and electronics used in vehicle body systems and their impact on comfort, convenience and safety. It will also assist you in developing a systematic approach to the diagnosis of complex system faults for which there is no prescribed method. This is achieved with hints and tips which will help you undertake a logical assessment of symptoms and then uses reasoning to reduce the possible number of options, before following a systematic approach to finding and fixing the root cause.

Contents

Information sources

Many of the diagnostic routines that are undertaken by a Master Technician are complex and will often have no prescribed method from which to work. This approach will involve you developing your own diagnostic strategies and requires you to have a good source of technical information and data (*see knowledge is power*). In order to conduct diagnosis on advanced body systems, you need to gather as much information as possible before you start and take it to the vehicle with you if possible.

Sources of information may include:

Table 4.1 Possible information sources

Verbal information from the driver (*see the interrogation room*)	Vehicle identification numbers
Service and repair history (*see the interrogation room*)	Warranty information
Vehicle handbook	Technical data manuals
Workshop manuals/Wiring diagrams	Safety recall sheets
Manufacturer specific information	Information bulletins
Technical helplines	Advice from other technicians/colleagues
Internet	Parts suppliers/catalogues
Jobcards	Diagnostic trouble codes
Oscilloscope waveforms	On vehicle warning labels/stickers
On vehicle displays	Temperature readings

Always compare the results of any inspection or testing to suitable sources of data. Remember that no matter which information or data source you use, it is important to evaluate how useful and reliable it will be to your diagnostic and repair routine.

Automotive Master Technician

Electronic and electrical safety

Working with any electrical system has its hazards. The main hazard is the possible risk of electric shock, but remember that the incorrect use of electrical diagnostic testing procedures can also cause irreparable damage to vehicle electronic systems. Where possible, isolate electrical systems before conducting the repair or replacement of components.

Always use the correct tools and equipment. Damage to components, tools or personal injury could occur if the wrong tool is used or misused. Check tools and equipment before each use.
If you are using electrical measuring equipment, always check that it is accurate and calibrated before you take any readings.

If you need to replace any electrical or electronic components, always check that the quality meets the original equipment manufacturer (OEM) specifications. (If the vehicle is under warranty, inferior parts or deliberate modification might make the warranty invalid. Also, if parts of an inferior quality are fitted, they might affect vehicle performance and safety). You should only carry out the replacement of electrical components if the parts comply with the legal requirements for road use.

Table 4.2 The operation of electrical and electronic systems

Electrical/Electronic system component	Purpose
ECU	The electronic control unit is designed to monitor the operation of vehicle systems. It processes the information received and operates actuators that control system functions. An ECU may also be known as an ECM (electronic control module).
Sensors	The sensors are mounted on various system components and they monitor the operation against set parameters. As the vehicle is driven, dynamic operation creates signals in the form of resistance changes or voltage which are sent to the ECU for processing.
Actuators	When actioned by the ECU, motors, solenoids, valves, etc. help to control the function of vehicle for correct operation.
Digital principles	Many vehicle sensors create analogue signals (a rising or falling voltage). The ECU is a computer and needs to have these signals converted into a digital format (on and off) before they can be processed. This can be done using a component called a pulse shaper or Schmitt trigger.
Duty cycle and PWM	Lots of electrical equipment and electronic actuators can be controlled by duty cycle or pulse width modulation (PWM). These work by switching

Advanced Light Vehicle Technology

Table 4.2 The operation of electrical and electronic systems

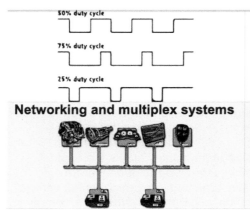 50% duty cycle / 75% duty cycle / 25% duty cycle	components on and off very quickly so that they only receive part of the current/voltage available. Depending on the reaction time of the component being switched and how long power is supplied, variable control is achieved. This is more efficient than using resistors to control the current/voltage in a circuit. Resistors waste electrical energy as heat, whereas duty cycle and PWM operate with almost no loss of power.
Networking and multiplex systems	Many modern vehicle systems are controlled using computer networking. In these systems a number of ECU's are linked together and communicate to share information in a standardised format. The most common network system is the Controller Area Network (CAN Bus).

Vehicle design and safety systems

Multiplexing and network systems

As the amount of technology on cars has increased, demand for faster computer operation and processing has also risen. Advances in vehicle management include:

- Engine management.
- Body control.
- Chassis systems.
- Transmission.

- Infotainment.
- Traction control.
- Safety systems.

ECU's were becoming bigger to cope with system requirements, and large amounts of wiring were needed to distribute electrical power around the car. These demands also generated a rise in the number of sensors required, leading to complication, extra weight and increases in the cost of manufacture.

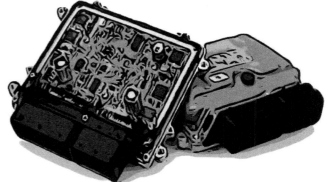

Figure 4.1 Electronic control unit ECU

Notes

The reason why ECU's were becoming larger and larger was due to the need for more connections and pins where they joined the wiring. There was a limit to how small these connections could be made, and how closely the wires in the loom could be bundled. Multiplexing has allowed a reduction in sensors and wiring, as it allows the ECU's to share information on a network.

To reduce the amount of sensors and wiring needed for system operation, **multiplexing** was introduced. Multiplexing simply means carrying out more than one operation at a time. Multiplex networking brings with it the benefits of efficient control of vehicle systems, it gives manufacturers the opportunity to configure vehicle options late in the manufacturing process and the ability to update software to overcome running issues and fix bugs once the vehicle has been launched.

Instead of a single large ECU in a vehicle, smaller ECU's were developed that managed individual systems. These single ECU's became known as **nodes**. The nodes are connected to each other by a communication wire, which allows information to be shared in a **network**. When one of the ECU's receives information from a sensor, it processes the signal and acts if required. It then passes on this information to the communication network wire linking the other ECU's, which then use that information if required, and once again pass it on. This means that signals from a single sensor can be shared across a number of different vehicle systems.

Multiplexing – a method of carrying out more than one operation at once.

Nodes – ECU's connected to a computer network (from the Latin word 'nodus', which means knot).

Network – a number of computers connected together so that they are able to communicate with each other.

CAN bus

There are a number of different network types and manufacturers available, but the name CAN bus has been adopted by many technicians to describe nearly all network systems.

Controller area network (CAN) was introduced by Robert Bosch in the 1980s and is an international standards organisation (ISO) standard for a serial multiplex communication protocol.

The advantages of CAN bus are:

- Transmission speeds are much faster than those used in conventional communication (up to 1 Mbps), allowing much more data to be sent.
- The system is very immune to interference (noise), and the data obtained from each error detection device is more reliable.
- Each ECU connected via the CAN, communicates independently, therefore if an ECU is damaged/faulty, communications can be continued in many cases.

The CAN bus line consists of two cables, known as CAN L and CAN H (CAN Low and CAN High, respectively). The CAN high and low wires are twisted together and this helps to cancel out noise which may be caused by electromagnetic interference from other vehicle electrical systems. At the ends of each CAN line are terminal resistors which help to dampen out voltage spikes (**back EMF**) which could be caused as the communication is triggered on and off. The CAN bus lines connecting two dominant ECU's are the main bus lines, and the CAN bus lines connecting each individual ECU are the sub-bus lines. Each ECU communicates with the CAN periodically sending information from several sensors. This information is circulated on the CAN bus as a data packet. Each ECU needing data on the CAN bus can receive these data frames sent from each ECU simultaneously. A single ECU transmits multiple data frames. When data packets conflict with one another (when more than one ECU transmits signals at the same time), data is prioritised for transmission by a process called **'mediation'**.

If mediation is required:

1. The data frame with high priority is transmitted first according to ID codes embedded in the data packet.
2. Transmission of low-priority data is suspended by the issuing ECU until the bus clears (when no transmission data exists on the CAN bus).
3. The ECU containing suspended data frames transmits when the bus becomes available.

NOTE: If the suspended state continues for a specified time, new data (data packet content) is created and sent.

Advanced Light Vehicle Technology

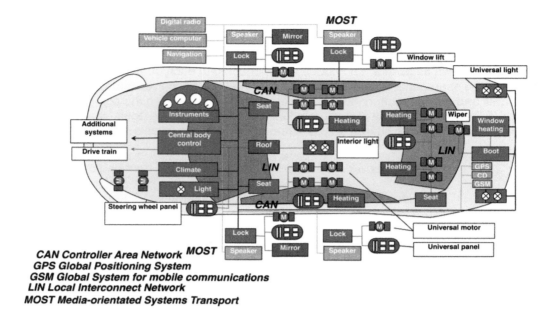

MOST

CAN Controller Area Network *MOST*
GPS Global Positioning System
GSM Global System for mobile communications
LIN Local Interconnect Network
MOST Media-orientated Systems Transport

Figure 4.2 Vehicle network layouts

The nodes are connected by a single communication line, which allows the exchange of multiple pieces of data. The communication line can link these ECU's in the following layouts:

- A large loop, known as a daisy chain.

- A star pattern, known as a server system.

- Connected in parallel to a single bus line.

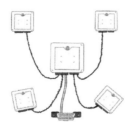

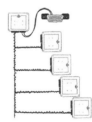

When a daisy chain layout is used, the data sent travels in both directions at once, which gives much greater reliability. If one wire is damaged or broken within the loop, the information can still arrive at the appropriate ECU as it comes from the other direction. The data is not only more reliable but this system also improves malfunction diagnosis.

Back EMF - this is a voltage spike (a reverse electromotive force) created when a circuit is switched off.

Mediation - a negotiation to resolve difficulties or conflict. In multiplex it is the rule that sets the priority of messages.

Automotive Master Technician

> **Gateway** - an ECU that joins two different system/speed busses together and translates/transmits the data packets into a format that can be used on the different part of the network.

Communication wires

The wiring for network communication can be constructed from three main methods:

- Twisted copper wiring.
- Coaxial wiring.
- Fibre optical.

It is possible to use a mixture of layout types and communication wires for different vehicle systems (i.e. powertrain, chassis, body and entertainment). These systems can then be connected to each other through individual electronic control units known as **'gateways'**. Gateways are used in order for busses of different speeds to co-operate. The gateway is used to link the busses and it is used to encode and decode messages. The gateway may be a separate unit or it may be incorporated in an ECU. The gateways allow the different networks to share information by translating between different bus communication speeds. When the networks are shown laid out on a diagram, it is known as the 'physical layer' or 'topology'.

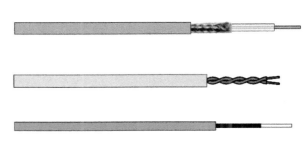

Figure 4.3 Communication wires

> The word 'bus' is used in various situations. One meaning of bus is a vehicle that collects you from one place and delivers you to another. This is very similar to its meaning within a communication network. Information is picked up at one point on the communication line, it then takes a route around the system and stops at various ECU's (like bus stops).

Communication data

When an ECU receives a signal from a vehicle sensor, it processes this and places the information on the network bus as a data packet. The data packet is usually made up of the following:

- A header, **SOF (Start of Frame)**: the equivalent of 'hello, I am transmitting a message'.
- The priority **ID (Identifier) region**: how important this message is, e.g. vehicle safety information will be more important than a bodywork communication such as a command to open an electric window.
- Data length **Control region**: this is so the receiver knows it has not lost or 'misheard' any of the information.
- Data type **Data region**: what type of information is contained, e.g. speed, temperature, etc.
- Data **Data region**: the actual sensor information itself.
- An error detection code **Cyclic Redundancy Check (CRC) region**: this says 'has all the information been received?'
- End of message **EOF (End of Frame)**: 'goodbye'.
- Finally, a request for a response from the receiving ECU **ACK (Acknowledge) region**: this says 'thank you, I got your message'.

Advanced Light Vehicle Technology

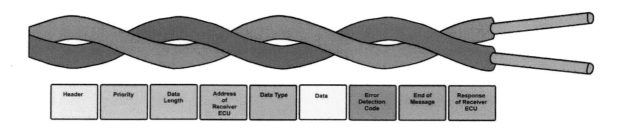

Figure 4.4 A communication data packet

To help reduce the possibility of data **corruption** caused by misinterpretation or external electromagnetic interference, a CAN bus system uses two communication wires instead of one, twisted over and over each other in a spiral. The same data is sent on both of these communication wires as an on and off voltage signal. One signal is sent as a positive switch and one is sent as a negative switch, providing a mirror image on each network wire, which are known as CAN high and CAN low. The **potential difference** between the voltages on the two lines produces a digital signal that can be processed into information.

> If the oscilloscope patterns from CAN high and low don't line up, check the channel speeds of your scope before assuming that there is a fault with the system.

DATA TRANSMISSION - High Speed

The transmitting ECU sends switched voltage through the CAN H and CAN L bus.
2.5 to 3.5 V signals to the CAN HIGH line.
2.5 to 1.5 V signals to the CAN LOW line.
The receiving ECU reads the data from the CAN lines as potential difference of between 3.5 and 1.5 volts.

In Figure 4.5, 'Recessive' refers to the state where both CAN H and CAN L are under the 2.5 V state, and 'Dominant, refers to the state where CAN H is under the 3.5 V state and CAN L is under the 1.5 V state. These values correspond to a binary value of either 1 or 0.
Recessive = Logic value of 1
Dominant = Logic value of 0

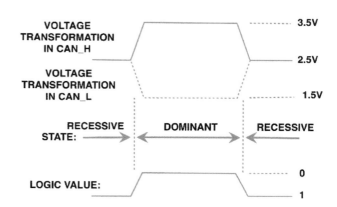

Figure 4.5 Data transmission high speed

Automotive Master Technician

DATA TRANSMISSION - Low Speed

The transmitting ECU sends switched voltage through the CAN H and CAN L bus.
0 to 4 V signals to the CAN HIGH line.
1 to 5 V signals to the CAN LOW line.
The receiving ECU reads the data from the CAN HIGH and CAN LOW as potential difference of between 5 and 0 volts.

In Figure 4.6 'Recessive' refers to the state where CAN H is at 0V and CAN L is at 5V
'Dominant' refers to the state where CAN H is at 4V state and CAN L is 1V.
Recessive = Logic value of 1
Dominant = Logic value of 0

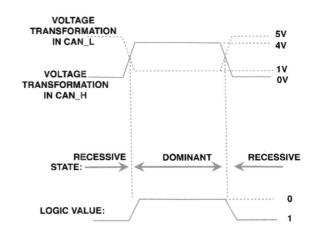

Figure 4.6 Data transmission low speed

The communication wires are exactly the same length, and as the data travels at the same speed, both versions of the data should arrive at the receiver at the same time. These messages can now be compared to help identify data corruption. The opposing voltage of the signals transmitted down the communication wires will also help cancel out electromagnetic interference from other systems.

Terminating resistances

Another factor that might affect a CAN bus system, causing data corruption, are network voltage spikes caused by back EMF during switching. A 120 ohm resistor is connected across the CAN high and CAN low circuits at each end which help to reduce voltage spikes and reduce the possibility of corruption; they act like shock absorbers for electricity.

Termination resistances can give a good indication of correct circuit operation.
If an ohmmeter is connected in parallel across CAN high and CAN low (using pins 6 and 14 of the data link connector for example) the total recorded resistance will be halved.

- If 60Ω is shown CAN High and CAN low should be OK.
- If O/L (Infinity is shown) an open circuit exists in both lines.
- If 0Ω is shown a dead short exists.
- If 120Ω is shown one CAN line may be at fault (confirm communication using Oscilloscope).

Corruption – a breakdown of integrity or communication.

Potential difference – the difference between two voltage values.

Infotainment - a combined information and entertainment system.

Hexadecimal - a method of counting in batches of sixteen that can be used as a computer programming language.

Bus speeds

There are three main CAN bus speeds:

- Low speed: used for instrumentation, body control and comfort, etc.
- It operates at a rate of 33,000 bits of information per second (33kbps).
- High speed: used for powertrain control and safety critical information, etc.
- It operates at a rate of 500,000 bits of information per second (500 kbps).
- Very high speed: used for high volumes of data transmission in **infotainment** systems (such as streaming video and music), etc.
- It will operate at a rate of 25,000,000 bits of information per second (25 mbps).

This is sometimes known as baud rate, but only if related to binary (1 & 0). Modern computer communication may have more than 2 symbol states (**hexadecimal** programming language) so baud rate can be misleading.

Notes

The data is sent as an on and off signal. It would be seen on the screen of an oscilloscope as a square waveform.

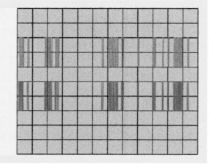

Tip

The main CAN Bus lines (high and low) can often be tested with an oscilloscope by connecting to pins 6 and 14 of the 16 pin diagnostic socket.

System reliability

The CAN bus system is more reliable than a standard wiring system due to the fact that a single open bus wire would not stop communication. Two open bus wires can stop communication, but as more ECU's are used for control only part of the system may fail.

Short circuits can have a catastrophic effect on network communication. A short to either positive or earth will disrupt the communication on the bus wire, as an on and off signal can no longer be transmitted. If viewed on and oscilloscope screen, this would be a flat line at either 0V or 5V. To avoid total failure of the system, bus-cut relays can be used. These are a type of circuit breaker that isolate part of the network, allowing the rest of the system to continue communicating.

Automotive Master Technician

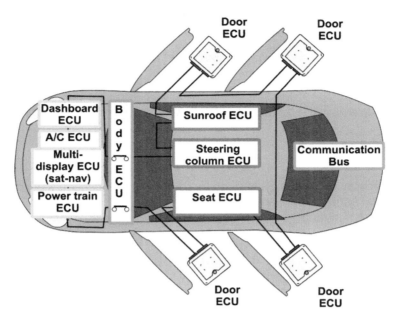

Figure 4.7 A daisy chain network layout with bus cut relays

Multiplex and networked diagnosis

If a critical network failure occurs, such as short to positive or earth, the vehicle may suffer a complete communication loss.

With a networked system, if communication is lost within a certain area, number of items will not work and numerous trouble codes may be generated. Having connected a scan tool and retrieved the diagnostic trouble codes, you should look for the code that is the root cause. Communication failures are normally an effect of the original fault (i.e. 'unable to communicate' or 'communication lost'). You should ask yourself, 'Is this the cause or an effect created by the fault?' CAN bus systems report communication faults as live data. As a result of this, once you have identified the cause trouble code, you should be able to conduct a diagnosis by disconnecting and isolating components or sections of wiring loom until communication is re-established.

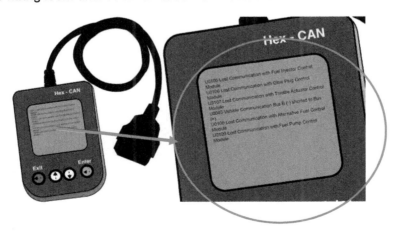

Figure 4.8 CAN Bus diagnostic trouble codes

Active and passive safety systems

Modern cars come equipped with advanced systems to try to protect the occupants in the event of an accident. There are two types of vehicle safety: active and passive.

Active safety attempts to prevent an accident occurring in the first place and includes systems such as:

- Anti-lock brakes.
- Traction control.

- Electronic stability programmes.
- Emergency brake assist.

Passive safety attempts to protect the occupants of a car in the event of an accident or impact and includes systems such as:

- Safety cells.
- Side impact protection.
- Crumple zones.
- Head restraints (sometimes called head rests).

- Airbags.
- Seatbelts.
- Pre-tensioner systems.

The term supplementary restraint systems (SRS) is applied to one or more individual systems that provide additional protection in the event of an accident. The most notable of these systems are airbags and seatbelt pre-tensioners.

Mechanical systems

Early systems for activating airbags and pre-tensioners were mechanical and used a triggering mechanism similar to a gun.

In the event of an impact, metal weights making use of the principle of **inertia**, were used to activate the airbag or **pre-tensioner** by:

- Igniting a small explosive which started a chemical reaction in the airbag, creating a gas to inflate it.
- Operating a piston in the pre-tensioner unit of the seat belt.

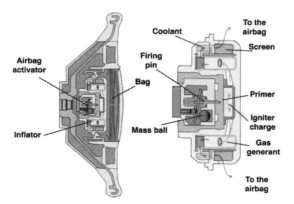

Figure 4.9 Mechanical type air bag system

Mechanical airbags and pre-tensioners were unstable and prone to accidental deployment. You must take care when working around a mechanical airbag as sudden shocks could cause the firing mechanism to go off. Always set any safety locks on the airbag units when removing and refitting them.

Pre-tensioner – a mechanism that slightly tightens a seatbelt in the event of an accident.

Inertia - the tendency of a body to resist acceleration; the tendency of a body at rest to remain at rest or of a body in straight line motion to stay in motion in a straight line unless acted on by an outside force.

Deployment – putting into action.

Electronic systems

Electronic SRS systems are now the most common, as they give precise control and increased levels of safety. The use of electronic sensing devices and processing computers means that different types of accident can be detected and the action of the SRS components tailored to provide the greatest amount of protection. Electronic SRS components include:

- **Driver's airbags**: mounted on the steering wheel, protecting the head and face in the event of a frontal impact.
- **Passenger's airbags**: mounted in the dashboard, protecting the head and face in the event of a frontal impact.
- **Door-mounted airbags**: mounted in the door panel to help protect the occupants in the event of a side impact.
- **Seat-mounted airbags**: mounted in the seat to help protect the occupants in the event of a side impact.
- **Curtain shield airbags**: mounted at the edges of the roof headlining, they deploy across the windows to help protect occupants from broken glass in the event of a side impact.
- **Roll over airbags**: mounted in the headlining to help protect the occupants in the event of the car rolling over during an accident.
- **Lower dash airbags**: mounted in the bottom section of the dashboard to help protect the driver and passenger's legs in the event of a frontal impact.
- **Seatbelt pre-tensioners and force limiters**: working in conjunction with the airbags, these are designed to put occupants in the correct position for safest contact with the airbags.

Airbag operation

Mounted around the car in strategic positions are a number of crash sensors. It is their job to send a signal to the SRS ECU in the event of an impact. An ECU, usually mounted centrally to the floor of the car, contains a second sensing mechanism that detects the rate of deceleration and the angle of impact. This second sensor is often referred to as the safing sensor.

Advanced Light Vehicle Technology

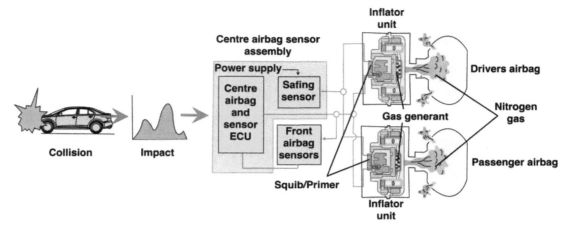

Figure 4.10 Electronic type air bag system

If an accident occurs, the crash sensor sends a signal to the ECU. If the safing sensor determines that the impact is happening at above a predetermined speed and within a certain impact angle, it will trigger the seatbelt pre-tensioners and deploy one or more airbags.

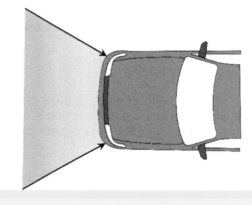

Figure 4.11 Angle of impact

Front airbags often need an impact of around 30 degrees to the centre line of the vehicle against an immovable object in order to activate. Many airbags are designed not to deploy if the impact occurs below a set speed, typically around 10–20 mph.

Airbag inflation

When an impact occurs which meets the predetermined criteria, the ECU will actuate a small igniter device in the airbag unit called a **squib**. The squib is an explosive detonator which heats up chemicals stored in a gas generator unit. A chemical reaction creates a large quantity of nitrogen and carbon dioxide gas, which is used to inflate the airbag. The nylon airbag will then burst out from its cover at over 200 mph.

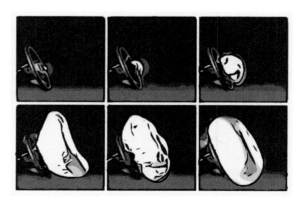

Figure 4.11 Air bag inflation

Automotive Master Technician

Squib – a small explosive detonator used to start the deployment of an airbag or seatbelt pre-tensioner.

The airbag will inflate to a pressure of around 0.5 bar and help cushion the impact on the occupant. As soon as the individual has collided with the airbag, large holes in the rear of the bag are designed to allow the gas to escape and the bag rapidly deflates. Passenger airbags normally function in a similar manner to the driver's airbag, although they are generally much larger, having a capacity up to five times that found on a steering wheel airbag. Frontal airbags are designed to work in conjunction with seatbelt mechanisms so that the occupant is in the correct position for their face to impact the bag in the middle.

The cover on an airbag is designed with a weak or perforated seam so that as the airbag is inflated it will tear open easily without obstructing deployment. A powder such as corn starch or talcum powder is used to help lubricate the nylon and keep the bag pliable. This can normally be seen as the airbag deploys, and is often mistaken for smoke created by the explosive or gas generator.

If an incorrect seating position is used or seatbelts are not worn, an airbag can cause serious injury or even death. During the vehicle design process, crash test dummies often have lipstick applied to their faces, so that when tested, outlines and marks are left on the airbag following a collision. This allows manufacturers to recommend the best seating positions and seatbelt use for safe operation.

Other forms of airbag deployment

Additional airbags, such as curtain shield and roll over, may be deployed using a pre-pressurised gas container. In the event of an impact of sufficient speed and direction, a signal from the ECU ignites a squib which punctures a gas container. The pressurised gas is allowed to escape into the nylon airbags, deploying them at high speed.

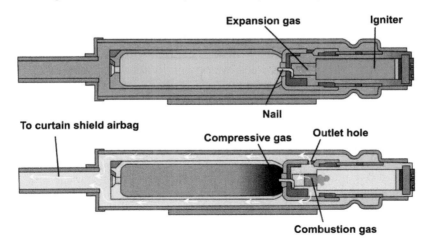

Figure 4.12 Pre-pressurised gas container

Advanced Light Vehicle Technology

Nylon airbags and gas generators degrade over time. As a result, airbags may be considered service items and many manufacturers recommend that they are replaced approximately every ten years.

Clock springs

In order to work correctly, the driver's airbag needs a reliable electrical connection which is able to rotate with steering wheel operation. Conventional wiring and connectors could create a possible problem with premature wear or breakage caused by the constant turning of the steering wheel. To overcome these issues, a special component called a clock spring is fitted between the wiring harness and the airbag module behind the steering wheel. A **Mylar** tape is wound in a similar manner to the spring found inside a clock and is able to wind up and unwind with steering wheel rotation. One end of the Mylar tape is connected to the fixed wiring on the steering column and the other is connected to the steering wheel. This allows a constant electrical connection to be maintained with the airbag unit at all times.

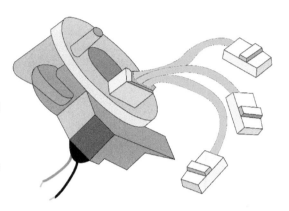

Figure 4.13 Clock spring spiral cable

Mylar – a form of polyester resin used to make heat-resistant plastic films and sheet.

SRS wiring

To help prevent accidental deployment of airbags and pyrotechnic pre-tensioners, the SRS system usually has its own wiring loom which is separate from the main vehicle wiring. It is connected in parallel to the main wiring and reduces the possibility of stray electrical signals from other systems activating airbag squibs. The SRS wiring is usually covered in bright yellow insulation in order to differentiate it from other vehicle wiring and to act as a warning that it should be handled/tested with caution.

SRS electronic control units

The SRS ECU receives signals from the crash sensors and safing sensors and processes the information. If the signals meet a pre-set criteria, the ECU will supply a voltage to the airbag squib and initiate deployment. It is common for the ECU to contain one or more of the sensors, and so its positioning when installed in the car is vital to correct operation. Many ECU's have markings on them which show how they should be fitted and orientated.

SRS ECU's are designed so that in the event of an accident and airbag deployment, codes are stored in the memory which cannot be erased. This safety feature means that a new unit must be fitted when repairing/resetting the system after an accident.

Figure 4.14 SRS electronic control unit

SRS systems contain a backup power source so that they can still deploy in an accident, even if the battery has been damaged. A number of **capacitors** can be charged and act as an independent power source for the airbags and pre-tensioners. If you are going to work on an SRS system, you must disconnect the battery and allow time for the capacitors to discharge before you start any work.

Sensors

There are a number of different crash and safing sensor types used in the construction of an SRS system. Although mechanical sensors are sometimes used, the most common type in modern SRS is electronic.

Mechanical sensors

- **Mass/inertia sensor**: this has a small weight in the form of a roller or ball which is housed inside an enclosure. During the rapid deceleration created by the impact of an accident, inertia causes the ball or roller to move against the force of a contact spring which will send an electric signal to the ECU.

- **Mercury sensor**: this has a small amount of mercury contained in a tube which is inclined upwards at an angle. It is mounted so that when an accident occurs, the mercury will rise up the tube and bridge two electrical contacts sending a signal to the ECU.

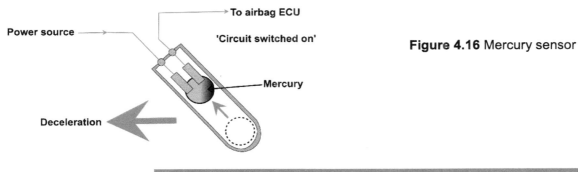

Figure 4.16 Mercury sensor

Mercury is toxic. If an SRS sensor is known to contain mercury it must be treated as hazardous waste.

Electronic sensors

Electronic sensors can work on the principle of **accelerometers**, or can be a strain gauge type device. This type of sensor uses a small plate which has a weighted flap that moves slightly when force is applied to it. A strain gauge and a small integrated circuit are also mounted on the plate.

During the rapid deceleration created in an accident, the flap is forced to move. This creates a signal in the integrated circuit, which can be transmitted to the ECU. This type of sensor is very accurate and it also produces a signal that is **proportional** to the force of deceleration. Because of this, the ECU is able to calculate the severity of the impact and apply the appropriate safety measures.

Advanced Light Vehicle Technology

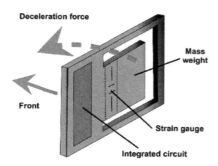

Deceleration force

Mass weight

Front

Strain gauge

Integrated circuit

Figure 4.15 Mass sensor strain gauge

Terms

Capacitor - an electronic component that acts as a temporary storage device for electricity.

Accelerometer – an instrument for measuring acceleration.

Proportional - corresponding in size or amount to something else, usually by ratio.

Cinching – tightening up and securing.

Pyrotechnic – involving the use of explosives, in a similar manner to fireworks.

Pre-tensioners and force limiters

During an accident, if the people involved are not in the correct position when airbags are deployed, injury or death may occur. Seatbelt pre-tensioners are designed to work in conjunction with the airbags by cinching up the belt components to remove any slack and bring the body and face into line with the airbag. The pre-tensioner mechanism can be in the inertia reel or the seatbelt stalk buckle mechanism. In a similar manner to airbags, many pre-tensioners use a squib and a gas generator to create pressure on a type of piston which is able to retract the seatbelt slightly when an accident occurs. Because these systems use small explosive charges, they are often called **pyrotechnic** pre-tensioners.

In the moments after deployment has occurred, the force tightening the belts is released as gas pressure is exhausted, in a similar manner to an airbag deflating after an accident. This force limiting reduces the pressures on internal organs which can rise rapidly following an impact.

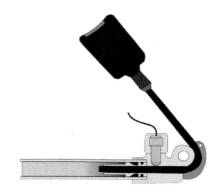

Figure 4.17 A seatbelt pre-tensioner system

Automotive Master Technician

Notes

Seatbelt pre-tensioners are designed to be used only once. Many pre-tensioner systems include an indicator device which is shown after the belt mechanism has deployed. This makes the occupants aware that the components must be replaced.

Timing of SRS operation

When an accident occurs, the pre-tensioners operate and the airbags will deploy from their housings at over 200 mph. Approximate timings of SRS operation are shown in Table 4.3.

Table 4.3 Approximate timings of SRS operation

Timing of SRS operation	Illustration
0 to 10 milliseconds **Crash sensors and safing sensors detect the impact and signal the SRS ECU.**	
10 to 15 milliseconds **The airbag is activated and starts to inflate. The seatbelt pre-tensioners are deployed, taking up the slack in the belt and pulling the occupant into an appropriate position.**	
20 to 30 milliseconds **Seatbelt tensioning is complete and the airbag is fully inflated. Inertia keeps the person moving and they come into contact with the airbag, cushioning the impact.**	
70 to 90 milliseconds **The main impact is over and the force limiters on the seatbelt slacken. As the occupant moves against the airbag, the pressure of gas inside is released through the holes in the back of the bag and it begins to deflate.**	
120 to 150 milliseconds **The airbag fully collapses and the inertia reel action of the seatbelt returns the occupant to an upright position.**	

SRS fault diagnosis

If a fault occurs with a supplementary restraint system, the malfunction warning light should illuminate on the dashboard. This warning light has a self-check procedure: when the ignition is first switched on, it should light up for around 5 to 10 seconds; if no fault exists, it should then go out.
If the light remains illuminated or flashes, this indicates that the system has detected a fault and the airbags and pre-tensioners are disabled and will not deploy in the event of an accident. This should not affect normal seatbelt operation.

Figure 4.18 SRS warning light

Use a diagnostic resistor to simulate an air bag component and then try and clear trouble codes. If the code can be reset, a new SRS component is required.

In order to diagnose the system, you need to have access to an appropriate diagnostic trouble code reader.
1. Connect the code reader to the vehicle's diagnostic socket.
2. Read all diagnostic trouble codes and record them.
3. Some SRS ECU's have no memory facility and only display faults that are present at the time of testing. Other ECU's are able to log trouble codes, and these will remain in the memory until cleared. To ensure that these codes have not been created during connection and disconnection or testing processes involved with your diagnosis, clear the memory using the code reader and then recheck.

Some early SRS system trouble codes could be accessed by counting 'blink codes'. A connection could be made at the diagnostic socket that allowed the warning light on the dashboard to flash or an LED could be inserted so that it also flashed. By counting the number of long or short flashes produced by the light, a two-digit code could be retrieved and compared to the manufacturer's trouble code list.

You should not use conventional electronic test equipment to check SRS components as they sometimes produce small electric currents which could make the SRS system deploy accidentally. The safest method of checking the condition of an SRS system is to use a scan tool that is able to produce live data, such as resistances, from the serial port.

1. Once you have retrieved the trouble codes, carefully inspect components and connections for damage.
2. If a code indicates that an SRS component is faulty, you can often disconnect it and replace it with a special diagnostic resistor, which simulates an operational part.
3. You can now clear the codes and, if the fault has disappeared, you should fit a new SRS component.
4. If the trouble code indicates that there is an issue with the wiring, disconnect all SRS components and isolate the wiring, then check with a multimeter.

Automotive Master Technician

Check under seats for loose connections if SRS light is on, as this can be a common issue.

Working on or around the components of a supplementary restraint system can be very dangerous. You must always take special care when handling components. Table 4.4 shows some precautions that you should observe, but you should also ensure that all health and safety practices are carried out during repair work. Always follow manufacturers recommended procedures.

Table 4.4 Precautions to take when working on or around SRS systems

Do's	Don'ts
Before undertaking any repairs to a vehicle, remove the airbag and pre-tensioner sensors if any shocks are likely to be applied to the sensors or in the vicinity. These shocks may not deploy the SRS, but they can damage the sensors so that they don't work in the event of an accident.	Never use memory keepers when working on SRS airbag systems. These special tools are designed to keep electrical components powered when the battery has been disconnected, so that radio codes, etc. are not lost during repair procedures. Using memory keepers means that the SRS system will remain 'live' and could deploy accidentally during repair work.
If you are undertaking any electric welding on a vehicle, always disconnect the airbags and seatbelt pre-tensioners. Electrical welding uses the vehicle body to conduct the welding current. The action of welding may cause accidental deployment of airbags and pre-tensioners.	Never use conventional electronic test tools on SRS system components. For example: • Ohmmeters create current flow in the circuits they measure. • Power probes are designed to supply electric current for test purposes. • Test lights connected in parallel draw extra electric current through a circuit. All of these tools may lead to accidental deployment of airbags or pre-tensioners.
Always check that all diagnostic trouble codes are clear after you have completed the repair work. Any codes that remain in the memory can mean that the system is disabled and will not deploy in the event of an accident.	Never reuse an SRS ECU if the vehicle has been involved in a collision which has resulted in deployment of airbags or pre-tensioners.
Always treat SRS ECU's as toxic waste when disposing of them if they are known to contain the mercury switch type safing sensor. Mercury is toxic and special controls are needed to ensure safe disposal.	Many SRS ECU's are connected to the vehicle body to provide a safe earthing method. Never disconnect or reconnect the SRS ECU wiring loom if it is not mounted to the vehicle's body. Static electricity could cause deployment of airbags or pre-tensioners.
If a vehicle has been involved in a collision, even if the airbags or pre-tensioners have not deployed, always inspect them thoroughly and check for system trouble codes.	Never use second-hand SRS components from a donor vehicle. They may be faulty, incompatible or beyond their service life and may not function correctly in the event of an accident. Always use new parts.
When disposing of airbag or pre-tensioner assemblies, they must always be deployed first. This includes	Never attempt to repair SRS assemblies, including: • sensors

Advanced Light Vehicle Technology

Table 4.4 Precautions to take when working on or around SRS systems

Do's	Don'ts
scrapping a vehicle – end of life (ELV) – with these assemblies still in place. You should only deploy airbags and pre-tensioners if you have been trained in how to do so safely. Special service tools are available to enable safe deployment, and you should always follow the manufacturer's recommended procedures.	• airbags • pre-tensioners • ECU • loom They may not function correctly in the event of an accident. Always replace the faulty part.
Always store removed airbags and pre-tensioner mechanisms in a specially designed explosives cabinet and ensure that others are aware of the contents.	Never expose airbag or pre-tensioner components to heat and never use grease or cleaning agents on SRS components.
Always follow manufacturer specific instructions when diagnosing SRS airbag faults.	Never place removed airbags deployment side down. If they deploy, they will take off like a rocket and this can be very dangerous.
Always wear appropriate PPE when working on SRS system components.	Do not hang diagnostic scan tools on steering wheels during test procedures. Electronic discharge through exposed contacts or electromagnetic forces may cause accidental deployment. This can result in the scan tool being propelled at high speed into the tester.

Pedestrian safety

Until recently, vehicle safety design features have focused on protecting the occupants in the event of an accident, but a large proportion of injuries and deaths are caused to pedestrians. Table 4.5 shows examples of the number of pedestrians killed or injured in the UK in 2010.

Table 4.5 Number of pedestrians killed or injured in the UK in 2010

	Fatalities	Serious Injuries	Slight Injuries	All Injuries
Children (0-15)	26	1,620	6,283	7,929
Adults (16-59)	224	2,475	11,019	13,718
Adults (60+)	155	1,020	2,472	3,602
All Pedestrians (including figures where no age was recorded)	405	5,200	20,240	25,845

Legislation is being introduced to ensure that all new vehicle designs take into account their possible effects on pedestrians in the event of an accident and offer some levels of protection. Vehicles are tested and rated by an organisation called Euro NCAP that have been conducting crash tests on cars for over 10 years. The cars tested are given star ratings for the level of occupant and pedestrian protection that they offer, and most manufacturers try to achieve the highest rating possible when designing their cars. (For further information on legislation and testing, visit http://www.euroncap.com).

Active pedestrian safety

Many of the pedestrian safety designs, used in new vehicles, focus on the absorption of impact created by the front bumper area and bonnet.

The bonnet of many cars is often made from sheet metal, which is a compliant energy absorbing structure and therefore poses a comparatively small threat to pedestrian injury. Most serious head injuries occur when there is not enough clearance between the bonnet and the underlying engine components. A gap of approximately 100 mm is usually enough to allow the pedestrian's head to have a controlled deceleration and a significantly reduced risk of death. Creating room under the bonnet is not always easy because usually there are other design constraints, such as aerodynamics and styling. Engineers have attempted to overcome this problem by using deformable mounts, and by developing more ambitious solutions such as airbags that are activated during the crash and cover the bonnet and lower windscreen area. Some manufacturers have managed to create a pop-up bonnet design, which can add extra clearance over the engine if the bumper senses a hit. Impact and crash sensors operate in a similar manner to those used in SRS systems and care should be exercised when working around these structures, as unexpected deployment may cause injury.

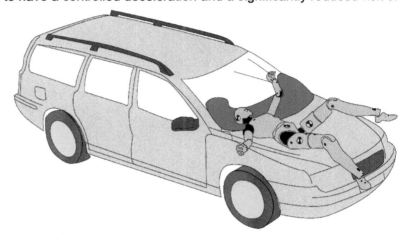

Figure 4.19 A pedestrian air bag system

Climate control

Air conditioning systems and components

Air conditioning is a convenience system designed to ventilate and lower passenger compartment temperature. It also dehumidifies and purifies the air. A well maintained air conditioning system provides the driver and passengers with an environment which is free from dust and dirt particles, moisture free and will help them to regulate their body temperature. This leads to a comfortable atmosphere in which the drivers and passengers are more alert, helping to improve road safety.

When the air conditioning is operating on full, it may take around 5 horse power from the engine to drive the refrigerant compressor. Many manufacturers now produce vehicles where the air conditioning compressor is switched on by default, but a manual override will be available to enable the driver to switch the system off.

The systems work on the principle of the refrigeration cycle.

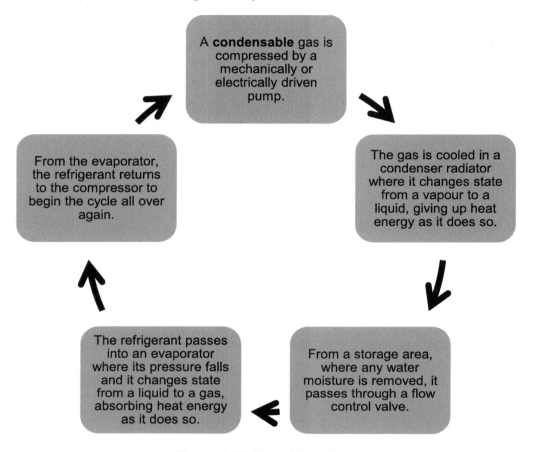

Figure 4.20 The refrigeration cycle

The refrigerant cycle is a process of heat transfer. It works by absorbing heat energy from the passenger compartment of the car, as air is circulated through the evaporator, and transferring it to the external atmosphere as it passes through the condenser radiator on the outside of the car. None of the heat energy is destroyed, it is simply taken from the inside of the car and moved to the outside.

Condensable - a substance that is able to change state from a gas/vapour to a liquid.

Automotive Master Technician

Air conditioning components

Table 4.6 lists the components found in air conditioning systems and an image to help you identify them.

Table 4.6 Identification of air conditioning system components

Air conditioning component and function	Example
Compressor - compresses the refrigerant	
Clutch assembly - disengages the compressor from the drive motor	
Condenser - radiates heat from the refrigerant	
Condenser fan – assists with cooling of the condenser	
Evaporator - absorbs heat from the passenger compartment	
Suction accumulator - stores liquid refrigerant before returning to the compressor	
Thermal expansion valve TXV - temperature controlled valve	

Advanced Light Vehicle Technology

Table 4.6 Identification of air conditioning system components

Fixed orifice tube FOT - an accurately sized restriction valve	
Receiver drier - stores liquid refrigerant before entering the evaporator and helps to remove any water moisture	
Hoses and pipes - connect various air conditioning components	
Connections - join various air conditioning components	
Service ports - allow the safe recovery and filling of refrigerants in an air conditioning system	
Mufflers - act as silencers to reduce any noise created by circulating refrigerant	
Pressure/Temperature switch - control the air conditioning operation due to pressure or temperature	
Relays - control system electric current	

Automotive Master Technician

The air conditioning pressure switch can be combined in one unit to turn off the compressor clutch if system pressures rise too high or falls too low. This is known as a binary switch and is designed to help protect the compressor form excessive loads caused by high or low pressures. High pressure in an air conditioning system may also be an indication of high condenser temperatures. The pressure switch can also sometimes operate the relay controlling the condenser cooling fan if a pre-set pressure value is reached. This type of switch is called a trinary switch as it performs three functions - pressure too high, pressure too low and condenser fan operation.

System construction

Air conditioning systems are sealed to prevent the escape of refrigerant gas to atmosphere. They are made using various materials such as:

- Steel - used in the construction of some receiver driers or suction accumulators.
- Rubber- used in the construction of system hoses.
- Aluminium - used in the construction of condensers and receiver driers.

To join the various system components, joints and connections can be:

- Flare type - the end of a metal pipe is spread out to seal against a mating surface when secured by a union nut.
- Ring type - the end of a metal pipe is sealed against a mating surface with a rubber 'O' ring and secured with a union nut.
- Block/pad type - a metal plate on the end of a pipe or component is squashed against a mating surface by an external clamp; the surfaces will often have a form of rubber gasket to assist with sealing.
- Spring lock type - a connection at the end of a pipe can be held against a mating surface by spring pressure; these connections are often quick release if the spring is compressed.

To help you identify the different pressure circuits on an air conditioning system, the high pressure section will normally have narrow pipe work with a large service connector. The low pressure section will normally have wider pipe work with a small service connector.

System layouts and operation

Figure 4.21 Air conditioning controls

Depending on how the vehicle is to be used, an air conditioning system may have a number of different layouts and configurations. On a standard system, the evaporator is mounded behind the dashboard, just before the heater matrix. Ventilation air can be drawn through the evaporator and heat, moisture and dust/dirt can be removed. The cooler, dry air can then be directed into the passenger compartment to lower the temperature. If the driver requires warm, dry/clean air, the air conditioning system can be adjusted so that having passed through the evaporator the ventilation air is then passed through the heater matrix before entering the passenger compartment. This allows the driver to use the air conditioning system all year round to maintain a comfortable environment and also assist with the de-misting of windows during winter months.

Advanced Light Vehicle Technology

Some systems, particularly in vans, may mount the evaporator overhead, allowing the cool clean air to pass down through the passenger compartment. Other systems may have two evaporators to enable different temperatures to be selected (for front and rear passengers for example). These are known as dual control/zone systems.

System operation

The refrigeration cycle of an air conditioning system is a closed loop, and as a result has no beginning or end. For the purpose of these descriptions we will start and finish at the compressor.

Thermal expansion valve type (TXV)

1. The compressor pump is operated by the engine or an electric motor. Internal pistons or vanes are used to raise the pressure of the refrigerant gas (compressing the refrigerant will also raise its temperature). This system pressure increase will raise the boiling point of the refrigerant gas.
2. The refrigerant leaves the compressor as a hot high pressure gas.
3. The refrigerant enters the top of the condenser radiator mounted on the outside of the vehicle. As the hot gas passes backwards and forwards across this radiator, air passing through the condenser will transfer some of the heat to the surrounding atmosphere. The fall in temperature of the refrigerant gas, as well as its high pressure, will allow the refrigerant gas to condense into a high pressure hot liquid (giving up a large amount of **latent heat** as it changes state).
4. From the condenser radiator, the hot, high pressure liquid refrigerant is transferred to a storage container known as a receiver drier. (Thermal expansion valve TXV system). The receiver drier also contains a silicone desiccant which will remove any water moisture.
5. From the receiver drier, a regulated amount of high pressure liquid refrigerant is allowed to pass through a valve into the evaporator unit, mounted inside the vehicles passenger compartment.
6. The evaporator is located in the low pressure/suction part of the air conditioning circuit. As the high pressure liquid refrigerant enters the low pressure space inside the evaporator, it instantly boils and turns into a low pressure gas (absorbing a large amount of latent heat as it changes state).
7. Passenger compartment air that is passed over the outside of the evaporator by an electrically driven fan has some of its heat energy removed and absorbed into the low pressure refrigerant gas.
8. The low pressure gas (with some heat energy from the passenger compartment) is then drawn back into the compressor where the refrigerant cycle begins all over again.

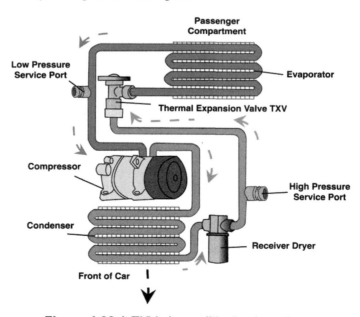

Figure 4.22 A TXV air conditioning layout

Automotive Master Technician

Latent heat - heat energy that is absorbed or given off as a substance changes state from a solid to a liquid, or a liquid to a gas.

Fixed orifice tube type (FOT)

1. The compressor pump is operated by the engine or an electric motor. Internal pistons or vanes are used to raise the pressure of the refrigerant gas (compressing the refrigerant will also raise its temperature). This system pressure increase will raise the boiling point of the refrigerant gas.

2. The refrigerant leaves the compressor as a hot high pressure gas.

3. The refrigerant enters the top of the condenser radiator mounted on the outside of the vehicle. As the hot gas passes backwards and forwards across this radiator, air passing through the condenser will transfer some of the heat to the surrounding atmosphere. The fall in temperature of the refrigerant gas, as well as its high pressure, will allow the refrigerant gas to condense into a high pressure hot liquid (giving up a large amount of latent heat as it changes state).

4. From the condenser radiator, the hot, high pressure liquid refrigerant is transferred via pipes to an accurately sized restriction known as a fixed orifice tube (FOT). The flow of refrigerant to the FOT is controlled by regulating the system pressure, normally by switching the compressor on and off using a clutch mechanism.

5. From the fixed orifice tube, a regulated amount of high pressure liquid refrigerant is allowed to pass through into the evaporator unit, mounted inside the vehicles passenger compartment.

6. The evaporator is located in the low pressure/suction part of the air conditioning circuit. As the high pressure liquid refrigerant enters the low pressure space inside the evaporator, it instantly boils and turns into a low pressure gas (absorbing a large amount of latent heat as it changes state).

7. Passenger compartment air that is passed over the outside of the evaporator by an electrically driven fan has some of its heat energy removed and absorbed into the low pressure refrigerant gas.

8. After the refrigerant leaves the evaporator, it enters a storage container known as a suction accumulator, which will allow time for any liquid refrigerant that remains to boil away before it returns to the compressor. The suction accumulator contains a silicone desiccant which will remove any water moisture.

9. The low pressure gas (with some heat energy from the passenger compartment) is then drawn back into the compressor where the refrigerant cycle begins all over again.

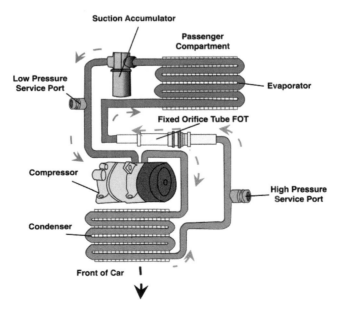

Figure 4.33 An FOT air conditioning layout

Advanced Light Vehicle Technology

Air distribution

An air conditioning system allows the driver to adjust the levels of ventilation and temperature in the passenger compartment. It also allows the selection of air vents from which the conditioned air is directed. The choice of air direction will often fall into three main areas:

- Up - towards the screens and windows.
- Out - towards the body and face.
- Down- towards the feet.

Control devices

Depending on whether the system is air conditioning (manually controlled) or climate control (automatically controlled) a number of options are available to manage the function and operation of the refrigerant cycle.
In an air conditioning or climate control system, the driver will normally have the following options:
Temperature control - in an air conditioning system this will usually be in the form of a dial or slide which can vary the electrical resistance in the temperature sensing electrical circuit of the air conditioner. By changing the electrical circuit resistance, the operation of the compressor can be controlled to regulate the evaporator temperature or move flaps in the ventilation circuit to control the flow of air through the evaporator and heater matrix.
With climate control the desired temperature is set by the driver using dashboard controls and the cabin temperature is then monitored by a sensor in the passenger compartment. The sensor sends information to an ECU which then operates actuators regulating the compressor and evaporator flow to ensure that the set temperature is maintained.

Air distribution - in an air conditioning system this will usually be in the form of a dial or slide which can operate flaps within the HVAC system to direct air flow to various outlets or vents. The options will normally include windscreen, face and feet but depending on vehicle design, other auxiliary vents may be included. The control will operate the flaps via cables, vacuum pipes and servos or small electric motors.
With climate control, the flaps will operate in a similar manner to those found in air conditioning but will normally only use electric servo motors to provide the positioning of the flaps. An ECU is able to manage the positioning of the flaps depending on how the driver has set the controls on the dashboard. During rapid warm up or cool down of the passenger compartment the ECU may override the drivers setting by initially directing warm air down (warm air rises) and cool air upwards (cold air falls). This way, the desired cabin temperature can be achieved in the quickest possible time before the flaps return to the driver settings.

Ventilation fan speed - in an air conditioning system this will usually be in the form of a dial or slide which controls the power to an electric motor driving a fan unit. Many air conditioning systems will use a **rheostat** which sends power through a set of dropping resistors connected in series with the electric motor to reduce the voltage available. Rheostats will often give the option of 3, 4 or 5 speeds. An example of a three speed fan operation is described below.
Off - the dial is in the off position, causing an open circuit and no current flows to the motor.
One - the electric motor feed is directed through a resistor in the rheostat which converts a large amount of the energy into heat. For example: if the resistor uses up 6 volts, then only 6 volts will be available at the motor giving slow speed.
Two- the electric motor feed is directed through a resistor in the rheostat which converts some of the energy into heat. For example: if the resistor uses up 3 volts, then 9 volts will be available at the motor giving medium speed.
Three - the electric motor feed bypasses the rheostat completely meaning that 12 volts is available at the motor giving full speed.

A good indication that the rheostat speed control unit has failed in the ventilation fan unit is that only full speed is available when the control is operated.

> **Rheostat** - an electrical component designed to control current flow by varying resistance.
>
> **Transistor**- an electronic component which acts like a switch with no moving parts.

With climate control it is often possible to set ventilation fan speed stages in a similar manner to air conditioning, but many systems will often have a greater selection of settings or a dial/slider which gives variable control. In these systems, speed control is adjusted via the ECU using duty cycle. Duty cycle is an electricity regulating method which uses **transistors** to rapidly switch a component on and off, which unlike using resistors can operate with very little loss of power. This rapid switching does not allow the fan motor to fully speed up or slow down and depending on the duty cycle's 'on' time power to the motor can be controlled between an upper and lower limit giving various speeds. Using an ECU and duty cycle allows the climate control system to manage the air flow to the passenger compartment, for example:

- During rapid warm up or cool down, a high fan speed can be used to get the cabin temperature to the desired level quickly.
- At varying road speeds ventilation can be regulated so that the forced air effect caused by the moving car can be compensated for. (i.e. at high road speeds the fan can be slowed down to allow for the air being rammed into the passenger compartment by momentum, and when the vehicle is stationary the fan speed can be increased to maintain desired ventilation).

Recirculation

Recirculation is another form of ventilation control. It is often operated by a different button or dial than the rest of the dashboard controls. When operated a flap acts to block off the outside air vents that allow ventilation in by the lower edge of the windscreen scuttle. This can be used to prevent unpleasant smells or fumes from entering the passenger compartment. In a climate control system, the recirculation flap can be automatically activated to assist with rapid warm up, cool down or demisting. The ECU can use information provided by an air purity sensor to control its operation in polluted situations.

Climate control sensors

For a climate control system to automatically manage the heating, ventilation and air conditioning (HVAC), it needs to gather information from various sensors. A description of some of these sensors is shown in Table 4.7.

Table 4.7 Climate control sensors

Sensor	Purpose and operation
Ambient air temperature sensor 	The ambient air temperature sensor is mounted in a position where it is able to take a measurement of the outside air temperature. It is a small **thermistor** which may be a **NTC** or **PTC** type. As outside temperature rises or falls, the resistance of the sensor changes and information is relayed to the ECU as a varying voltage. The ECU will then compare the information provided by this sensor with values of cabin air temperature and adjust the loading on the climate control system to compensate. This ensures that the most efficient operation of the heating, ventilation and air conditioning is achieved.

Advanced Light Vehicle Technology

Table 4.7 Climate control sensors

Sensor	Purpose and operation
Cabin air temperature sensor 	The cabin air temperature sensor is mounted in a position where it is able to take a measurement of the passenger compartment temperature. It is often located inside a small pipe, called an aspirator tube that can be connected to the **plenum chamber** of the ventilation fan. This aspirator tube is then able to create a low pressure which helps draw cabin air across the sensor. The sensor is a small thermistor which may be a NTC or PTC type. As inside temperature rises or falls, the resistance of the sensor changes and information is relayed to the ECU as a varying voltage. The ECU will then compare the information provided by this sensor with values of ambient air temperature and the desired temperature set by the driver. It will then adjust the loading on the climate control system to compensate. This ensures that the most efficient operation of the heating, ventilation and air conditioning is achieved while maintaining a constant temperature inside the cabin.
Coolant temperature sensor 	Climate control systems will blend warm air that has passed through the heater matrix with purified/dehumidified air that has passed through the evaporator. In order for the ECU to accurately control the temperature which has been set by the driver, it needs to know the temperature of the coolant in the heater matrix. This measurement can be obtained from the engine management coolant temperature sensor, which is often a form of NTC thermistor and then shared with the climate control ECU via network communication.
Evaporator temperature sensor 	The evaporator temperature sensor is mounted in a position where it is able to take a measurement of the evaporator core temperature. It is a small thermistor which may be a NTC or PTC type. As evaporator temperature rises or falls, the resistance of the sensor changes and information is relayed to the ECU as a varying voltage. The ECU will then compare the information provided by this sensor with values of cabin air temperature and adjust the loading on the climate control system to compensate. The refrigerant cycle is carefully controlled from information provided by this temperature sensor in order to keep the evaporator core at just above 0°C. If this is not done, it would be possible for the condensed water moisture from the air to freeze on the outside of the evaporator core, preventing air flow and reducing overall efficiency of the system
Condenser temperature sensor 	To ensure correct operation of the refrigerant cycle, the condenser must efficiently dissipate heat to the surrounding air. This means that the condenser temperature must be kept within certain limits. At times of high system load and when ambient temperature is also high, extra air flow may be required in order for the condenser to radiate heat. The sensor will be mounted in a position where it is able to measure the temperature of the condenser core and operate a cooling fan if it becomes too hot.
Sun load sensor 	The sun load sensor is a small photoelectric cell. It is normally mounted on the dashboard of the car behind the windscreen. In direct sunlight you will feel hotter than the actual ambient cabin temperature due to the radiation produced by the sun's rays. It is the job of the sun load sensor to pick up this added solar radiation and adjust the loading on the climate control system to compensate, regardless of passenger compartment temperature.

Automotive Master Technician

Table 4.7 Climate control sensors

Sensor	Purpose and operation
Air quality sensor 	The air quality sensor is mounted in the main ventilation air intake system. It uses a chemical reaction to indicate the presence of carbon monoxide or nitrogen dioxide which changes the resistance of the sensor. The resistance change is used by the ECU to detect pollution and regulate the position of the recirculation flap controlling the exchange of incoming air to the passenger compartment.
Engine speed sensor 	Information from the engine management systems engine speed sensor is often shared with the climate control systems ECU via network communication. The engine speed is normally registered by the crankshaft position sensor which may be inductive or hall effect. The information from this sensor can be processed so that the compressor clutch of the climate control system does not cut in immediately after start-up, but instead waits until the engine has been running for a short period of time to ensure that excessive loads are not placed on the engine. This provides a more stable start-up/warm-up period, reduces fuel consumption and emissions and can increase engine performance during acceleration.
Vehicle speed sensor 	Information from the cars dynamic control system vehicle speed sensor is often shared with the climate control ECU via network communication. The vehicle speed is normally registered by the transmission speed sensor or a wheel speed sensor which may be inductive, MRE or hall effect. The information from this sensor can be processed so that the speed of the ventilation fan can be regulated. When the car is moving fast, more air is forced into the passenger compartment and the ventilation fan slows down. When the car is moving slowly or is stopped, less air enters the passenger compartment so the ventilation fan is speeded up.

Thermistor - a temperature sensitive resistor (thermal resistor).

NTC - negative temperature coefficient. When used with thermistors, as temperature rises, resistance falls.

PTC - positive temperature coefficient. When used with thermistors, as temperature rises, resistance rises.

Plenum chamber - an air storage chamber.

Advanced Light Vehicle Technology

Other comfort, convenience and safety systems

Some of the other comfort, convenience and safety systems you will come across are shown in Table 4.8.

Table 4.8 Other comfort, convenience and safety systems

Rear view camera systems 	Many modern vehicles include a form of visual display unit as part of their infotainment system. Because of this it is then a very small step to add additional features to this visual display which will help with comfort and convenience. A camera can be mounted at the rear of the car and connected to a screen found on a TV or DVD player for example. It can then be linked to the reversing light system so that when backing up, additional vision is available for safety.
Lane change warning/control 	A series of sensors can be used that identify road lane markings. The information supplied to an ECU can inform the driver if they begin to drift across lanes by mistake. A vibration or buzzer attached to the seat can be used to alert the driver, while some systems may make small corrections to the steering via the electronic power-assisted steering (EPS) if the indicators have not been used or a positive input to the steering has been made.
Self-parking 	The addition of a Doppler parking radar means that manufacturers are able to combine engine, transmission and electronic power steering control to allow a car to park itself in a parallel parking space. The driver just needs to find a parking space of suitable size, press a button and the car will automatically reverse into the space. (Some systems require that the driver controls the vehicle speed with the throttle and brake during the automatic parking procedure).
Anti-collision 	Many manufacturers are now beginning to include active safety systems which use sensor information to apply the brakes and avoid a collision. How the system operates normally depends on a number of parameters, including speed and direction.
Driver and passenger seating 	To improve vehicle comfort, many manufacturers offer the option of heated and sometimes cooled driver and passenger seating. In the heated mode, a thin electrical element is placed just below the seat covers. When activated by the driver, electric current is driven through the element where it is converted to heat. During cooling mode, ventilated air from the climate control system is directed to vents into the seat frame where it is distributed to the through the cushion and backrest which can lower the temperature by 3°C to 5°C and also reduce humidity by approximately 20%. The amount of heating or cooling may be controlled automatically via the vehicles ambient temperature thermistor from the climate control system, or it may have its own temperature sensitive thermistor. Seating position can also be controlled to provide extra comfort and convenience. Electric motors may be attached to the seat frame which can adjust its position according to driver/passenger requirements. Position adjustment may include: • Fore and aft adjustment of the entire seat frame • Up and down height adjustment of the seat base • Tilt of the seat back • In and out of lumbar support • And height adjustment of the head rest Once the desired position has been achieved, these settings can be stored in the memory of an ECU, for recall by the driver. Small variable resistors are

Table 4.8 Other comfort, convenience and safety systems

	attached to the frame and motors which send information about the seating position in the form of voltage or resistance values. This way, if the seat is moved and needs to be reset, it becomes a one touch switch operation. When activated the ECU is able to drive the seat into the desired position using motors on the frame according to the values shown on the potentiometers. This is known as memory seating.
Doppler radar 	The Doppler effect is a scientific principle which detects the frequency of waves. These may be sound waves (sonar) or radio waves (radar). The principle of the Doppler effect can be described by listening to the noise of an approaching car (a racing car is a good example to use). The sound of the racing car as it approaches you, is high pitched as the sound waves are compressed by its forwards motion. As the car passes you, the noise emitted is stretched out and its tone lowered as the sound waves are allowed to expand when following the movement of the car. This principle can be applied to a number of convenience systems such as: Anti-theft: A Doppler system is able to detect proximity and is capable of giving an audible warning when someone is too close. It works by sending out an electromagnetic radar wave and analysing the signal that is bounced back to the receivers. If the frequency of the signal bounced back increases, someone is moving towards the car. If the frequency of the signal bounced back reduces, someone is moving away from the car. Parking aids: mounted in the rear bumper (and increasingly also in the front bumper), small Doppler **transponders** are able to detect objects and their proximity to the front or rear bumpers. The radar frequency given off and received by these sensors can then be used to give the driver an indication of the distance of objects in front or behind by beeping with increased intensity as the distance between the object and the bumpers reduces. (Due to the sharing of information via the vehicles multiplex network system, these Doppler sensors are increasingly being used to provide information to passive pedestrian safety systems, such as external air bags or pop-up bonnets).
Head up displays 	Head up display (HUD) is a system that allows certain data to be projected onto the driver's windscreen in a readable but transparent format. This way the driver is able to make use of the information provided without having to take their eyes off the road. Head up display systems have been in use with fighter pilots for a number of years but this system is likely to only be offered as an optional extra on cars as many drivers may feel that it is a distraction and would prefer to have a clear view of the road ahead. Information that can be provided by a head up display includes: • Current speed • Turn by turn navigation • Lights or turn signals are activated • RPM • Lane deviation • Radio station • Collision avoidance Research and development is underway by a number of manufacturers to enable the HUD to **augment** the driver's vision. With the addition of special cameras and radar, objects in front of the driver can be highlighted on the windscreen when vision is poor. Examples could include: • Highlighting pedestrians/cyclists in the dark • Highlighting the edge of the road when foggy

Advanced Light Vehicle Technology

Table 4.8 Other comfort, convenience and safety systems

Automatic windscreen wiper control 	Some cars are now being produced with rain sensing windscreen wiper operation. A small area of the front windscreen glass, usually located on the outside of the vehicle, opposite the rear-view mirror, is monitored by an optical sensor. The sensor is designed to project infrared light at the windscreen at an angle and then read the amount of light that is reflected back. A clean windscreen will reflect nearly all of the inferred light back, while a wet or dirty windscreen will cause the light to scatter. The optical sensor can determine the necessary frequency and speed of the windscreen wipers by monitoring the amount of light reflected back into the sensor. As a safety precaution, and to prevent damage to the wiper mechanism, nearly all rain-sensing wipers must be activated each time they are used (initially switched on by the driver). The activation process prevents the system from automatically wiping a frozen windscreen, or triggering while the vehicle is in a car wash; both instances could damage the blades or electric motor powering the wipers.
Advanced front lighting systems 	A recent development in lighting systems is the ability to move the headlight beam in response not only to vehicle steering and suspension dynamics, but also to weather and visibility conditions, vehicle speed, and road curvature or contour. This development is known as advanced front-lighting systems (AFS). This system of moving headlamps has been around for many years, but early operation relied on mechanical rods and linkages. In contrast, AFS uses dynamic vehicle information from various chassis, speed and acceleration sensors that is available on the vehicle communication network to determine the most effective use of headlamp aim and angle. The headlamps can then react to vehicle movement and operation using actuators and motors built into headlamp units. Some manufacturers have been producing vehicles equipped with AFS as standard since 2002. These auxiliary systems may be switched on and off as the vehicle and operating conditions call for light or darkness at the angles covered by the beam. Developments now mean that AFS systems can use GPS signals to anticipate changes in road curvature, making them truly intelligent. Many manufacturers are now offering an option to have the vehicles lighting system automatically switch on and off depending on ambient light conditions. A light sensitive resistor is mounted at the front of the car (usually in the area of the front scuttle or windscreen). The driver must select the automatic lighting option to override any other light combination choices. This way, when the ambient light levels fall to a pre-set value, a standard lighting arrangement, including dip-beam headlamps is automatically switched on. The automatic light system may be left on permanently and will cut the lights when the engine is switched off and the ignition key is removed, but if the driver requires a different lighting pattern, fog or spotlights for example, the automatic option must be switched off manually.

Transponder - a radio, radar or sonar receiver that automatically transmits a signal on reception of a designated incoming signal.

Augment - to add to, extend or increase.

Telemetry and vehicle monitoring

Telemetry is a system of automatic transmission and measurement of data from remote sources by wire, radio or other means. Although the term most commonly refers to wireless data transfer mechanisms, such as radio, Bluetooth or infrared, it also includes information transferred using other methods such as a telephone, computer network, optical link or other wired systems. Many modern telematics can take advantage of the low cost mobile phone networks by using SMS to receive and transmit data.

Radio transmission

Radio signals are broadcast by creating a waveform in the **electromagnetic spectrum** which are invisible and completely undetectable by humans.
Radio waves are created by switching electrical circuits on and off, and if produced at the right frequency and optimised for distance, can be transmitted over vast areas at the speed of light. If a receiver is then tuned to the correct frequency of the broadcast signal, it can be decoded and converted in to information/data to provide wireless communication.

Telemetry - the automatic transmission and measurement of data from remote sources by wire, radio or other means.

Electromagnetic spectrum - the range of electromagnetic radiation, laid out in order of frequency and wavelength.

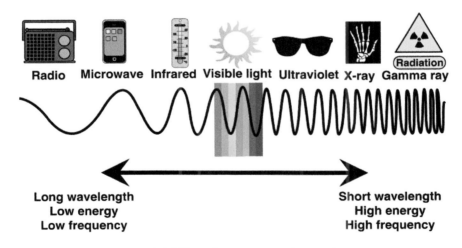

Figure 4.34 The electromagnetic spectrum

Telemetry has been in use for many years, but its development for motor racing applications has advanced its usefulness to a point where manufacturers are including systems on most road cars. Telemetry may be used for various options including:

- Satellite navigation.
- Communication.
- Vehicle diagnostic monitoring and data logging.

- Vehicle position monitoring, including anti-theft.
- Two-way telemetry, including preventative maintenance.

Advanced Light Vehicle Technology

Bluetooth is a system that uses standardised short range radio communication to connect two devices. It is very useful for transmitting data between two devices that are in close proximity and is now commonly used for transmitting vehicle management data wirelessly to a scan tool or vehicle communication interface (VCI) for example.

Satellite navigation

Driver navigation aids can be manufactured as part of the vehicle integrated design, or fitted by the owner as auxiliary aftermarket equipment. Auxiliary navigation equipment relies on Global Positioning System (GPS) satellites, whereas integrated systems can use dead reckoning, GPS or a combination of the two.

Figure 4.35 Navigation satellites

Dead reckoning is a system that makes use of vehicle sensor data such as vehicle speed, steering angle and wheel speed, and matches this to maps that are pre-programmed into the navigation system. If dead reckoning is used without GPS combination, it has to be calibrated by setting a start position and, as the car is driven along the road, its location is calculated from the information supplied by the vehicle sensors

Navigation using GPS relies on radio signals sent from satellites in space. The system is based on a series of American military satellites, but other countries are now beginning to launch their own satellites for navigation purposes. 24 satellites are used for GPS at any one time, with a number of spares also in orbit in case one goes wrong. Each of these solar-powered satellites circles the earth at a distance of about 12,000 miles (19,300 km), making two complete rotations every day. The orbits are arranged so that at anytime, anywhere on Earth, there are at least four satellites 'visible' in the sky.

The job of the GPS receiver, mounted in the car, is to locate four satellites and work out its distance from them. It can then use this information to establish its own position and map this against information in its memory. This process is called trilateration. Once the position of the GPS receiver has been determined, software programmed into the unit is able to calculate the route to a desired destination. Many systems are equipped with voice commands which give timely instructions to the driver, allowing them to reach their destination without having to concentrate on a map.

Trilateration

Each satellite sends out radio signals containing time, position and climate information. When the GPS unit in the car receives a signal from a satellite, it can work out its distance from the satellite by the length of time it has taken the signal to travel. (Radio signals sent from the satellites travel at the speed of light: 299,792,458 metres per second, 1,079,251,985 kilometres per hour or 186,282 miles per second). When the distance from four satellites is combined your exact position can be accurately calculated and matched to a map stored in the receiver's memory.

Automotive Master Technician

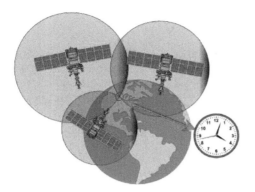

Figure 4.36 Trilateration

Advantages and disadvantages of satellite navigation

The advantages and disadvantages of satellite navigation are shown in Table 4.9.

Table 4.9 Advantages and disadvantages of satellite navigation

Advantages	Disadvantages
Guides the user to unfamiliar destinations	Limited use for the majority of journeys
Avoids the use of maps when driving	May take driver down unsuitable roads
Gives accurate and timely instructions	Some systems may not be aware of current traffic problems

Safety

• You must not mount the sat-nav receiver where it will obscure the view of the road ahead.

• You must not update the system while the vehicle is moving.

Communication

The use of mobile phones when driving is considered a distraction and so legislation has been enacted to restrict the use of hand-held phones while driving. A number of methods have been produced that facilitate the use of mobile telephone systems while driving, including:

Figure 4.37 In-car telephone communication

- **Built-in infotainment:** With built-in equipment, car manufacturers design a telephone system which is directly integrated with the entertainment and information system. No external telephone is required as the microphone and speaker are mounted inside the passenger compartment. Many systems have a method for placing or receiving a call by voice control or buttons mounted on the steering wheel.

- **Auxiliary phone connections:** Usually a cradle system is mounted inside the car which the telephone is attached to while in use. The cradle may be connected to the in-car entertainment system to make and receive calls, or can have its own speaker and microphone system.

Advanced Light Vehicle Technology

- **Bluetooth:** Many mobile phones have a Bluetooth capability. Bluetooth uses standardised short range radio communication to connect two devices. With the Bluetooth switched on, the phone is able to act as a remote transmitter and receiver for telecommunication. An appropriate Bluetooth speaker and microphone device is required to use this system, which can be stand-alone or built into the in-car entertainment system.

Vehicle diagnostic monitoring and data logging

On-board diagnostics (OBD) has been in use on vehicles since the early 1980's. Originally each manufacturer designed and used their own methods for connecting dedicated diagnostic equipment and extracting diagnostic trouble codes (DTC). By January 2000, legislation dictated that all vehicles designed for sale and use in Europe had to become E-OBD compliant, standardising the socket for the data link connector (DLC) and using a generic set of codes for malfunctions that affect engine management systems linked to emissions.

Although many scan tools used to conduct on-board diagnosis are still physically connected to the DLC by a wired system, more and more equipment manufacturers are producing Bluetooth or Wi-Fi transmitters which when plugged into the data link connector can transmit information wirelessly. This gives the technician the flexibility to move around the vehicle with the scan tool and conduct diagnostic routines while not restricted to the passenger compartment.

An extension of this technology is to bypass the data link connector completely and transmit information wirelessly from a management system to a receiving host (telemetry). This way, vehicles can communicate directly with the manufacturer or workshop, providing data before the vehicle even arrives for repair. If the ECU has a **data logging** capability, the amount of information supplied can be increased to take into account intermittent faults or glitches.

Data logging - a method of recording diagnostic information from a vehicle management system so that it can be played back and studied at a later time.

Transceiver - a device that can both receive and transmit data.

Transducer - a device that converts a signal in one form of energy to another form of energy.

Vehicle position monitoring, including anti-theft

Wireless communication has enabled manufacturers to improve vehicle security. Immobilisers have used this technology for a number of years and nearly every vehicle produced today will have some form of anti-theft device, where a transponder chip embedded in a key will send a coded signal to a **transceiver** unit in the car. When the correct coded signal is received, the immobiliser system is the disabled and the vehicle can be started in the normal manner. An advancement of this system is keyless entry. Some cars have a central locking system that is triggered if a **transducer** type advanced key is within a certain distance of the car. Sometimes called hands-free or smart-key, this system allows the driver to unlock their vehicle without having to physically push a button on the key fob. They are also able to start or stop the engine without physically having to insert the key and turn the ignition.

As the driver approaches the car, the vehicle senses that the key is approaching (located in a pocket for example). When inside the car's required distance, there are two methods typically used by manufacturers to unlock the doors:

1. Once the driver is within the required distance for the key to be recognised, the car automatically unlocks the driver's door.
2. Once the driver is within the required distance for the key to be recognised, the car doesn't unlock the door until the key holder touches a sensor located behind one of the door handles.

Automotive Master Technician

If passengers attempt to gain entry, the system senses that the driver is within the required proximity and as they touch the sensors behind their door handles, the car unlocks their door.

Although these passive security systems make it more difficult to steal a car, they are not always sufficient to prevent a determined thief. GPS has given manufacturers an added security benefit in the form of vehicle tracking. If the vehicle is equipped with a Global Positioning System, then this can often be used to transmit its location, if stolen, using the mobile phone network. This design, although not actually preventing the theft of the vehicle in the first place, is helping to recover more cars once they have been stolen.

Two-way telemetry, including preventative maintenance

Two-way telemetry was originally developed for use in motor racing applications. Not only could the car transmit information and data about its operation, but technicians could then make electronic adjustments to the management systems wirelessly from the pits.

This technology can also be applied to road cars, allowing a workshop to receive information about the operating conditions of a car and then update the systems remotely. The communication could be over a short range wireless system, or use the mobile phone network to conduct updates anywhere in the world with a signal.

Chapter 5 Alternative Fuel Vehicles

This chapter will help you develop an understanding of the construction and operation of alternative fuel vehicles. It will give an overview of the types and configurations of alternative fuel vehicles available and cover the relationship between cost, efficiency and ecological impact. It will also assist you in developing a systematic approach to the diagnosis of complex system faults for which there is no prescribed method. This is achieved with hints and tips which will help you undertake a logical assessment of symptoms and then uses reasoning to reduce the possible number of options, before following a systematic approach to finding and fixing the root cause.

Contents

Information sources

Many of the diagnostic routines that are undertaken by a Master Technician are complex and will often have no prescribed method from which to work. This approach will involve you developing your own diagnostic strategies and requires you to have a good source of technical information and data (*see knowledge is power*). In order to conduct diagnosis on advanced alternative fuel systems, you need to gather as much information as possible before you start and take it to the vehicle with you if possible.
Sources of information may include:

Table 5.1 Possible information sources

Verbal information from the driver (*see the interrogation room*)	Vehicle identification numbers
Service and repair history (*see the interrogation room*)	Warranty information
Vehicle handbook	Technical data manuals
Workshop manuals/Wiring diagrams	Safety recall sheets
Manufacturer specific information	Information bulletins
Technical helplines	Advice from other technicians/colleagues
Internet	Parts suppliers/catalogues
Jobcards	Diagnostic trouble codes
Oscilloscope waveforms	On vehicle warning labels/stickers
On vehicle displays	Temperature readings

Automotive Master Technician

Always compare the results of any inspection or testing to suitable sources of data. Remember that no matter which information or data source you use, it is important to evaluate how useful and reliable it will be to your diagnostic and repair routine.

Electronic and electrical safety

Working with any electrical system has its hazards. The main hazard is the possible risk of electric shock, but remember that the incorrect use of electrical diagnostic testing procedures can also cause irreparable damage to vehicle electronic systems. Where possible, isolate electrical systems before conducting the repair or replacement of components.

When diagnosing faults on hybrid and all-electric drive systems, take care when working around the high voltage components. The high voltage system can normally be identified by its reinforced insulation and shielding which is often coloured bright orange. These systems carry voltages that can cause severe injury or death.

Always use the correct tools and equipment. Damage to components, tools or personal injury could occur if the wrong tool is used or misused. Check tools and equipment before each use.
If you are using electrical measuring equipment, always check that it is accurate and calibrated before you take any readings.

If you need to replace any electrical or electronic components, always check that the quality meets the original equipment manufacturer (OEM) specifications. (If the vehicle is under warranty, inferior parts or deliberate modification might make the warranty invalid. Also, if parts of an inferior quality are fitted, they might affect vehicle performance and safety). You should only carry out the replacement of electrical components if the parts comply with the legal requirements for road use.

Table 5.2 The operation of electrical and electronic systems

Electrical/Electronic system component	Purpose
ECU	The electronic control unit is designed to monitor the operation of vehicle systems. It processes the information received and operates actuators that control system functions. An ECU may also be known as an ECM (electronic control module).
Sensors	The sensors are mounted on various system components and they monitor the operation against set parameters. As the vehicle is driven, dynamic operation creates signals in the form of resistance changes or voltage which are sent to the ECU for processing.
Actuators	When actioned by the ECU, motors, solenoids, valves, etc. help to control the function of vehicle for correct operation.

Table 5.2 The operation of electrical and electronic systems

Digital principles	Many vehicle sensors create analogue signals (a rising or falling voltage). The ECU is a computer and needs to have these signals converted into a digital format (on and off) before they can be processed. This can be done using a component called a pulse shaper or Schmitt trigger.
Duty cycle and PWM	Lots of electrical equipment and electronic actuators can be controlled by duty cycle or pulse width modulation (PWM). These work by switching components on and off very quickly so that they only receive part of the current/voltage available. Depending on the reaction time of the component being switched and how long power is supplied, variable control is achieved. This is more efficient than using resistors to control the current/voltage in a circuit. Resistors waste electrical energy as heat, whereas duty cycle and PWM operate with almost no loss of power.
Networking and multiplex systems	Many modern vehicle systems are controlled using computer networking. In these systems a number of ECU's are linked together and communicate to share information in a standardised format. The most common network system is the Controller Area Network (CAN Bus). *See Chapter 1.*

Alternative fuel vehicles

Petrol and Diesel are gradually becoming more expensive to produce. One of the main reasons for this is that the demand for these fuels has increased. To overcome the issues created by **'peak oil'** and environmental pollution many manufacturers are researching and designing cars that will run on alternative fuels.

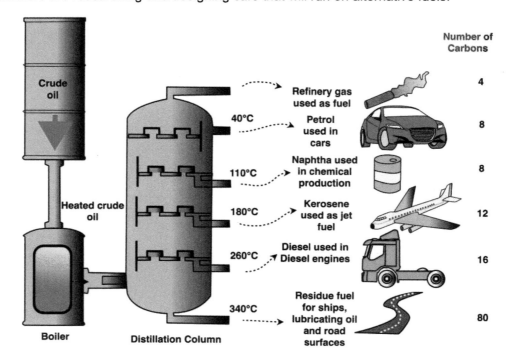

Figure 5.1 Refining of crude oil

Automotive Master Technician

An alternative fuel vehicle is one that runs on a fuel other than traditional petroleum-based fuels (i.e. petrol or Diesel). Between 2008 and 2009 there were around 33 million alternative fuel and advanced propulsion technology vehicles on the world's roads. This represented around five per cent of the world's vehicles. This number is increasing every year.

Peak oil - a situation where the demand for crude oil exceeds the supply.

Retrofitting – aftermarket vehicle conversion.

Liquefied petroleum gas (LPG)

LPG, sometimes known as 'autogas' is produced during the normal refining process of crude oil. Liquefied petroleum gas (LPG) is a low pressure liquefied gas mixture made up of mainly propane and butane, but essentially it is petrol gas. This fuel can be burned in a conventional internal combustion engine and produces less CO_2 than petrol. The LPG is stored under pressure as a liquid because in this state it is 250 times denser than in its gaseous form. A standard petrol car can often be converted to run on LPG (this is known as **retrofitting**) by adding a second tank and fuel system for the LPG. Because the petrol tank and petrol fuel operating system stay, the car becomes 'dual fuel'.

Benefits of using liquefied petroleum gas LPG include:

- ✓ Smoother, quieter and cleaner running engines than those using conventional petrol.
- ✓ The life-span of the engine can be extended by as much as 30%.
- ✓ Lower emissions output, leading to less environmental pollution.
- ✓ Much lower fuel costs.
- ✓ As the engine runs cleaner, servicing costs are reduced.
- ✓ If the vehicle is designed to be a dual fuel car, (one that operates on either petrol or LPG) its range is increased.
- ✓ Resale values may be increased because the car is cheaper to run than its competitors.
- ✓ The storage container for LPG in cars is normally of stronger construction than that of petrol tanks and as a result, crash damage safety is improved.
- ✓ LPG has a high ignition temperature, around twice that of petrol, making it less likely to spontaneously combust and reducing the risk if fire.

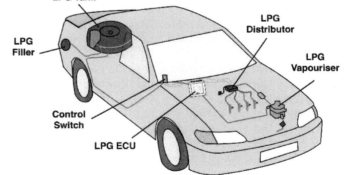

Figure 5.2 An LPG equipped car

Advanced Light Vehicle Technology

Safety

Remember that an LPG tank will never really be empty. As fuel is used and pressures fall, eventually the propane left over will remain as a gas.
An LPG tank should also never be completely filled. It is recommended that tanks are only taken to around 80% capacity to allow for expansion and contraction of the propane due to different ambient temperatures.

Unfortunately, LPG has a lower energy density than petrol and therefore fuel consumption is increased, but a lower duty of tax will often offset this higher consumption giving a lower overall operating cost.
When compared to conventional petrol or Diesel engined vehicles, the emissions produced are far less harmful to health and the environment. Table 5.3 shows how petrol and Diesel compare to liquefied petroleum gas LPG.

Table 5.3 emissions comparison between petrol, Diesel and LPG

LPG compared to petrol	LPG compared to Diesel
75% less carbon monoxide	60% less carbon monoxide
40% less oxides of nitrogen	90% less oxides of nitrogen
87% less potential of forming ground level ozone	70% less potential of forming ground level ozone
85% less hydrocarbons	90% less particulate matter (soot)
10% less carbon dioxide	

To combust properly inside an engine, petrol and Diesel need to be vaporised, as it is the fumes that burn not the liquid fuel. Complicated methods are needed to introduce petrol or Diesel, such as a carburettor or fuel injection, and turn it into a vapour. An advantage of using LPG as alternative fuel source to power vehicles is its very low boiling point of between - 0.6 degrees Celsius for butane and -42 degrees Celsius for propane. Because of its very low boiling point, most LPG used for vehicles is based on propane. With LPG you only require some form of simple nozzle to introduce the fuel and it will vaporise easily.

Converting a standard car to run on LPG

Although LPG is a very safe fuel to use for vehicle propulsion, care must be taken when converting a car to run on propane, otherwise safety may be compromised.
If the car is to be operated as a dual fuel vehicle, a second tank must be fitted to store the LPG. The tank may be cylindrical in shape and need a mounting space in the boot, or it can be round and take the place of the spare wheel. A filling hose will be required and an external filling point will need to be added to the vehicle body, normally close to the location of the petrol filling point. LPG from the tank can then be transferred to the engine via copper piping routed along the underside of the car.

A solenoid valve and filter are mounted in the fuel delivery line in order to remove dirt particles and prevent the flow of LPG when a dual fuel engine is being run on petrol.

A small regulator unit is then mounted on the engine's intake system, which warms the propane with heat from the engine cooling system, turning it into gas.

A mixer/distributor takes information from various engine sensors or ECU and introduces a controlled amount of gas to the intake manifold via injectors. The regulator and mixer unit will normally include a safety circuit which will cut the flow of propane gas to the inlet manifold if the engine should stall or cut out.

Figure 5.3 An LPG fuel tank

Special electronic circuitry is then required so that:

- A functioning fuel gauge can be created for the LPG system.
- An automatic or manual switching system can be created to allow the engine to swap between petrol and LPG.
- Petrol fuel injection can be simulated electronically to the ECU to avoid the engine management system storing diagnostic trouble codes when the engine is running on LPG; this is achieved using a component known as an **'emulator'**.

Emulator - a component that copies the behaviour of something else.

Compressed natural gas (CNG)

An alternative to LPG is high pressure compressed natural gas (CNG). CNG can be produced from a number of sources and is mainly composed of methane. When crude oil is extracted from the ground, around 3% is natural methane gas.

It can be used to fuel normal internal combustion engines instead of petrol. The combustion of methane produces the least amount of CO_2 of all fossil fuels. Petrol cars can be retrofitted to CNG and become bi-fuel natural gas vehicles (NGV) in a similar way to LPG conversion. Because the original petrol tank and petrol fuel operating system stay, the car becomes 'dual fuel'.

The construction and operation of a CNG system is similar to that used in LPG, but instead of the fuel being stored in a liquid state, it is stored as a gas. If methane was to be stored as a liquid it would not only require high pressures, but also cooling to extremely low temperatures.

A disadvantage of this system compared to LPG is the density of the stored fuel. As a result, CNG vehicles will require a larger fuel tank in comparison to LPG in order to store similar quantities.

An advantage of this system however, is that compressed natural gas can be stored at a far lower pressure, in a form known as absorbed natural gas (ANG), and vehicles can be refuelled from the normal natural gas network without any further compression. This improves the logistics for making this fuel widely available to customers, as an infrastructure already exists which could be adapted for vehicle use.

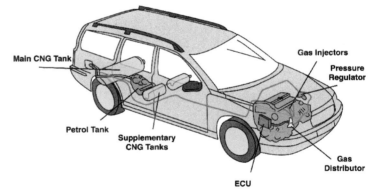

Figure 5.4 A CNG equipped car

Biogas

Biogas normally refers to a gas produced by the biological breakdown of **organic matter** in the absence of oxygen. Biogas normally comes from **biogenic** material, such as dead plant and animal matter, and is a type of renewable biofuel. A biogas generator plant will normally contain a unit called an **anaerobic** digester where plant material, animal matter and sewage are allowed to decompose and form methane, carbon dioxide and small amounts of hydrogen sulphide. After purification of the raw gas, compressed biogas can be used to power normal internal combustion engines in a similar manner to compressed natural gas (CNG).

Advanced Light Vehicle Technology

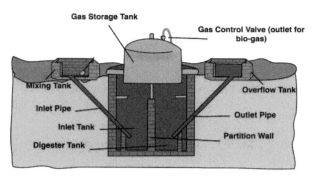

Figure 5.5 Anaerobic digester

Anaerobic - without oxygen.

Organic matter – something that has come from a once living organism that is capable of decay.

Biogenic – produced or brought about by living organisms.

Biodiesel

Biodiesel (fatty acid methyl ester) is a way of making a form of Diesel fuel from a very wide range of oil-producing plants, such as:

Algae oil	Jojoba oil	Radish oil
Artichoke oil	Karanj oil	Rapeseed oil
Canola oil	Kukui nut oil	Rice bran oil
Castor oil	Milk bush	Safflower oil
Coconut oil	Pencil bush oil	Sesame oil
Corn	Mustard oil	Soybean oil
Cottonseed oil	Neem oil	Sunflower oil
Flaxseed oil	Olive oil	Tung oil
Hemp oil	Palm oil	Waste vegetable oil (WVO)
Jatropha oil	Peanut oil	

Figure 5.6 Oil producing plants

Automotive Master Technician

Waste vegetable oil

Waste vegetable oil can often be used as a direct substitute for normal Diesel, although it can cause engine and fuel system problems. If the oil is unfiltered, the relatively high **viscosity** and any debris, may cause damage to the fuel pumps and block the fuel filter. The viscosity also leads to poor atomisation of the fuel at the injectors and inefficient combustion. Poor combustion leads to high carbon deposits inside the cylinder and around the injectors and valves. Incomplete combustion may also cause contamination of the engines lubricating oil. The best way to prevent these issues is to refine the oil through a process known as **transesterification**.

Once treated the waste vegetable oil will have:

- A lower viscosity.
- A lower boiling point.

- A lower flash point.
- Damaging glycerides removed.

Viscosity - a fluids resistance to flow.

Transesterification - the process of exchanging the alcohol group of an ester compound with another alcohol.

Transesterification

Oils produced from plants or animal fat will need fully refining through a process known as transesterification before they can be used as a biodiesel. This procedure involves the separation of ester from the glycerides found in these oils.
To do this, the fat/oil is reacted with alcohol using a chemical catalyst such as potassium or sodium hydroxide. The mixture is often heated in a sealed container which is kept just above the boiling point of the alcohol. The heavier glycerol settles out and can then be removed, recycled and sold on to the pharmaceutical industry.

In order for the separated ester to be used as fuel, the alcohol must then be removed, which is done by a process of flash evaporation or distillation. When complete the biodiesel is then checked to ensure that:

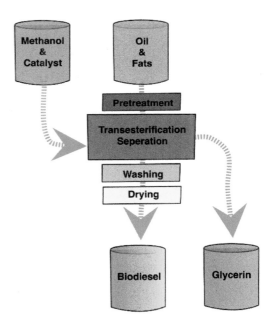

Figure 5.7 Transesterification

- ✓ All of the glycerine has been removed.
- ✓ All of the alcohol has been removed.

- ✓ All of the catalyst chemical has been removed.
- ✓ The fatty acids have been removed.

Once this is done, the resulting biodiesel should act as an adequate fuel for use in internal combustion engines.

Advanced Light Vehicle Technology

Table 5.4 shows some of the advantages and disadvantages of biodiesel when compared to fossil Diesel.

Table 5.4 The advantages and disadvantages of biodiesel

Advantages of biodiesel	Disadvantages of biodiesel
Because biodiesel is mainly produced from plants or crops, it is considered a renewable energy source when compared to fossil fuels.	Biodiesel has a lower energy density than fossil Diesel fuel. This means that biodiesel vehicles are not quite able to keep up with the fuel economy of a normal fossil-fuelled Diesel vehicle.
Biodiesel is often considered carbon neutral when used as a vehicle fuel. This is because the carbon dioxide released during combustion is equal to the carbon dioxide absorbed from the atmosphere during the plants growing period and life-span.	The production of biodiesel can be expensive and use a large amount of energy. The energy needed often comes from non-renewable sources and therefore creates carbon dioxide from production.
Biodiesel is completely biodegradable and non-toxic, as a result fuel spillages are less of a risk to environmental pollution than fossil fuels.	Some biodiesel, especially those produced from waste vegetable oil, contain chemicals which can damage natural rubber leading to leaks from fuel system seals.
Biodiesel has a higher flash point than fossil Diesel, reducing the risk of fire in the event of an accident.	Large areas of land are required to grow the oil producing crops. This often leads to reduced area for the production of food crops or deforestation.

Flash point – the temperature at which a fuel forms an ignitable mixture in air.

Biodiesel from algae

Biodiesel can also be created from algae. Algae can be grown in open ponds or sealed systems known as 'photo bioreactors'. Algae is one of the most efficient organisms on earth, due to its very rapid growth rate and can be harvested in areas which do not affect the production of food crops. Once dried and pressed to remove the oil, algae produces around 300 times more oil than standard biofuel crops and can be harvested on a 1 to 10 day cycle rather than yearly.

Bioalcohol/Ethanol

Different types of alcohol are often able to be used as a fuel source for internal combustion engines. The two main types of alcohol used are methanol and ethanol. Methanol is often created from natural gas and is therefore not normally considered a biofuel (although it can be produced from biogas, but this is often a more complex and expensive procedure).

Ethanol is mainly produced from biological material through a process of fermentation, which is similar to that used in the production of alcohol for drinking. Ethanol can also be produced from petroleum based products, but would be very dangerous to drink; causing blindness or death.

The alcohols used for fuel have a lower energy density than conventional petrol or Diesel, with around one and a half, to two times the volume required to produce the same amount of heat energy. However, alcohol has a naturally high **octane**, meaning that it can be run at higher compression pressures inside the engine which then produce similar results in performance and fuel economy. There is also a reduction in tail pipe emissions when compared to petrol or Diesel.

Automotive Master Technician

A disadvantage of alcohol is that it tends to be corrosive or promote corrosion in fuel system components. Over a period of time it is possible that fuel systems may become excessively corroded or blocked. In recent years many manufactures have been designing cars with fuel systems that are able to tolerate around 10% ethanol which is often now blended with standard petrol, but to run solely on alcohol, fuel system materials need to be adapted and engine management systems reprogrammed to take into account its use.

Diesel engines are also able to operate on alcohol, but as the fuel has a low **cetane** value, additives such as glycol need to be combined with the fuel to improve ignition.

Octane - a colourless flammable hydrocarbon with the ability to suppress detonation.

Cetane - a measure of a fuels ignition delay (the time period from injection until it starts to burn).

Other alcohols that can be used as a fuel include butanol and propanol. These are more expensive and difficult to produce than methanol and ethanol making them less viable for fuel production. If a method can be found to produce butanol economically, then fuel economy and performance could be increased, as it has a very similar energy density to petrol.

Ammonia Green

Ammonia Green (NH3) is being used with success by some vehicle manufacturers because it can run in spark ignition or compression ignition engines with only minor modifications.

Using ammonia to power internal combustion engines, provides some advantages over other fuel sources:

- It has a very high energy density when compared to other non-petroleum based fuels.
- It contains no carbon, and therefore releases no carbon dioxide during combustion and generates no particulate matter.
- As ammonia can be manufactured relatively easily, it won't run out like fossil fuels.

Ammonia is manufactured by reacting hydrogen with the nitrogen in air to produce NH3, meaning the raw materials needed to produce ammonia are water and air.

The most common method of creating hydrogen for the production of ammonia comes from natural gas in a process which can cause a significant amount of greenhouse gasses to be released to atmosphere. This is known as brown ammonia.

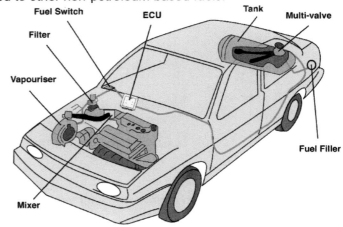

Figure 5.8 An Ammonia green equipped car

If the hydrogen for the production of ammonia is created through electrolysis, then green non-polluting methods can be used, these include:

- Wind power
- Solar
- Hydroelectric
- Wave power

Advanced Light Vehicle Technology

If the ammonia is produced through electrolysis from renewable energy sources, it is known as ammonia green. Liquid ammonia has half the density of petrol or Diesel, this means that it can be easily carried in sufficient quantities in vehicles, and unlike some other alternative fuels, produce a good driving range. On combustion it produces no emissions other than nitrogen and water vapour, making the exhaust non-polluting.

Ammonia is considered to be very toxic, but if stored and handled correctly, it is no more dangerous than petrol or LPG.

Hydrogen internal combustion engines (HICE)

The gas hydrogen can be used to directly power internal combustion engines with a few modifications. The advantages of running an engine on hydrogen is that the only emission caused by combustion should be water vapour, and the fuel has the potential to provide a higher power output than petrol or Diesel. Producing an engine that runs on hydrogen is not difficult, but getting an engine that runs well on hydrogen is a different matter.

To allow an engine to run on hydrogen will normally require the adaption of the fuel delivery system. Hydrogen can be introduced with three main methods:

- Single-point injection.
- Multi-point injection.
- Direct injection.

Single-point and multi-point require the least amount of engine adaption and hydrogen becomes a gas easily inside the intake manifold. Direct injection may need the engine to be considerably modified but does offer the best range of operation when running on hydrogen as the air fuel mixture can adjusted in far greater detail.
The ignition system on a hydrogen internal combustion engine can be similar to a standard petrol ignition system, although cold running spark plugs will be required to prevent **pre-ignition**. Also platinum electrode spark plugs cannot be used as this metal acts as a catalyst and reacts with hydrogen causing it to oxidise with air.

Pre-ignition - a condition where the fuel in the combustion chamber ignites before the spark has been produced.

Although hydrogen is a good source of fuel for vehicles, its properties have advantages and disadvantages when used in internal combustion engines.

.Hydrogen is highly flammable. If an accidental leak in a hydrogen system should occur, ensure that all sources of ignition are removed.

Automotive Master Technician

Table 5.5 describes some advantages and disadvantages of using hydrogen in internal combustion engines.

Table 5.5 The advantages and disadvantages of hydrogen when used in HICE

Property	Advantages	Disadvantages
Its wide range of flammability.	Hydrogen will burn when used with extremely weak mixtures that are well below the recommended stoichiometric values. This has the advantage of easy engine starting and a more complete combustion overall.	There is a limit to how weak an air fuel mixture can become before the combustion temperature falls to a point where power output from the engine is reduced.
Its low ignition energy requirements.	The amount of energy needed to ignite hydrogen is far less than that needed for petrol. This means that a standard spark ignition system can be used to ignite the fuel, even with extremely weak mixtures.	Because of hydrogens low ignition energy requirements, hot spots inside the cylinder may cause pre-ignition, leading to a misfire. Sources of combustion cylinder hot spots could be carbon build-up from burnt lubrication oil, or overheated spark plug electrodes.
Its ability to resist quenching.	Unlike petrol which will sometimes be extinguished during its combustion as it approaches the cooler cylinder walls, hydrogen is less likely to be quenched and therefore use up more of its available energy.	The small quenching distance can increase the risk of a backfire as the combusting hydrogen is able to burn closer to a nearly closed inlet valve. If the induction system uses single-point or multi-point injection in the manifold, it is possible that fuel here will ignite and cause a fire.
Its high auto ignition temperature.	A high auto ignition temperature is important to the stability of the fuel, meaning that it resists detonation solely due to air temperature. This stability means that it can run at far higher compression ratios than a standard petrol engine and as compression ratios rise, so does the performance output.	The high auto ignition temperature makes hydrogen unsuitable for use in compression ignition engines.
Its ability to spread out due to its low density.	As hydrogen is a gas even at very low temperatures. If a leak in the system occurs, it will spread out and disperse very quickly reducing the risk of an accidental fire.	The low density of hydrogen means that it takes up a lot more space in an engine cylinder than vaporised petrol. This means that the energy density available for combustion is much lower, reducing the engines performance output.

The ideal are fuel ratios for a petrol engine are 14.7:1 (by mass) compared to 34:1 (by mass) for hydrogen. Although this is the recommended air fuel ratio, hydrogen internal combustion engines will often run far weaker than this, down to values as low as 180:1.

Advanced Light Vehicle Technology

Hydrogen fuel storage

Unlike petrol and Diesel, which attach their hydrogen atoms to carbon so the fuel can be stored in a liquid form, hydrogen is a gas at normal temperatures. This means that to store pure hydrogen on board a vehicle as a liquid would require extremely high pressures and low temperatures.

Crankcase ventilation

As with a normal internal combustion engine, blow-by of combustion mixtures from a hydrogen internal combustion engine can leak past the piston rings and into the crankcase. Crankcase ventilation needs to be correctly controlled for two main reasons.

1. A build-up of hydrogen in the crankcase could lead to an explosion, igniting other flammable materials such as oil.
2. A by-product of hydrogen combustion with air is water. Water can contaminate the engine oil, reducing its lubrication properties.

Figure 5.9 Water contamination of engine oil

Alternative propulsion comparisons

Table 5.6 shows comparisons in efficiency for different fuel and vehicle drive types.

Table 5.6 System drive types and efficiency

Fuel/drive system type	Efficiency
Steam produced by coal	In practice, a steam engine exhausting the steam to atmosphere will typically have an efficiency (including the boiler) in the range of 1-10%; this means that it is around 90% inefficient. With the addition of a condenser and multiple expansion, this figure can be improved to 25% or better.
Petrol	The efficiency of a petrol-powered car is surprisingly low. All of the heat that comes out as exhaust or goes into the radiator is wasted energy. The engine also uses a lot of energy turning the various pumps, fans and generators that keep it going. So the overall efficiency of a petrol engine is about 20%. That is, only about 20% of the thermal-energy content of the petrol is converted into mechanical work.
Diesel	Diesels, are able to reach an efficiency of about 40% in the engine speed range of idle to about 1800 rpm. After this speed, efficiency begins to decline due to air pumping losses within the engine.
Hybrid	Depending on the hybrid drive type: Series Parallel Combination Efficiency values will vary. If you combine the efficiency of a petrol engine of around 20%, with the efficiency of electric motors of around 90%, then in theory you should get an overall efficiency of about 70%. Unfortunately, due to construction and design, heat losses will reduce this figure to around 55% to 60%, putting them in the same sort of efficiency as a modern common rail Diesel. These figures though do not take into account the large reduction in emissions produced by hybrids when compared to Diesel engines.

Automotive Master Technician

Table 5.6 System drive types and efficiency

Fuel/drive system type	Efficiency
Plug in electric	A battery-powered electric car has a fairly high efficiency. The battery is about 90% efficient (most batteries generate some heat, or require heating), and the electric motor/inverter is about 80% efficient. This gives an overall efficiency of about 72%. But that is not the whole story. The electricity used to power the car had to be generated somewhere. If it was generated at a power plant that used a combustion process (rather than nuclear, hydroelectric, solar or wind), then only about 40 percent of the fuel required by the power plant was converted into electricity. The process of charging the car requires the conversion of alternating current (AC) power to direct current (DC) power. This process has an efficiency of about 90 percent. So, if we look at the whole cycle, the efficiency of an electric car is 72% for the car, 40% for the power plant and 90% for charging the car. That gives an overall efficiency of 26%. The overall efficiency varies considerably depending on what sort of power plant is used. If the electricity for the car is generated by a hydroelectric plant for instance, then it is basically free (no fuel was burned to generate the electricity), and the efficiency of the electric car is about 65%.
Solar	Because photovoltaic cells (solar panels) are unable to convert all of the energy in the electromagnetic spectrum produced by sunlight, they are only currently around 10% efficient. This means that a 1 square meter solar panel that is capable of receiving 1 kilo watt of energy from the sun can only convert this into 100 watts of energy. This is around 1.3 horse power (UK). This very low power output makes solar panels unsuitable for powering cars but they can be used to extend the range of plug-in electric vehicles or be used in the electrolysis processes used to create hydrogen or ammonia which can then be used in vehicle propulsion.
Hydrogen powered ICE	When compared to a standard internal combustion engine running on petrol, a hydrogen engine has the potential to produce a greater power output because of the higher compression ratios that can be used. Unfortunately, due to the space required inside the engine cylinder by the volume of hydrogen in its gaseous state, air/fuel ratios tend to run extremely weak leading to a low efficiency overall of around 25%.
Hydrogen fuel cell	If the fuel cell is powered with pure hydrogen, it has the potential to be up to 80% efficient. That is, it converts 80% of the energy content of the hydrogen into electrical energy. However, we still need to convert the electrical energy into mechanical work. This is accomplished by the electric motor and inverter. A reasonable number for the efficiency of the motor/inverter is about 80%. So we have 80% efficiency in generating electricity, and 80% efficiency converting it to mechanical power. That gives an overall efficiency of about 64%.
LPG, CNG and biogas	Although liquefied petroleum gas LPG is currently cheaper to buy than the equivalent quantity of petrol or Diesel, it is less fuel efficient than its alternatives. A vehicle fitted with an LPG system will use more fuel than an equivalent vehicle fitted with a petrol engine and will therefore get less miles per gallon (Mpg). A factor of approximately a 15% efficiency reduction should be taken into consideration for any fuel economy calculations when compared to the equivalent petrol.
Ammonia green NH3	When compared with petrol, ammonia green NH3 has a very similar overall energy content. Due to the way it is manufactured, only around 70% of the energy put into the production process is converted into hydrogen and therefore, when burnt in a standard internal combustion engine it will produce an efficiency of approximately 28%.

Advanced Light Vehicle Technology

Table 5.6 System drive types and efficiency

Fuel/drive system type	Efficiency
Bioalcohol/Ethanol	Bioalcohol or ethanol produced from crops is approximately 34% less efficient than the equivalent petrol. A standard engine running on ethanol would therefore not produce the same fuel economy as a comparable petrol engine. As ethanol has a higher octane rating than petrol, it does however increase the ability of engine designers to raise compression ratios or use forced air induction and therefore regain some overall efficiency.

Cost savings of alternative fuel vehicles

In order to work out if it is financially viable to convert a vehicle to an alternative fuel, you will need to know certain information.

- The total cost of the conversion.
- Annual mileage.
- On average how much you spend on petrol or Diesel each week.
- Average fuel economy of your car (Mpg).
- The current cost of petrol or Diesel per litre.
- The current cost of the alternative fuel per litre.

These values can then be substituted in the following calculations to work out if the conversion is economically viable.

Miles per litre = miles per gallon/4.546

Average mileage per week = total annual mileage/52

Average fuel consumption per week = weekly mileage x miles per litre

Weekly fuel cost = fuel cost per litre x total fuel consumption per week (litres)

Total current fuel cost per year = average cost of fuel per week (petrol or Diesel) x 52

Total cost of alternative fuel per year = average cost of alternative fuel per week x 52

Annual fuel cost difference (saving) between current fuel and alternative fuel = current cost of fuel per year - cost of alternative fuel per year

Time required (in months) needed to pay off the conversion = total cost of the conversion/annual fuel cost saving x 12

Note: these calculations do not take into account any efficiency differential between petrol/Diesel and the alternative fuel type chosen, *see Table 5.6*. (For example an approximate efficiency loss of around 15% should be factored into these calculations when using LPG as the alternative fuel source).

Environmental issues and low carbon technologies

As demand for natural resources rise, the impact on our environment also increases. One of the largest environmental pollutants is carbon dioxide, and anthropogenic (manmade) carbon dioxide is creating climate change.

Nearly everything we do in life creates carbon dioxide, and the amount of carbon dioxide produced through individual consumption and use is known as your carbon footprint. As the world's population increases, our requirements will also continue to rise and unless care is taken, we may cause irreparable damage to our planet. A large amount of carbon is generated through transport, and many manufacturers are exploring different technologies which can reduce overall CO_2 output (known as low carbon technologies), but there are also things that we can do as individuals when travelling that can help including:

- Ensuring that our vehicles are maintained and operate within recommended specifications.
- Not making unnecessary journeys.
- For short journeys, walk or cycle.
- Using public transport where appropriate.
- Sharing journeys with others (car sharing for example).
- Modify driving styles to ensure the most efficient operation of vehicles.
- Use vehicles with alternative methods of propulsion.

Anthropogenic - environmental pollution and pollutants originating from human activity.

Carbon footprint - the amount of carbon dioxide or other carbon compounds emitted into the atmosphere by an individual, company or country.

Fossil fuels - a hydrocarbon deposit, such as petroleum, coal, or natural gas, derived from living matter of a previous geologic time and used for fuel.

The need for alternative propulsion

In recent years a number of issues have arisen from the use of **fossil fuels** used for vehicle propulsion systems. These include:

- Limited quantities and reserves of crude oil.
- Peak oil production.
- Hazardous exhaust emissions.
- Environmental pollution.

Crude oil, which is the basis for all fossil fuels used in the propulsion systems of modern vehicles, is created by the biological breakdown of organic materials, from plant and animal life, underground. This process requires extreme heat and pressure, and takes place over millions of years. Crude oil then forms a rich source of hydrocarbons (HC) which can be extracted from the ground, and refined into fuels.

Figure 5.10 Crude oil pumping

We are all aware that the sources of fossil fuel are limited, but there are currently known reserves for some years to come, there is however another problem. Peak oil production is an issue where demand is beginning to outpace the ability to supply the quantities of fuel needed. As our use of fossil fuels increases, the ability to extract crude oil form the ground and process it into fuels quickly enough is beginning to lead to shortages. As a result, alternative fuel sources are being sought that do not solely rely on fossil fuels. One alternative to this is the use of electricity.

Electric vehicle design

Electric cars are not a new idea. The invention of electric vehicles is often attributed to various people, but can mainly be traced back to a Hungarian inventor called Ányos Jedlik in 1828, who managed to make a small model car fitted with an electric motor that he designed. By 1838, a Scotsman named Robert Davidson had built an electric locomotive that managed a top speed of 4 mph. These designs predate the first internal combustion engine cars by nearly 50 years.

Recent improvements in engineering, motor and battery design have now made electricity a viable alternative source of propulsion to fossil fuels.

Advanced Light Vehicle Technology

All electric vehicles

An all-electric vehicle is one where its main source of power for propulsion comes from high voltage batteries or high quantity capacitors and motors. These types of vehicles are sometimes known as battery electric vehicles (BEV's) and because they are often charged from mains electricity they are also called 'plug-in'.

There are also other types of all electric vehicle that get their power source from methods other than mains charging including solar and fuel cell.

Figure 5.11 All electric car

Hybrid vehicles

A hybrid vehicle is one where the main source of power for propulsion is provided by a combination of internal combustion engine and electric motor. The engine and electric motor can be connected in three main formats to provide drive:

- Series
- Parallel
- Combination

Many modern hybrid vehicles are of the parallel or combination type and are made up of the following components:

- Batteries
- Motors
- Cabling
- Control units
- Circuit protection

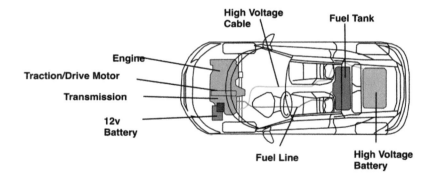

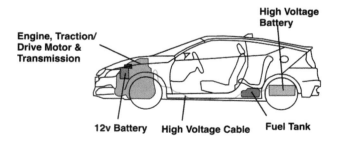

Figure 5.12 Hybrid system layout

Automotive Master Technician

Some hybrid drive vehicles are able to supplement the charge in their high voltage batteries by connecting to mains electricity when not in use. These types are known as 'plug-in hybrids'.

Hybrid drive

A hybrid vehicle is one which combines an internal combustion engine with an electric motor to provide drive. This gives the flexibility of a petrol or Diesel engine with the fuel economy and low pollution characteristics of electric motors. There are three main types of hybrid drive:

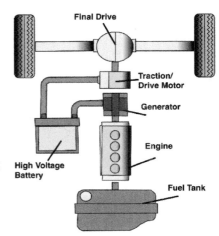

- **Series hybrid**: A small capacity internal combustion engine is used to act as a generator. This then charges batteries that are used to power the electric traction motors that drive the wheels. There is no direct connection between the engine and the wheels, meaning that a gearbox is not required. The advantage of this system is that no driving loads are placed on the engine and it can run at a constant speed. This reduces fuel consumption, emission output and engine wear.

Figure 5.13 Series hybrid

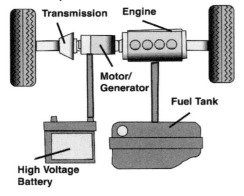

- **Parallel hybrid**: An integrated electric motor is used to support or boost the performance of a small capacity internal combustion engine. When not required, the electric motor can be converted into a generator to recharge the high voltage electric batteries.

Figure 5.14 Parallel hybrid

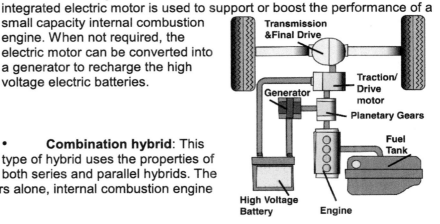

- **Combination hybrid**: This type of hybrid uses the properties of both series and parallel hybrids. The car can operate on electric motors alone, internal combustion engine alone or a combination of both.

Figure 5.15 Combination hybrid

Vehicle designers are continuously creating unusual types of hybrid drive systems. This not only improves efficiency and operation, but also overcomes **copyright** issues.
Some manufacturers have been known to use the terms **series** and **parallel** in a slightly different context when describing the operation of their hybrid drive systems. Make sure that you have fully studied and understood the manufacturer's description of their system before you make any assumptions about how they operate. This will help reduce confusion and misunderstanding.

Advanced Light Vehicle Technology

> **Copyright** – an exclusive legal right of design ownership and use.
>
> **Series** – connected in a line (one after another).
>
> **Parallel** – connected side by side.

Two common hybrid systems are described in the next section.

Collaborative Motor Drive

A **Collaborative** Motor Drive system has two separate motors that run parallel and are capable of producing drive: an internal combustion engine and an electric motor. When required they work together to provide a smooth drive system.

Start
- When pulling away, the petrol engine is not used and the electric motor provides the drive.

Climb
- When going up a hill, the engine and electric motor both power the vehicle to provide maximum performance.

Slow
- When decelerating or braking, the system recycles the kinetic energy to recharge the batteries.

Overtake
- When overtaking, the engine and electric motor both power the vehicle to provide maximum performance..

Stop
- When the vehicle is stationary, both the engine and the electric motor automatically switch off to save power.

Automotive Master Technician

Integrated Motor System

An Integrated Motor System has a compact electric motor, sandwiched between an internal combustion engine and the gearbox, in a similar position to the flywheel. When needed this electric motor is able to boost, the performance from the engine.

Start
- When pulling away, the electric motor provides maximum torque to assist the engine for strong acceleration and reduced fuel consumption.

Low Speed
- When the vehicle is cruising at low speed, the engine is stopped (intake and exhaust valves remain closed) and the electric motor is used by itself for drive.

High Speed
- When the vehicle is accelerating gently or cruising at high speed, the vehicle runs on engine power alone and the electric motor is switched off.

Overtake
- When accelerating rapidly, the engine and electric motor both power the vehicle to provide maximum performance.

Slow
- When decelerating or braking, the engine is stopped (intake and exhaust valves remain closed) and the electric motor acts as a generator to recharge the batteries.

Stop
- When the vehicle is stationary, the engine and electric motor automatically switch off to save power.

Collaboration – working together in co-operation (this is also known as synergy).

Dangers involved with working on high voltage electric vehicles

There are many dangers involved with the maintenance and repair of high voltage electric vehicles, these can include:

Table 5.7 The hazards associated with high voltage electrical systems found on vehicles

Electric shock	Electric shock is caused when electricity passes through the human body leading to injury or death. The risk of electric shock increases as electrical voltage rises. Voltages higher than 60v DC or 30v AC RMS values increase the risks posed.
Burns	A common effect of electric shock on the human body is burning. This is due to the fact that the electrical energy is converted to heat as it is discharged. Burns caused by electrical discharge through the body may be internal and cause tissue damage that might not be immediately apparent. If an accidental short circuit is created during the connection or disconnection of a high voltage electrical circuit, the heat given off can cause external burning to the skin.
Arc flash	Arc flash is caused by a sudden, accidental discharge of electrical energy and creates an arc similar to that of welding. The temperatures created when this occurs can be around 20,000 degrees Celsius. This temperature is enough to vaporise clothing and human flesh. Arc flash is most likely to occur at voltages above 480v, which are not uncommon on electric vehicle drive systems.
Arc blast	Arc blast is an explosion caused by a short circuit in a high voltage system. It has many of the same effects as arc flash, causing severe burns, but also has the potential to cause injury at a distance. Added dangers may involve flying particles from the system, and a sound pressure wave that might lead to permanent hearing damage.
Fire	A rapid discharge of electrical energy will create large amounts of heat. This heat will then have a high potential of igniting flammable materials. When working on high voltage systems it is advisable to have a suitable fire extinguisher to hand (one that will not conduct electricity - class E).
Explosion	High voltage systems often combine electricity with chemicals, gasses and fumes that pose the risk of explosion. Care should always be taken to try and remove any sources of ignition and correctly isolate electrics before working on these systems.

Levels of current and voltage that may present hazards

An electric shock is usually painful and can be lethal. The level of voltage is not a direct guide to the level of injury or danger of death. Physical effects and damage are generally determined by the amount of current and the duration of electric shock. Even a low voltage causing a current of extended duration can be fatal.
The type of damage caused by electric shock can be different depending on the voltage, duration, current and path taken.

- Current entering the hand has a threshold of around 5 to 10 milliamps (mA) for DC and about 1 to 10 mA for AC.
- At Voltages of between 110 to 220 volts AC, current traveling through the chest for a fraction of a second can cause a heart attack with currents as low as 60mA.
- With DC, 300 to 500 mA is required.

Automotive Master Technician

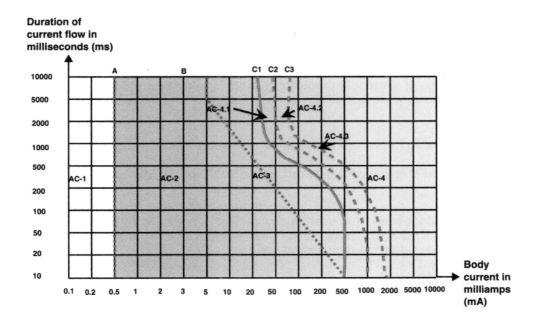

Figure 5.16 Levels of current and voltage that may present a hazard

For further guidance on the dangers posed by electric current, *see Figure 5.16.*

AC-1 zone: Imperceptible

AC-2 zone: Perceptible

AC-3 zone : Reversible effects: muscular contraction

AC-4 zone: Possibility of irreversible effects

AC-4-1 zone: Up to 5% probability of **heart fibrillation**

AC-4-2 zone: Up to 50% probability of heart fibrillation

AC-4-3 zone: More than 50% probability of heart fibrillation

A curve: Threshold of perception of current

B curve: Threshold of muscular reactions

C1 curve: Threshold of 0% probability of **ventricular fibrillation**

C2 curve: Threshold of 5% probability of ventricular fibrillation

C3 curve: Threshold of 50% probability of ventricular fibrillation

Advanced Light Vehicle Technology

Safety precautions to be taken before carrying out any maintenance and repair procedures on hybrid vehicles

Typical voltages used for a range of electrically propelled and hybrid vehicles is 100 to 650V. Working with high voltage systems found on hybrid vehicles and all electric vehicles will require the use of specific personal protective equipment.

Table 5.8 shows the recommended PPE and reasons why it should be used.

Table 5.8 PPE required when working on vehicle high voltage systems

PPE	Recommendations
Overalls 	It is recommended that overalls should have non-electrically conductive fasteners. Metal fasteners may also create issues when working around the extremely strong magnets used in hybrid system motor/generators.
Gloves 	Rubber insulating gloves are one of the most important pieces of personal protective clothing for working on or around high voltage vehicle electrical systems. Gloves should be rated at least Class 0—Maximum use voltage of 1,000 volts AC, proof tested to 5,000 volts AC, and be in good condition with no rips or tears. Before use you should blow into the open wrist section and then roll the glove up from that point to see if it is able to hold the air. If any leaks are present the gloves should be replaced.
Protective footwear 	Safety footwear is important when working on any vehicle system as reinforced toe caps provide protection from crush injury. When working with high voltage systems there may be occasions where insulated rubber boots might be necessary. (When a vehicle has been submerged in water for example, as this adds to the risk of electric shock). Insulating rubber boots or insulating rubber over-boots are available which will help provide extra protection.
Goggles 	As with any electrical system, sparks created by accidental discharge through a short circuit have the potential to cause damage to the eyes. Safety glasses will help provide some defence from this, but goggles and face shields that fully enclose the eyes will give the most protection.

The effects of electric current and the standards needed to protect operators' form the dangers they pose are regulated by the International Electrotechnical Commission - ICE 60479 and ICE 479-2.

ECE R100 (relating to vehicle regulations) paragraph 2.14 clearly defines high voltage:

"High Voltage means the classification of an electric component or circuit, if its working voltage is > 60 V and ≤ 1500 V DC or > 30 V and ≤ 1000 V AC root mean square (RMS)."

NOTE: This is different to definitions in commercial and domestic use which are:
- Extra Low Voltage <50 V RMS AC and <120 V DC
- Low Voltage 50-1000 V RMS AC and 120-1500 V DC
- High Voltage >1000 V RMS AC and >1500 V DC

Other safety precautions that should be observed when working with vehicle high voltage systems are shown in Table 5.9.

Table 5.9 Safety precautions when working on high voltage systems

Hazard	Precautions
Using electrical equipment with high voltage systems	Wear appropriate PPE. Where possible, use electrical hand tools that are specifically designed for use with high voltage systems. When using a multimeter, ensure that your fingers are behind the insulating finger guards of the test probes. If the multimeter screen shows a low battery indicator, replace the battery immediately. A low battery in the multimeter, may lead to incorrect readings being taken which could lead to injury or death.
Disposal of waste materials	Under the Environmental Protection Act 1990 (EPA), you must treat old batteries (lead acid and metal hydride) as hazardous waste and dispose of them in the correct manner. They should be safely stored in a clearly marked container until they are collected by a licensed recycling company. This company should give you a waste transfer note as the receipt of collection.
Dealing with leakage	Wear appropriate PPE. If battery leakage occurs, cover the spill with a neutralising agent such as sodium carbonate or baking soda mixed with equal parts water. After neutralising, rinse the contaminated area with clean water. If the spill involves a large amount of electrolyte, call the fire services and allow them to handle it. This may help prevent you getting seriously hurt and reduce any environmental issues.
High voltage electrical system isolation	When working on or around the vehicles high voltage components, you should correctly isolate and insulate the system, this should involve: ✓ Wear appropriate PPE. ✓ Make others aware that the high voltage system is being worked on. ✓ Remove the ignition key (if the ignition key cannot be removed, due to damage for example, take out all of the fuses in the fuse box). ✓ If the ignition key is a 'smart key' which doesn't require insertion into a key slot, it should be safely stored more than thirty feet away from the vehicle. ✓ Disconnect the low voltage (12v) auxiliary battery. ✓ Following manufacturer's instructions, remove the high voltage service plug or switch and lock the high voltage isolator button so that they cannot be accidentally reconnected. ✓ Allow time for any system capacitors to discharge and check the voltage has fallen to a safe level using a multimeter. ✓ Do not cut any orange high voltage cable. ✓ Cover any disconnected terminals with insulating tape.

Table 5.9 Safety precautions when working on high voltage systems

Hazard	Precautions
Submerged vehicle safety	To handle a hybrid or electric vehicle that has been partially or fully submerged in water, the high voltage system and air bags will need to be safely isolated. Wear fully insulating electrical PPE (as described in Table 5.8). Immobilise vehicle and remove ignition key (if the key is a 'smart key', ensure it is kept beyond its range of operation). Where possible allow any water to drain/dry. Remove all system fuses and isolate the high voltage system as described above.
Highly magnetic components	The magnets used in hybrid electric motors are around 10 to 15 times stronger than a standard iron magnet. The naturally high magnetic field produced can affect the correct operation of heart pacemakers. Care should be taken to remove all metal jewellery and avoid the use of delicate electronic equipment such as mobile phones while working on these systems. If the rotor of a hybrid drive motor needs to be removed/fitted, it will be necessary to use a special tool, so that the magnetic attraction to other metal components does not affect its fitment and create damage to the motor or cause personal injury.
Medical conditions that may be affected by high voltage or magnetic fields	Existing medical conditions such as heart conditions can be affected by both the very strong magnetic fields produced in hybrid drive motors and the high voltage systems. It is not recommended that people with heart pacemakers work on these systems.
Checking voltage prior to working near or on high voltage systems	Before any work is started on a high voltage system, the electrical circuits should be isolated and capacitors allowed to discharge. The system voltage should be checked with a multimeter to see that it has fallen low enough to begin any work. Voltages higher than 60v DC or 30v AC RMS are likely to cause electric shock leading to injury or even death.

Safety

Do not attempt to repair any high voltage electric wiring or components. If any wiring or components become faulty or damaged, they should be replaced with new parts

Electric vehicle design

The efficiency of a hybrid or electric vehicle can be improved with the addition of special engineering design features.

Aerodynamics and body shape

A vehicles efficiency can be improved with the use of **aerodynamic** body styling.
Any vehicle moving through air will encounter resistance caused by **drag** which often result from body shape/design and friction created by the material of the skin. As a car is propelled forwards two forces oppose the movement. A high air pressure is created at the front of the car and a low air pressure is created at the rear. If the body is aerodynamically designed, using shapes that reduce these two opposing pressures, overall efficiency is improved, this is why many hybrid and electric vehicles may have unusual **aesthetics**.

Certain design features of cars will also create drag as the air moves over the surface of the vehicle while in motion. By integrating body components such as bumpers and mirrors, and enclosing wheels inside bodywork, drag can be reduced further still.

Aerodynamic - having a shape that reduces drag as air moves past.

Drag - a retarding force which tries to slow something down.

Aesthetics - how nice something looks.

Many hybrid or electric vehicles use specially designed tyres that reduce their overall rolling resistance. This will also help reduce drag on other systems and improve efficiency.

Transmission

A hybrid vehicle using a combination of a petrol engine and an electric motor produces a combination of torque ranges.

- An electric motor gives its greatest amount of torque when starting from rest.
- A petrol engine gives its greatest amount of torque at speed.

Many manufacturers use a continuously variable transmission (CVT) with hybrid vehicles which will deliver the most efficient amount of torque and speed to the road wheels no matter which motor is driving at the time.

Continuously variable transmission (CVT)

A continuously variable transmission (CVT) gearbox is a form of automatic transmission. Although the design has been around for over 100 years, many manufacturers are now offering this type of gearbox with hybrid vehicles as an option because of its efficient delivery of torque and power. For a description of CVT design and operation, *see Chapter 3.*

Hybrid and electric vehicle air-conditioning and climate control

The air conditioning or climate control of a hybrid or electric drive vehicle works in the same way as a conventional system (*see chapter 4*), with one main exception. Because the compressor in a conventional system is driven by the internal combustion engine, it cannot be operated when used with all electric drive or during stop/start on hybrid motors. Instead the compressor is driven by an electric motor, powered by the high voltage batteries. Because the compressor uses the high voltage system, a special lubricating PAG (polyalkylene glycol) oil is needed that will not conduct electricity. Care must be taken when servicing the air conditioning systems of an all-electric or hybrid drive vehicle that only recommended oil is used.

Advanced Light Vehicle Technology

The refrigerants used in air-conditioning systems are environmental pollutants. R12 creates ozone depletion and R134a contributes to the greenhouse effect. International agreements and protocols have led to legislation which restricts how refrigerants are used, and controls their release to atmosphere. In order for a technician to handle the refrigerants used in air conditioning systems during maintenance and repair, they must be suitably qualified.

The operation of electric motors used to drive vehicles

Two main types of electric motor are in common use on hybrid and electric drive vehicles Direct Current motors (DC) and Alternating Current motors (AC).

Direct current motors

A simple direct current (DC) motor can be made by passing an electric current through a coiled wire that is wound around a central shaft called the **armature** – this creates an electromagnet.

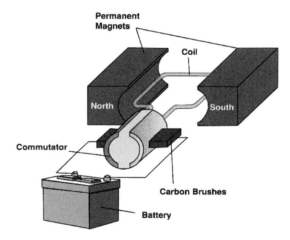

- The electric current produces an invisible magnetic field, which is repelled (pushed away) or attracted (pulled towards) by the permanent magnets surrounding it. This causes the armature to turn.
- Once the armature has turned out of the magnetic field, it would normally stop.
- To keep it rotating, the **polarity** of the electricity passing through the electromagnet mounted on the armature must be changed. This is done by a component called a **commutator**.
- Two spring-loaded electrical contacts called **brushes** are mounted on the end of the armature to maintain electrical connection with the commutator as the shaft rotates.
- When electric current is switched off, the motor will stop.

Figure 5.17 A simple electric motor

Armature – the central shaft of an electric motor.

Polarity - the positive and negative connections of an electric circuit or the north and south poles of a magnet.

Commutator – a segmented electrical contact mounted on the end of a motor armature, designed to change electrical polarity as the motor turns.

Brushes – spring-loaded electrical contacts that transfer current to the rotating armature.

Automotive Master Technician

> **Field coil** – copper wiring wrapped around magnets – this can increase the magnetic field produced when supplied with electricity.

The stronger the magnetic forces inside the motor, the more power it will produce. The magnets inside the motor casing that surround the armature can produce stronger magnetic fields if they are wrapped in wire coils and electric current is passed through them. These external magnet wires are called **field coils**.

Alternating current motors AC

Another type of electric motor that can be made using magnetic fields is the alternating current (AC) motor. Unlike a direct current (DC) motor where polarity has to be continuously changed through the use of a commutator, alternating current through its very nature changes direction as it operates and this can be used to swap polarity and keep the motor rotating.

AC motors can be split up into different designs including:

- Brush type.
- Synchronous type.
- Induction type.
- Three phase type.

Brush type AC motors

A simple brush type AC motor design uses two permanent magnets placed either side a rotating wire coil in a similar manner to that described for DC motors. The magnets are lined up so that the coil faces the north pole of one magnet and the south pole of the other. Conducting brushes touch two slip ring connectors, which feed the inner coil of the armature with electric current. When fed with current, magnetic forces make the armature coil rotate so that the south pole turns to the north pole of the permanent magnet, and vice versa. When the supply current alternates (changes direction), the magnetic south and north of the coil swap places and the motor continues to turn. The frequency of the alternating current will help determine the motor speed.

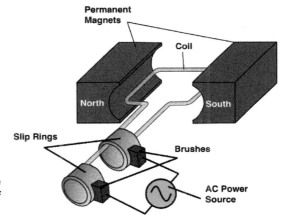

Figure 5.18 A brush type motor

Synchronous type AC motors

A synchronous type AC motor produces a very accurate motor speed. This design has a set of coils surrounding a rotor, but instead of a wire coil for the rotor, it is a permanent magnet. The electric coils are arranged as opposing pairs in a stationary housing around the edge of the motor known as the stator. The north-south pairs attract the north-south poles of the permanent magnet on the rotor, turning it. As the alternating current cycles backwards and forwards, the motor will rotate with a very accurate speed depending on the current frequency and the number of coils used. This type of system is known to create large quantities of heat when used in the design of hybrid electric vehicles which often requires it to have its own dedicated cooling system.

Advanced Light Vehicle Technology

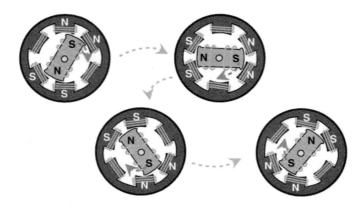

Figure 5.19 Synchronous type AC motor

The permanent magnets used in hybrid electric motors are very strong "rare earth" magnets. Despite their name, rare earth magnets can be made from fairly common compounds and are often in the form of neodymium iron boron (NIB) or samarium cobalt. These rare earth magnets are around 10 to 15 times stronger than a standard iron magnet making them ideal for use in powerful electric motors.

Rare earth magnets that are used in hybrid electric motors have an extremely strong magnetic attraction to each other or metal surfaces.
• They are strong enough to cause injuries to body parts pinched between two magnets, or a magnet and a metal surface, even causing broken bones or severed fingers.
• NIB magnets are very brittle and if allowed to get too near each other can strike together with enough force to chip and shatter the brittle material; the flying chips can cause injuries.
• The strength of the magnetic fields created by rare earth magnets is enough to disrupt electronic equipment such as heart pacemakers.
Care must always be taken when working on or around these magnets, and always follow manufactures instructions, including the use of any specialist tooling.

Induction type AC motors

An induction type AC motor doesn't use brushes or conducting slip rings. Instead, it uses an effect called induction, where a changing magnetic field produces an electric current in a similar way to an ignition coil. When an alternating current is passed through a series of conductor coils around the edge of the motor, they produce an invisible magnetic field. This magnetic field will induce a current in a winding attached to the armature of the motor, creating its own magnetic field, causing it to turn. This type of motor does not suffer with the wear or sparking produced by brush type motors.

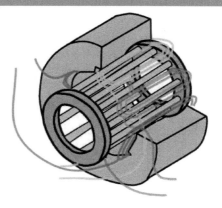

Figure 5.20 Induction type AC motor

Automotive Master Technician

Because the inner conductor coils on the armature of an induction type motor are shaped in a round, cage-like grid, engineers often refer to this design as a "squirrel-cage" motor.

Three phase type AC motors

Many hybrid and electric vehicles are able to use an **inverter**, which takes the high voltage DC stored in the batteries and converts it into an AC current in three separate phases. Each **phase** is 120 degrees apart from the next, which will form a complete 360 degree cycle. Powerful hybrid electric traction motors are often wired for this three-phase electricity. These motors have coils spaced 120 degrees apart, each coil being driven by one phase of the electricity. This arrangement produces a rotating magnetic field simply and efficiently, which is able to turn a permanent magnet rotor. An advantage of this design is that it can often be air cooled, reducing weight and design costs.

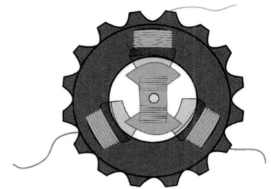

Figure 5.21 Three phase AC motor

Inverter - an electronic component that turns direct current DC into alternating current AC.

Phase - the relationship between each cycle of different alternating currents.

Some manufacturers are beginning to produce multi-phase (more than three) electric motors, in an effort to increase the efficiency and output of their designs.

The power delivery of current hybrid electric motors is between 10Kw and 50Kw; that's around 14 to 69 horse power (UK).

Battery technology

Low voltage lead acid

A standard lead-acid battery, the type often found in cars, contains a number of sections called cells (*see Figure 5.22*). Each cell is capable of producing approximately 2.1 volts.
A standard battery contains six cells linked together, creating a battery with a voltage of 12.6 volts. This is often rounded down, so we say that the battery has a voltage of 12 volts.

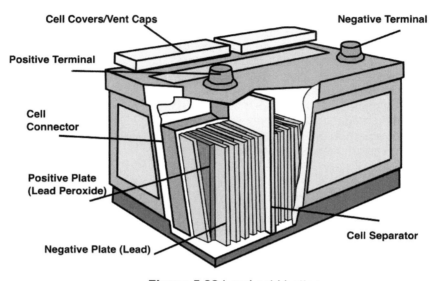

Figure 5.22 Lead acid battery

Each cell contains a number of lead plates, which are chemically different:

- The negative plate is made of lead.
- The positive plate is made of lead peroxide.

To prevent the plates from touching each other and causing a short circuit, thin sheets of material called separators are inserted between them.
Lead peroxide contains extra oxygen compared with normal lead. This means that the positive plate is chemically different from the negative plate, and it would like to share its electrons with the negative plate.
If connected in a circuit, the electrons are allowed to move from the negative plate to the positive plate. This creates electric current and provides the energy to power components in the vehicle.
The first half of the circuit is made with an electrolyte which consists of sulphuric acid and deionised water. The liquid covers the plates and allows electrons to move from one plate to another through the electrolyte. The top of each plate is then connected to the rest of the circuit. The circuit must contain a consumer to use up the electrical potential energy.
When the circuit is complete, the electrons combine with the electrolyte and move from one plate to another as a chemical reaction, creating current.

Automotive Master Technician

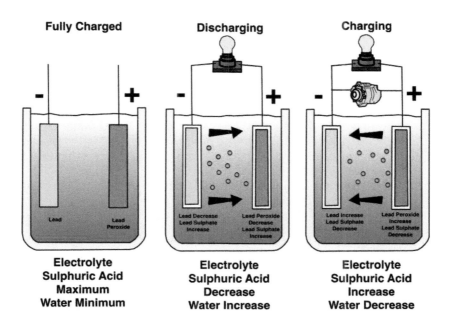

Fully Charged	Discharging	Charging

Electrolyte
Sulphuric Acid
Maximum
Water Minimum

Electrolyte
Sulphuric Acid
Decrease
Water Increase

Electrolyte
Sulphuric Acid
Increase
Water Decrease

Figure 5.23 Chemical reactions inside a lead acid battery

Why batteries need to be recharged

Batteries work with electricity flowing in one direction only (direct current). This means that eventually both plates in a low voltage lead acid battery will become chemically the same (they change to a substance called lead sulphate) and current will stop.

When this happens, no more electricity flows through the circuit, meaning that the battery is flat. When the battery is flat, it needs to be recharged. To do this, a generator (which can be thought of as a pump for electricity) is connected to the engine.

To recharge the battery, an electric voltage is supplied to the battery circuit with a pressure that is higher than the **EMF** (approximately 12.6 volts with all electrical consumers switched off). This reverses the chemical reaction – it forces electrons back through the **electrolyte** to their original positions in the lead and lead peroxide plates and recharges the battery.

In a hybrid car the high voltage batteries will also need recharging. In these systems, the generator is often the electric drive motor turned in reverse.

The hybrid drive motor will first charge the high voltage batteries which serve the hybrid drive system. The voltage is then stepped down through a DC to DC converter, where it is applied to the low voltage battery circuit.

EMF - electromotive force or the open circuit voltage with no current flowing.

Electrolyte - an electrically conducting material that is made as a liquid or gel.

Advanced Light Vehicle Technology

Table 5.10 explains some terms associated with lead acid batteries and some of their classifications/ratings.

Table 5.10 Battery terms and ratings

Term/Rating	Description
Amp hours (Ah)	A measurement of the electrical current that a battery can deliver. This quantity is one indicator of the total amount of charge that a battery is able to store and deliver at its rated voltage. The amp hours value is the total of the discharge-current (in amperes), multiplied by the duration (in hours) for which this discharge-current can be sustained by the battery. For example, a car battery could be rated as 100 Ah, which should contain enough electricity to provide: 100 amps for one hour 1 amp for 100 hours 10 amps for 10 hours Any other combination that multiply together to make 100 (e.g. 25 amps for 4 hours). The amp hour rating, is required by law in Europe, to be shown on a battery.
Cranking amps (CA)	A number that represents the amount of current a lead-acid battery can provide at 0°C (32°F) for 30 seconds and maintain at least 1.2 volts per cell (7.2 volts for a 12 volt battery).
Cold cranking amps (CCA)	A number that represents the amount of current a lead-acid battery can provide at −18°C (0°F) for 30 seconds and maintain at least 1.2 volts per cell (7.2 volts for a 12 volt battery). This test is more demanding than those conducted at higher temperatures.
Hot cranking amps (HCA)	A number that represents the amount of current a lead-acid battery can provide at 27°C (80°F) for 30 seconds and maintain at least 1.2 volts per cell (7.2 volts for a 12 volt battery).
Reserve capacity minutes (RCM), also known as reserve capacity (RC)	A lead-acid battery's ability to sustain a minimum stated electrical load. It is defined as the time (in minutes) that the battery at 27°C (80°F) will continuously deliver 25 amperes before its voltage drops below 10.5 volts.

Absorbed Glass Matt (AGM) low voltage battery

Some manufacturers use a form of lead acid battery to power the low voltage vehicle systems which is known as an Absorbed Glass Matt (AGM) battery. In this type the electrolyte is held in a glass fibre mesh which acts as separators between the plates. By containing the electrolyte in the glass matt, the amount of hydrogen gas given off during charging is considerably reduced. The battery is vented to external air to reduce any excessive gas build-up during operation and ensure that any internal pressures are kept within acceptable limits.

Because the electrolyte of an AGM battery is not in a free liquid state, it cannot be topped up, so the battery is sealed for life in a maintenance free (MF) form.

Safety

Due to their construction, AGM batteries need to have their charging carefully controlled as excess voltage and current can easily damage the internal components. If external charging is required, always use an approved battery charger.

Automotive Master Technician

Nickel cadmium battery

An alternative to the lead-acid battery is the nickel cadmium (NiCad) type, which is found in some cars. It works in a similar way to a standard lead acid battery but requires less maintenance and cannot be overcharged. This type of battery is made of the following materials:

- Positive plates: nickel hydrate.
- Negative plates: cadmium.
- Electrolyte: potassium hydroxide and water.

These batteries tend to be larger and more expensive than normal lead-acid batteries. However, they are better at coping with the extreme loads placed on them by modern electrical systems, especially in hybrid or battery electric vehicles.

Nickel-Metal Hydride battery (Ni-MH)

Nickel-Metal Hydride batteries are becoming the most popular for use with hybrid drive vehicles. They are very similar in construction to a nickel cadmium battery, but use a metal hydride (hydrogen atoms stored in metal) as the negative plate. A Ni-MH battery can have two or three times the capacity of an equivalent sized nickel cadmium battery, meaning physical size is reduced but electrical storage is increased. A single cell of a Nickel-Metal Hydride battery will produce 1.2 volts and are connected in series to each other in groups of six to form packs known as modules. The modules are then connected in series to form the complete high voltage battery.
Current hybrid drive systems (depending on manufacturer) are using batteries with a high voltage potential somewhere between 100V and 300V.

A Nickel-Metal hydride battery has the following advantages over other battery types:

- They have a high electrolyte conductivity, which allows them to be used in high power applications (such as hybrid drive and electric vehicles).
- The battery system can be sealed, which minimizes maintenance and leakage issues.
- They operate over a very wide temperature range.
- They have very long life characteristics when compared with other battery types – this offsets their higher initial cost.
- They have a higher energy density and lower cost per watt than other battery types.

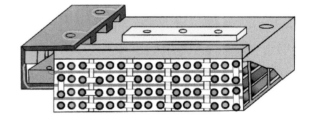

Figure 5.24 Nickel-Metal Hydride (Ni-MH) battery

Because of the characteristics of a Nickel-Metal hydride battery, the charging and discharging has to be very carefully monitored and controlled. If the battery pack is allowed to charge too quickly, overheating and damage can occur. The hybrid drive generator system will maintain a constant current to **trickle charge** the high voltage battery to try and maintain a set **state of charge (SOC)**. An ideal state of charge is around 60% of battery capacity, allowing the battery to work well within its capabilities.
Battery temperature is carefully monitored because as the cells become fully charged any excess energy will be converted into heat. **Thermistors** are used to measure battery temperature and if required an electronically controlled fan is able to draw air through ducting surrounding the high voltage battery unit and assist with cooling.

Figure 5.25 Battery cooling

Trickle charge - a slow charging method that is equal to or very slightly above the batteries natural discharge rate.

State of charge SOC - a rating that shows how much electricity is contained in the battery compared to its capacity.

Thermistors - temperature sensitive resistors.

The measurement of temperature in the high voltage battery system is also a good indication of the battery's state of charge (SOC). By monitoring temperature, the battery system ECU is able to send signals to the generator regulation control unit and alter the amount of charge supplied.

Lithium Ion Battery

For future hybrid and electric cars, a better battery type capable of delivering and storing high voltage electricity is needed.
Lithium ion (Li-ion) batteries are important because they have a higher energy density, (the amount of energy they hold by volume, or by weight) than any of the other currently used batteries. Generally, the Li-ion battery cells hold roughly four times as much energy as the Nickel-Metal Hydride (NiMH) batteries which are used in current hybrids.

High voltage system capacitors

When the hybrid or electric drive system requires electricity from the high voltage battery, its delivery of power must be supplied gradually to ensure that a sudden surge of electricity doesn't damage system components. Some manufacturers use capacitors to smooth out the delivery of electricity to and from the high voltage battery units.

Figure 5.26 High voltage system capacitors

High voltage battery module relays

A series of relays are often used to control the flow of electricity from the high voltage battery. These relays will connect and disconnect both the positive and negative high voltage electrical circuits. In addition to the main connection, a further relay and resistor unit are used when initially connecting power to the system. This introduces a controlled amount of power to the system before the main relay takes over, stepping up to full voltage.

Hybrid vehicle regenerative braking

A standard braking system uses friction to convert the kinetic (movement) energy of a vehicle into heat. Hybrid vehicles often use a different method for braking and slowing down, which is called **regenerative** braking. This is a highly efficient process that turns some of the vehicle's kinetic energy into electricity that can be used to help charge the high voltage battery system.

A hybrid vehicle uses a combination of an internal combustion engine and electric motor to provide drive. If the electric motor is driven mechanically, during braking for example, it can be converted into an electrical generator. The conversion of kinetic energy into electricity actually slows the vehicle down. Any extra deceleration required by the driver that is not achieved by the regenerative braking is handled by a brake-by-wire system which operates brake callipers and pads against discs. A sophisticated electronic control system is used to calculate the amount of braking required and splits the operation between the generator and the brakes.

Limitations of regenerative braking include:

- The regenerative braking effect drops off at lower speeds.
- Most road vehicles with regenerative braking only have power on some wheels, for example a two-wheel drive car. The regenerative braking power only applies to the drive wheels because they are the only wheels linked to the drive motor. In order to provide controlled braking under difficult conditions, such as wet roads, friction-based braking is necessary on the other wheels.
- If the batteries or capacitors are fully charged, no regenerative braking takes place.

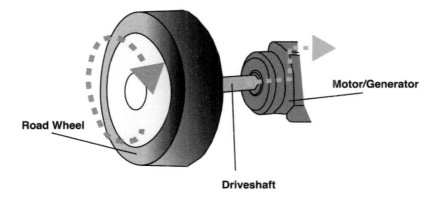

Figure 5.27 Regenerative braking

Regenerative – a method of braking in which energy is extracted from the parts braked, to be stored and reused.

A hybrid vehicle using regenerative braking will have reduced wear on the friction materials for the driven axles than a standard car. This is due to some of the retardation effort coming from the motor generators instead of the standard brakes. This can however sometimes lead to seizure/failure of the standard braking components when the braking is underutilised by over-cautious drivers.

Advanced Light Vehicle Technology

Brake by wire

The regenerative braking used in hybrid vehicles means that only some of the energy needed to slow the vehicle down comes from the hydraulic system. To ensure that the deceleration of the car is accurately controlled, manufacturers incorporate brake-by-wire systems in their vehicle design. Instead of the master cylinder applying hydraulic pressure directly to the brake calliper system, it operates as a pressure measurement sensor (also known as a stroke simulator). This sensor simulates pedal pressure so that brake operation feels normal to the driver. The signal from the brake pressure sensor is processed by an ECU, which operates a secondary master cylinder unit. In conjunction with the wheel speed sensors, the secondary master cylinder applies the appropriate brake force required to slow the wheels without allowing them to skid. If the regenerative braking and hydraulic brake-by-wire fails, the system enters conventional brake operation, which will allow the vehicle to be slowed down but the driver may have to push the pedal slightly harder and overall stopping distances may be increased.

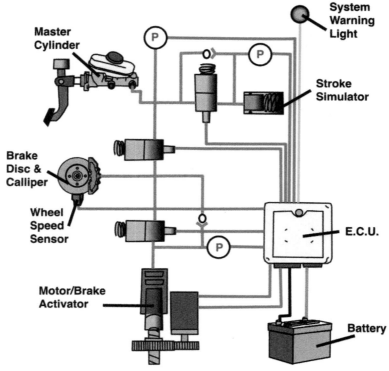

Figure 5.28 Brake by wire system

Other types of electric vehicle

Electric motors are a very effective method of providing a source of propulsion for cars and they produce no emissions while in use. Unfortunately, many of the methods used to create the electricity needed to drive these motors are not very efficient or can be polluting to the environment. Currently, renewable sources of electricity such as wind, solar and hydro-electricity are unable to supply the demands needed to make fully electric powered vehicles truly non-polluting.

Environmentally friendly methods used to create electricity for powering electric vehicles mainly comes from:

- Solar power
- Mains supply
- Hydrogen fuel cells
- Hybrid drive

Automotive Master Technician

Propulsion – the action of driving or pushing forward.

Efficient – how well something works.

Semiconductor - a material with the ability to switch its properties between an insulator and a conductor.

Solar

A solar car is an electric vehicle, powered by energy from the sun, which is obtained from solar panels on the car. A solar panel converts light energy into electricity that can be used as a source of power. The sun gives off approximately 1000 watts of energy for every square metre of the earth's surface. Solar panels that are sometimes found on cars, are also known as photovoltaic cells.

This comes from the words "photo" meaning light and "voltaic" meaning electricity. Photovoltaic cells, are made of **semiconductor** material such as silicon. When sunlight strikes the semiconductor material, some of the energy is absorbed and is converted into moving electrons within the material to create an electric current. The flow of electrons, is in one direction (direct current). If electrical contacts are placed above and below the photovoltaic cell, the electricity produced by converting photons can be used to power electrical circuits.

The silicone used in a photovoltaic cell works in this way because its atoms are not completely filled up with electrons in their outer orbit or shell. If impurities are added to the material, such as phosphorus, small amounts of extra electrons are available making it negative. On the other hand, if boron is added to the silicone, less electrons are available, making it positive. These two sections of silicone can then be joined, and if supplied with light from the sun, share the electrons between the positive and negative sections creating electric current.

Unfortunately not all of the energy provided by the sun can effectively be converted into electricity. This is because light comes in different wavelengths and photovoltaic cells can only use certain areas of the electromagnetic spectrum. Currently solar panels cannot be used to directly supply a car with enough power to provide drive to electric motors, but they can be used to charge batteries or extend the range of 'plug in' electric vehicles.

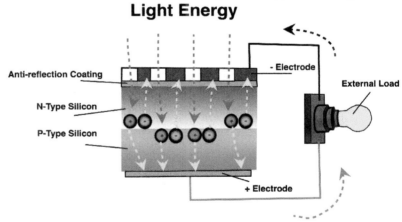

Figure 5.29 Photovoltaic cell

Mains supply

A number of manufacturers are producing a range of mains charged electric cars. Instead of an internal combustion engine, these vehicles (often known as 'plug in') are powered from high capacity batteries that drive electric traction motors. Although these vehicles produce no emissions when they are driven, mains generated electricity is often created using fossil fuel (which creates pollution) or atomic energy (which is dangerous and radioactive).

There are a number of limitations involved with owning and using a plug-in fully electric car including:

Range, this is the distance that a fully electric car can travel on a single charge.
Charging time, this is the amount of time that it takes to re-charge the batteries (which may be several hours).
Due to the expensive nature of the batteries, many manufacturers rent the battery packs to the car owners, and this figure must be factored into the running cost of these vehicle types.

Advanced Light Vehicle Technology

Household Charging System

14 hours @ 100 volts or 7 hours @ 200 volts

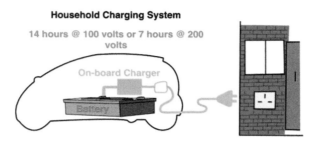

Quick Charging System

30 minutes @ 200 volts (approx 80% charge)

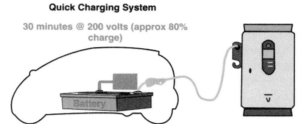

Figure 5.30 Plug-in electric cars

The owner of a plug-in all electric vehicle needs to meet certain criteria in order for it to be a viable option.
Do they have access to off road parking, such as a driveway or garage, where the vehicle can be re-charged over night?
Is their daily mileage under 100 miles? (The average maximum range for many current plug-in electric cars).

Unless official permission is granted by the local authority, it is not advisable to trail an electric cable across pavements or other public areas to connect a plug-in electric car parked on-street with a private/household electricity supply.

Currently, the infrastructure for charging electric vehicles is limited, with the majority of public charging points in the major cities, but as demand for plug-in cars rises the amount of charging points will also increase. Three types of charging point are currently used in the UK:

- 'Slow' points use a standard 13 Amp supply (6-8 hours for full charge).
- 'Fast' points use single or three-phase 32 Amps supply (3-4 hours).
- 'Rapid' points provide direct current supply (typically 80% charge in 30 minutes).

Automotive Master Technician

Hydrogen

Hydrogen is the most abundant gas in the universe and it makes an extremely good fuel for operating electric vehicles. Hydrogen is extremely flammable and produces very little in the way of harmful emissions (the main by-product being water). Unfortunately, hydrogen does not occur naturally on earth so must be manufactured.

The process of making hydrogen is fairly straightforward. It normally involves separating the hydrogen and oxygen in water by a process of **electrolysis**. In general terms, it takes about three times as much energy to make the hydrogen as can be obtained from the hydrogen itself. This makes hydrogen an inefficient fuel source in many ways. Also, hydrogen **molecules** are so small that they will leak through almost any container. This means that storage can be a problem. For example, if you filled up a standard fuel tank with hydrogen, even if you didn't use the vehicle, the tank would be empty in a few days.

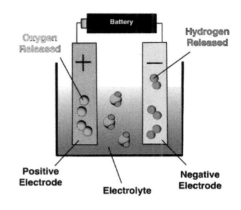

Figure 5.31 Making hydrogen from water

Electrolysis – a chemical decomposition produced by passing an electric current through a liquid.

Molecules – the smallest component of a chemical element.

Hydrogen fuel cell

Another method that can be used to power vehicles using hydrogen, other than burning it inside an engine, is a fuel cell. Some manufacturers are now producing cars with a hydrogen fuel cell, but because of their complexity they are expensive. The fuel cell uses hydrogen to create electricity that can then be used to power electric motors.

A standard vehicle battery, including those found in hybrid and electric vehicles, stores all of its chemical energy inside a casing which it uses to create electricity through a chemical reaction. Once the battery has used up all of its chemical energy, the battery is flat. Many car batteries can reverse this process by supplying an electric current with a voltage potential higher than that coming out of the battery. This is normally achieved using an alternator or generator.

A fuel cell is similar to a battery, but it doesn't store its own internal electricity in the form of a charge. With a fuel cell, as long as the cell is kept supplied with fuel, in this case hydrogen and oxygen, then it works like a battery that never goes flat. Hydrogen is stored in a separate container/fuel tank, and then mixed with the oxygen inside fuel cell to create electricity.

The hydrogen fuel cell is not a new idea. Sir William Grove invented the first fuel cell in 1839. He knew that water could be separated into hydrogen and oxygen if electric current was passed through it by a process known as electrolysis. Grove found that if the process was reversed, he could create electricity by recombining the hydrogen and oxygen and creating water. He then went on to create a very basic type of fuel cell which he called a gas voltaic battery.

Advanced Light Vehicle Technology

Fuel cell construction

The most common type of hydrogen fuel cell is made using a component called a **proton** exchange **membrane** (PEM). This is a material that separates the two sides of the fuel cell.

- One side of the fuel cell is fed with oxygen from the surrounding air.

- The other side is fed with hydrogen from a fuel tank.

Fuel cell operation

As hydrogen enters the fuel cell, a reaction takes place that strips the protons from the hydrogen atom and moves them through the membrane towards the oxygen on the other side. This leaves the **electrons** from the hydrogen atoms, which travel through a different circuit and create electric current. After the energy from the hydrogen has been converted into current and powered the vehicles electric circuit, the electrons reattach themselves to the protons and the hydrogen atom combines with the oxygen to form H2O. This means that the only emission from the fuel cell is water, making it clean and non-polluting.

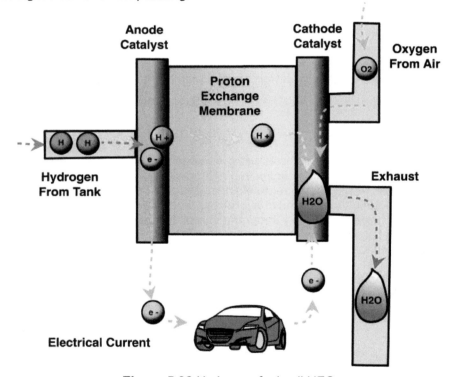

Figure 5.32 Hydrogen fuel cell HFC

Automotive Master Technician

The typical output from a single fuel cell is approximately 0.8V. This means that a number of fuel cells have to be combined (known as a fuel cell stack) to create a useable amount of voltage to drive electric motors.

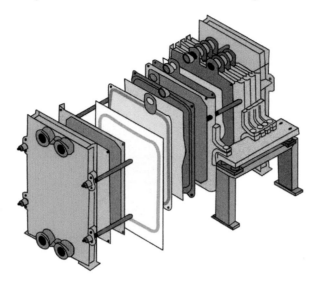

Figure 5.33 Hydrogen fuel cell stack

Protons – the positively charged particles of an atom.

Membrane – a thin layer of material that is used to separate two connected areas.

Electrons – the negatively charged particles of an atom.

Advanced Light Vehicle Technology

Chapter 6 Providing Technical Support and Advice to Colleagues in Motor Vehicle Environments

This chapter will help you develop an understanding of the abilities and procedures required to provide technical support and advice to colleagues in motor vehicle environments. It will give an overview of effective techniques that can be employed during a training or coaching session where technical information needs to be delivered. It will also assist you in developing a systematic approach to any training needs of colleagues within a working environment.

Contents

Technical information sources

In order for you to provide technical support to colleagues in a motor vehicle environment, you will need to have access to accurate information sources. This information can then be used to help support or train your associates in a reliable and consistent manner. Sources of information may include:

Table 6.1 Technical information sources

Verbal information from the driver	Vehicle identification numbers
Service and repair history	Warranty information
Vehicle handbook	Technical data manuals
Workshop manuals/Wiring diagrams	Safety recall sheets
Manufacturer specific information	Information bulletins
Technical helplines	Advice from other technicians/colleagues
Internet	Parts suppliers/catalogues
Jobcards	Diagnostic trouble codes
Oscilloscope waveforms	On vehicle warning labels/stickers
On vehicle displays	Temperature readings

Remember that no matter which information or data source you use, it is important to evaluate how useful and reliable it will be when providing personal and technical support to colleagues.

How to store technical information effectively

The access to technical information and data is vital in your role as a Master Technician, in order for you to support colleagues in their job roles and assist with any training or development needs. Once this information has been acquired, it should be stored for future reference in an effective manner that will allow it to be utilised in a productive way. No matter what type of technical information or data you wish to store, it should be tidy, logical and secure. This could be in electronic or hard copy format.

Hard copy - paper based materials can be indexed and catalogued in order that they can easily be found and used without excessive effort. Where possible, the format should be concise and to the point. If information in paper-based form is difficult to use, it will often be ignored in favour of personal knowledge.

Automotive Master Technician

When undertaking jobs in a workshop situation, information that is not used can lead to wasted time and incorrect repairs or processes taking place.

Electronic - a paperless system can reduce the space required for information storage and if correctly organised, can be easily found and used when required. Electronic information and data has traditionally been stored and accessed locally (i.e. on a workshop based machine) but **cloud storage** is now slowly taking over as the preferred method where data is held on a remote server. The advantages of cloud storage means that information can often be remotely accessed via an Internet connection, collaborated on, and updated by others. It is important that naming conventions are established for work files, so that a search can easily find what is required. It is also vital that when information is stored electronically, that regular backups are taken to reduce the possibility of data loss due to corruption or system crash.

Cloud storage - a method of storing and retrieving information on a remote server system via the Internet.

When technical information is stored, whether electronically or in paper-based format, you need to ensure that it complies with company health and safety, data protection and copyright policies.

For information on the data protection act, *see Chapter 8.*

The delivery of technical information in a training or support situation

Presentation and demonstration methods

It has been shown through research, that if information and guidance is delivered using tried and tested methods, the effectiveness of any training is enhanced and the retention of knowledge is improved. Presentations and demonstrations may be may be conducted in a formal or informal situation and could be to an individual or a group. No matter which type of presentation or demonstration you are asked to conduct, ensure that you have selected a suitable location, such as the workshop or a training area where all those involved will be able to see and hear what is going on, and participate where appropriate.

Presentation methods

If you are giving a formal presentation on technical matters to colleagues in a work situation, it is vital that you prepare effectively beforehand and choose the most appropriate style for the surroundings, task and resources available. You need to have a clear focus in mind as to why you are conducting the presentation and what you want the audience to learn. Presenting the information from a script or outline has pros and cons:

Table 6.2 Pros and cons of using a script

Format	Pros	Cons
Reading from a script	Helps ensure you remember the details. Gives you better control over the length of your presentation.	Can impede natural and spontaneous interactions with your audience.
Delivering from an outline	Lets you have more eye contact with your audience, thereby promoting responsiveness.	Requires you to rely heavily on your memory. Leaves more uncertainty about exactly how long it will take you to deliver the presentation.

Advanced Light Vehicle Technology

Whether you choose to deliver your presentation from a script or an outline, if you are inexperienced, you should write out your presentation word for word. You will need to go through several drafts, experimenting with what to include, what to exclude, how to express your ideas, and how to organise them. Once you have a final draft and have read through it many times, you can decide if you would like to deliver it from the script or from an outline.

Figure 6.1 Presentation methods

Visual aids

There are two main options for visual aids: paper hand-outs; or Power Point slides. No matter which type of visual aid you use, ensure that you give the audience the time to read and digest the information shown. You will have to carefully consider what sorts of visual aids would work the best for your particular project. In general, here are some pros and cons:

Table 6.3 Pros and cons of visual aids

Type of aid	Pros	Cons
Hand-outs	Helps listeners follow your main points. Can help you deal with long but important quotes or data sets. Gives people a space for taking notes.	Can tempt audience members to read ahead and stop paying attention to what you're saying. Can irritate your audience if you don't bring enough for everyone.
Power Point	Helps listeners follow your main points. Can help you deal with long but important quotes or data sets. Allows you to control when and how long people see the information, which helps make sure that it doesn't distract from your speaking. Allows everyone to see the visual aid without your having to worry about how many people to expect.	Can cause anxiety if you're not comfortable using the equipment or if the equipment is not present or fails to work properly. Can be awkward to speak from, by tempting you to turn your back to your audience as you look at the projection screen. If not correctly handled, can lead to boredom in a situation known as 'Death by Power Point'.

Whichever type of visual aid you choose, remember that visual aids should complement your presentation, and not distract from it. Use hand-outs or slides with a clear purpose in mind, not to "dress up" your presentation. Remember, with visual aids, often less is more.

Automotive Master Technician

Demonstration methods

The key to conducting a good demonstration is preparation.
It is understandable that in many real world coaching situations designed to support colleagues with their work and technical knowledge, these may be 'off the cuff' and no preparation time may be available. It is still important however:

- To ensure that any safety aspects are clearly pointed out and observed.
- The demonstration is engaging.
- The task is broken down into logical and systematic steps that are easy to follow.
- The demonstration is clear and your colleague has the opportunity to ask questions or practice the task for themselves.
- That you confirm knowledge, understanding and ability or competence.
- That you recap and summarise what has been covered at the end.

If the demonstration is one that has been planned as part of a colleague's ongoing professional development, then this should give you the opportunity for preparation.
For one to one training, demonstration and assessment activities, it is important to know your colleagues previous experience with the task and their preferred learning style. If you are unfamiliar with your colleague's background and abilities, you should consider having an informal interview with them, to ensure that any training or demonstration can be tailored so that it delivered in the most effective manner for the individual.

Learning styles

Figure 6.2 Demonstration methods

Individuals will all have a preferred method of learning; these are often known as a learning style. There are several different learning styles, but in their simplest form they can be broken down into three main categories:

- **Visual** - learns best when presented with ideas, concepts, data and other information that they can see.
- **Aural** - learns best when they can listen to ideas, concepts, data and other information and may have difficulty with instructions that are written.
- **Kinaesthetic** - learns best when carrying out a physical activity rather than listening to a presentation or watching a demonstration.

Although most people will have a mixture of these learning styles, it is often found that one will be stronger than the others and therefore a more effective training session can be produced if the presenter is aware of the trainees preference. In order to help you discover a preferred learning style, questionnaires are available for free on the Internet, which may be undertaken online, or printed off in hard copy format, which will give an indication of the most dominant style.

Notes

Although learning styles can be a useful tool when designing a presentation or demonstration, your colleagues will also benefit from strengthening any weaker learning preferences by including some aspects of their styles in any training. Good learners tend to have a balanced set of learning styles.

Advanced Light Vehicle Technology

To help you decide the best methods for designing a presentation or demonstration to support preferred learning styles Table 6.4 gives some examples of how they could be included.

Table 6.4 Learning styles

Preferred learning style	Possible strategy for the delivery of a presentation or demonstration
Visual learning style 	There are two important things to remember demonstrating for visual learners. 1. The first is to make sure you are showing the movements as completely and clearly as possible. If you're demonstrating a technique and part of the movement is hidden from view, you will need to find a way to rearrange things; always ensure that you position yourself so that everything you're doing is available for viewing. 2. Always take the time to show the technique from a number of different angles and encourage your colleagues to move around and find the best viewing angles. Give your demonstrations toward the middle of the floor, not near a wall. That way they can get all the way around you.
Auditory learning style 	When demonstrating or presenting to an auditory learner, let them ask questions as soon as they have them. Making them wait until the end or for permission to speak can waste the moment when your colleague is keen to learn. Auditory learners will often try to do exactly what you say. You need to speak clearly and completely or they're going to head off in the wrong direction and may do the wrong thing. Assuming you've got reasonable speaking skills, the thing to pay most attention to is giving a detailed verbal description of what you're doing. Saying "do it like this" is not enough. It's talking, but not actually describing anything. "Do it like this" means: Ignore what I'm saying and watch instead. Examples could include: • Instead of saying "put your hand here." Say "put your hand on the inside." • Instead of saying "push hard," say "push hard enough to hold it down." • Instead of saying, "move over here," say "move over next to the component." Avoid saying things that assume your colleague can see what you're talking about. Allowing your colleague to talk through ideas out loud, helps a lot of auditory learners. When they can both hear something and then say it for themselves it helps them process the information. Most auditory learners like to ask questions, so give them the chance. You can often get things started by asking some questions of your own, but don't make them feel like they're being tested by putting them on the spot.
Kinaesthetic learning style 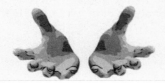	Kinaesthetic Learners like to get hands on and practice the skills that you are demonstrating. (This is often the most common preferred learning style of motor vehicle technicians). You'll often see the kinaesthetic types following along as you demonstrate - moving their arms in imitation of what you're doing. Moving is so fundamental to kinaesthetic learners that they often just fidget, it helps them concentrate better.

Automotive Master Technician

Table 6.4 Learning styles

Preferred learning style	Possible strategy for the delivery of a presentation or demonstration
	If you talk for more than ten minutes during a technical demonstration you've gone on for too long. Kinaesthetic learners need to get into the action as soon as possible. Even visual and auditory learners can't keep track of 10 minutes worth of non-stop details. If you can't demonstrate something in under three minutes break it down into smaller chunks. Say what you need say, don't say anything else and then get to work. The word Kinaesthetic simply refers to an awareness of changes in pressure, momentum, balance and body position in general. It's all about feeling what they are doing as they do it.

Demonstration and assessment planners

To make sure that you accomplish an effective demonstration and are able to assess what has been learnt, it is often a good idea to write out what you intend to do and how you will check that it has been achieved.
This method can be used for group training or in a one-to-one situation and will make sure that you are being inclusive and have thought about any individual support needs.
It will provide you with a structure to follow, and you will know that nothing has been missed. When any demonstration, presentation or training has been completed, it is important for you to evaluate how effective it was from both the colleague's point of view and your own. An example of a demonstration/training planner is shown in the next section that could be adapted to your particular situation.

Demonstration and assessment planner		
Demonstrator/assessor:		Date:
Candidate:		Task:
Aim of the training/demonstration	This is where you summarise what you intend to do during the demonstration or training.	
Learning Outcomes These should be Specific Measurable Achievable Realistic Timed	**By the end of the training the candidate will be able to:** This is where you state what you want the candidate to be able to achieve by the end of the demonstration or training. It is not realistic to expect any candidate(s) to be fully competent with any task until they have practiced it a number of times, so your learning outcomes should reflect this fact. Choose two or three key areas that are vital to the task and ensure these are done correctly and assessed. Make these targets SMART (see left hand column).	
Learner overviews Please include any preferred learning styles, previous experience and any targeted strategies. This section allows space to plan for any delivery strategies that can be specifically targeted towards the individual undertaking the task or demonstration. If possible you should conduct a brief interview/discussion with the candidate to find out if they have any previous experience with this particular task and how best they like to learn. This way you can plan an individualised method of demonstration, training and assessment.		
Safety precautions, PPE, VPE and hazards relevant to task: This section allows you to plan and ensure that the demonstration and task can be conducted appropriately and safely with the minimum risk of injury or damage. Where possible it would be advisable to conduct a risk assessment and this will also help you comply with legislation and any organisation or company requirements.		
Tools and equipment required for the task: In order for you to conduct an effective demonstration, training session and assessment, you will need to ensure that you have gathered all of the required resources before you begin and checked them for correct function and operation. This area could also be used to list any parts or consumables that would be required when undertaking the task. Preparing the resources beforehand will allow you to conduct your activities without interruption, leading to a more efficient and productive outcome.		

Information and data required to complete task:
Many demonstration, training and assessment activities will require you to have a good source of technical information and data. In order to conduct an effective session you need to gather as much information as possible before you start and have it available to you if possible. Examples of information sources are described at the beginning of the technical chapters in this book. Remember that no matter which information or data source you use, it is important to evaluate how useful and reliable it will be to your demonstration, training and assessment routine.

Evaluation: (how effective was the demonstration/assessment plan, delivery, content, resources and individual candidate achievement?)
Training and technical support will form an ongoing part of any good organisation's progression plans and help form part of the individual continuing professional development (CPD) of all staff. This section will allow you to record your evaluation and assessment of how well you feel the demonstration, and training went. The evaluation should cover yourself, the individual undertaking the training and also be used as a tool to enhance any future demonstrations and assessments. This can then be stored for future reference.

Approx. Timing or Stages	Trainer Activity Training delivery, including presentation & demonstration.	Candidate Activity Candidate practical activities in a systematic and logical order	Assessment of Learning
This section allows you to plan a systematic and logical order to your demonstration training and assessment. It could be used to provide a numbered structure or give an approximation of time that will be needed to deliver an effective procedure.	This section allows you to set out what you need to show during your presentation/demonstration. You should ensure that you include any key points so that they are fully explained and demonstrated. **Example:** Show the candidate the specific PPE required for the task and explain why it should be used.	This section allows you to set out what you expect the candidate to be doing while you are giving your presentation/demonstration. **Example:** Listening and asking questions about why the specific PPE is needed. Examining and using appropriate PPE.	This section allows you to set out the performance objectives that need to be evidenced giving an indication of competence. These performance objectives should be shared with the trainee to ensure that they know what is expected. **Example:** Candidate uses the correct PPE throughout the task.

Figure 6.3 Demonstration and assessment planner

Effective techniques for delivering instructions to colleagues

It is wise to have a checklist of items to ensure that you will provide the most effective instruction through presentation or demonstration that could include:

- Ensure that you are aware of any previous experience and current abilities to help determine individual support needs.
- No matter what the level of difficulty of the task being demonstrated, be supportive towards all individuals.
- Plan the location and timing of the presentation or demonstration so that interruptions and distractions are kept to a minimum.
- Never single out someone and put them on the spot. Avoid embarrassment of individuals.
- Break the presentation or demonstration down into a logical step by step process. To ensure that learning is taking pace, check and assess your colleague's understanding throughout.

Automotive Master Technician

Methods for checking colleagues work

When checking a colleague's work, it is important to have a clear set of standards which set out the **range** of work for the task being conducted. If the training is formal, this might come from the approved National Occupational Standards (NOS), but if it is informal the assessment criteria and performance objectives may be set by the organisation. The performance objectives should prove that the candidate has the ability to apply knowledge, understanding and practical thinking skills to achieve an effective standard, while being sufficiently flexible to meet any changing demands. It is best to make the candidate aware of any expectations before any assessment, and if any performance criteria is not directly observed, strategic questioning should be used to ensure that all outcomes are covered.

Figure 6.4 Checking a colleague's work

Range - the difference between two limits. In connection with performance outcomes, it is the amount of things that should be covered within a particular task.

Direct observation - this is where you are actually watching the assessed task being conducted.

Indirect observation - this is where you are examining evidence of the task once it has been completed; the end product.

Direct and indirect observation of colleagues work

Direct observation of a colleagues work should be done by prior arrangement so that the individual has had time to prepare themselves for the task being conducted, this will also involve being made aware of any assessment expectations and outcomes. It is your responsibility to put the candidate at ease to ensure that they can perform to the best of their ability, without any feelings of excessive pressure or intimidation.

Once the task has commenced, the direct observation should be conducted as unobtrusively as possible without interference or interruption, unless you recognise any health and safety issues or incorrect procedures/practices. If you feel at this point, that an observation of your colleagues work cannot continue, it may be possible to turn the activity into a training or coaching session to further support their abilities, but the assessment and observation of the task should be rescheduled.

During an assessment, if any of the agreed performance criteria are not directly observed, you should question the candidate fully to ensure that all of your objectives have been covered. A good questioning technique, to ensure an acceptable standard of technical understanding, is to develop a probing style. Ask open questions (these are ones without a simple yes or no answer), but if your colleague is having difficulty, re-phrase or adapt your questioning so that you can accurately assess their knowledge and understanding. Remember that it is possible that they know the answer, but do not understand the question.

Indirect observation will normally involve examining the end product of a task or evidence, showing the way your colleague carried out activities. This could be the finished vehicle repair for example, or a report on what has been done, i.e. job card/invoice with a full description of any work that has been conducted. This gives the opportunity to assess an activity using a diverse range of evidence.

Indirect observation will require that you will also use supporting evidence, such as questioning to provide confirmation that all of the performance objectives have been met. Other forms of supporting evidence may come via third party statements and testimonies from supervisors, colleagues or customers.

Advanced Light Vehicle Technology

No matter what type of diverse evidence is used for the assessment purpose, it is important to ensure that it is:

- Relevant to the task being assessed.
- Authentic, i.e. produced by the candidate.

- Sufficient, so that it covers the full range of performance criteria agreed.
- Meets the specified standard.

Action to take when your colleagues work is not in line with the agreed requirements

Any inconsistencies in a candidate's work or evidence should be clarified by using question and answer and then the questions and candidate responses should be recorded. If a colleague's work does not meet the required criteria it should be noted as to where performance objectives have fallen short and then discussed while giving specific constructive feedback. If a direct observation is being conducted and serious errors occur which could develop into a health and safety issue or potentially cause damage, you should intervene and bring the task to a stop. At this point the candidate should be supported so that they are able to complete the task appropriately. This may require coaching or further demonstration, but must be done in a way that does not undermine their confidence.

Make sure that when giving feedback on your colleague's performance that there is time to do so in a suitable location without distractions. It is often a good idea to do this in a quiet area away from others, so that it remains private.

An action plan can then be developed that will support the candidate with any extra training that is required and set a realistic timescale to be completed. At this point it may be possible to plan a date and time when a reassessment could take place ensuring that the candidate will be confident that they will be able to pass.

Figure 6.5 Giving feedback on performance

Constructive feedback

When giving feedback, be interactive, encouraging and constructive. Always start and finish any feedback on a positive note, with any negative feedback sandwiched in the middle. Feedback is a two way process, so make sure that you get feedback from the candidate to see how they feel.

Be aware of your colleagues' background and assess without prejudice, making sure that you adhere to equal opportunities.

Try to give positive assessment feedback (where possible), only with regard to area/evidence being assessed; on performance, essential knowledge, performance objectives and the range of work covered.

Constrictive feedback is not just praise or criticism but should clearly describe how your colleague can improve their performance in the future. Constructive feedback can be broken down into the steps shown next.

Step 1: State the purpose of your feedback.

First, briefly state your purpose by indicating what you'd like to cover and why it's important. This is particularly important if your colleagues work has not met the required standard. Your statement provides focus and gives the other person a 'heads up' about the overall outcome. If the task has met the required standard, your colleague may request feedback about a specific area; a focusing statement will make sure that you direct your feedback toward what the person needs. Remember to be clear and straight-to-the-point, but without being condescending.

Automotive Master Technician

You could begin your feedback with: 'I have a concern about.' 'I feel I need to let you know.' 'I want to discuss.' 'I have some thoughts about', for example.

Step 2: Describe specifically what you have observed.

Have a certain procedure or action in mind and be able to say when and how it happened, and what the results were. Stick to what you directly observed and don't try to make assumptions about things that you didn't see. Avoid describing what your colleague 'normally' or 'usually' does.

For example: 'When you finished the task, I noticed that you didn't reset the torque wrench to zero to preserve its calibration.'

Step 3: Describe the result of their actions.

Explain what the consequences of your colleagues approach and procedures could have on the task being conducted. Then give examples of the outcomes of these actions and how it may affect the operation of the systems being worked on. Describing the possible outcomes or consequences allows the other person to see and understand the impact their actions.

For example: 'By not conducting your diagnosis in a systematic and logical order, the task took longer than needed.' 'This means that the extra cost involved in completing the job will have to be passed on to the customer or absorbed by the company.'

Step 4: Give the other person an opportunity to respond.

Feedback should be a two-way street, don't forget to listen to what the other person has to say. Remain silent and meet the other person's eye, indicating that you are waiting for answer. If the person hesitates to respond, ask an open question (one without a simple yes or no answer).

For example: 'What do you think?' 'What is your view of this situation?' 'What is your reaction to this?' 'Tell me, what are your thoughts?'

Step 5: Offer specific suggestions.

Whenever possible make your suggestions helpful by including practical, feasible examples. Offering suggestions shows that you have thought past your evaluations and moved on to how to improve the situation. Constructive feedback is centred around continuing development and how to make things better next time.

For example: 'Next time you conduct a diagnosis, make notes as you go along, that way you can refer to them at any point and you'll have details of where you left your routine if you are interrupted.' or 'Remember to calibrate your electrical test equipment before you take a reading, this way you will know that its accurate.'

Step 6: Summarise and express your support.

At the end of the assessment and feedback, it's always important to review the major points you discussed. Summarise the action items, not the negative points of the other person's actions. For corrective feedback, stress the main things you've discussed that the person could do differently and how to develop their skills. It's important to always end on a positive note by expressing confidence in the person's ability to improve the situation.

For example: 'As I said, the way that you used the technical data to trace the fault to the malfunctioning sensor was very good. You've really followed a systematic routine to find a tough problem. Please keep using a logical approach on problems like that.'

Advanced Light Vehicle Technology

By summarising, you can avoid misunderstandings and check to make sure that your communication is clear. The summary is also an opportunity to show your support for the other person and an effective way to conclude even a negative feedback situation on a positive note.

Be careful how you conduct yourself when letting a candidate know that their work is not up to the required standard. If a colleague's confidence is undermined when giving feedback or corrections of their practical work, this will discourage them during any future assessments. If feedback is not given correctly, this could damage the relationship between the assessor and trainee. A colleague lacking confidence will be nervous and may make mistakes during subsequent assessment activities.

Throughout the training, assessment and feedback process, it is important that any managers or supervisors are kept informed of the process. Ultimately, the manager is in charge and will make decisions on any future training and assessments so they need to be aware of the candidate's performance and outcome. Also, it must be remembered that a workplace is a business, and the manager will have a responsibility to ensure that training and assessments are effective while having minimal impact on workshop productivity.

Alternative training and development solutions

The training and assessment of a colleague in technical matters may not always be directly your responsibility, but you might be involved in helping choose the most appropriate route for staff development.
Other sources of technical training may include:

- Further education/technical colleges.
- Independent training providers.
- Vehicle manufacturer training.
- Equipment manufacturer training.
- Specialist diagnostic training.

Recognising training needs

The role of a Master Technician will frequently include certain supervisory responsibilities. You will often have a vested interest in supporting the companies' vision of prosperity and growth by monitoring the overall performance of the workshop productivity.
This supervisory role will incorporate the monitoring of staff within your care and recognising any deficiencies in their knowledge or performance which reduces efficiency. These inefficiencies should then be brought to the attention of your manager so that action can be taken to support the staff with any training and development needs.

Recognising training needs can be formal or informal, but it is important that when performance issues are identified, they are recorded so that this can be discussed during a staff appraisal. The recorded issues can then be used to create an action plan which will help the individual develop within and beyond their particular job role. Recognising training needs can be achieved by:

- Analysing individual workplace performance through observations (both formally and informally).
- Analysing how regularly your colleagues ask for assistance or advice.
- Analysing a colleague's inability to use relevant equipment.
- Analysing a colleague's difficulty or inability to diagnose system faults.
- Analysing failure to carry out work to recognised company/industrial standards.
- The colleague's failure to complete tasks in required time.
- Analysing customer complaints which reflect an individual's performance.

Figure 6.6 Training and development solutions

Automotive Master Technician

SWOT analysis

To help you identify any further training or development needs, a SWOT analysis is a useful structured planning approach used to evaluate the current position, and any methods that could be utilised to assist in development.

SWOT stands for, Strengths, Weaknesses, Opportunities, and Threats.

A SWOT analysis can be carried out for a product, place or person and all of these factors will have a bearing on a colleague's abilities and development. The process of conducting a SWOT analysis involves specifying the objective of the business or individual and identifying any internal or external factors that are favourable and unfavourable to achieving that objective. Setting a final objective should be done after the SWOT analysis has been performed. This would allow achievable outcomes or goals to be set.

Strengths include: characteristics of the business or individual that give it an advantage over others.
Weaknesses include: characteristics that place the individual or team at a disadvantage relative to others.
Opportunities include: elements of ability that the company or individual could exploit to their advantage.
Threats include: elements in the environment or lack of skills that could cause difficulties for the business or individual.

SWOT ANALYSIS

	Helpful	Harmful
Internal origin	**Strengths** S	**Weaknesses** W
External origin	**Opportunities** O	**Threats** T

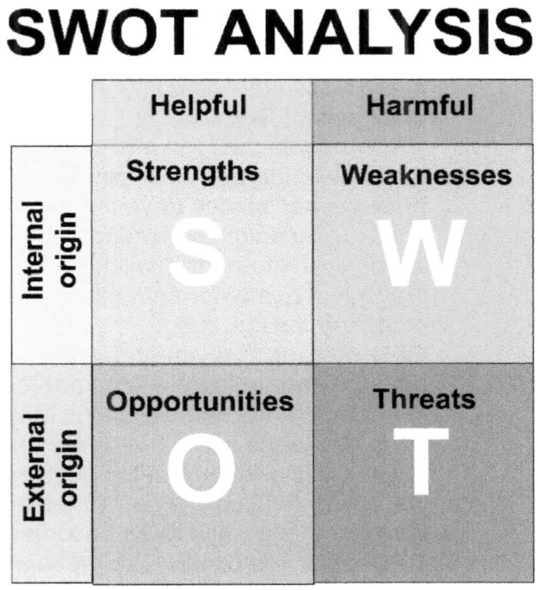

Figure 6.7 A SWOT analysis matrix

The identification of SWOTs are important because they can inform any later steps in planning to achieve the objective.

When using SWOT analysis, you need to ask and answer questions that generate meaningful information for each category (strengths, weaknesses, opportunities, and threats) to make the analysis useful and find any possible advantages.

A SWOT analysis can be laid out in a matrix as shown in Figure 6.7, and then collaboration should be used to complete the chart before deciding the way forward and creating an action plan.

The importance of continuous development and learning

The importance of continuing professional development or CPD should not be underestimated when it comes to an individual's future. Training and development should be seen as a career-long investment and obligation for those wishing to become a professional in their industry.

Well planned and facilitated continuing professional development is vital because it delivers benefits to the individual, their profession and the public.

Advanced Light Vehicle Technology

- CPD ensures your capabilities keep pace with the current standards and any technological advances within your industry. This will help maintain or improve overall task time including that spent within the diagnosis of faults.

- CPD ensures that you maintain and enhance/improve the knowledge and skills you need to deliver a professional service to your customers. As your practical techniques develop, a better job will be accomplished and this will be reflected in a reduced number of customer complaints and improved customer satisfaction.

- CPD ensures that you and your knowledge stay relevant and up to date. You are more aware of the changing trends and directions in your profession. The pace of change in the motor industry is rapid, and this is an ongoing feature of the sector we that work in. If you stand still you will get left behind as the currency of your knowledge and skills becomes out-dated.

Figure 6.8 Continuing professional development CPD

- CPD helps you continue to make a meaningful contribution to your team. You become more effective in the workplace. This assists you to advance in your career and move into new positions where you can lead, manage, influence, coach and mentor others.

- CPD helps you to stay interested and interesting. Experience is a good teacher, but it does mean that we get stuck in a rut and tend to do what we have done before. Focused CPD opens you up to new possibilities, new knowledge and new skill areas.

- CPD helps advance the body of knowledge and technology within your profession.

- CPD can lead to increased public confidence in individual professionals and their profession as a whole.

- CPD contributes to an improved protection and quality of life, the environment, sustainability, property and the economy.

- CPD helps to raise self-esteem and improve the overall image of the industry.

Many schemes and national bodies provide support with continuing professional development and set standards which individuals can peruse. This helps to foster aspirational ideals and increase enthusiasm to the point where CPD can become a self-sustaining process.

Automotive Master Technician

Recording information and making suitable recommendations/giving feedback

At all stages of planning, demonstrating/presenting, assessment and feedback, you should record information and make suitable recommendations. Some of this recorded information will help you plan and set-up for demonstration and training, but much of it could be used to help provide opportunities for assessment and constructive feedback. The Table below shows some examples of how to do this.

Table 6.5 Recording information and making suitable recommendations/giving feedback

Stage	Information	Recommendations
Before you start	Conduct a SWOT analysis.	Using an analysis matrix record any perceived strengths, weaknesses, opportunities and threats.
	Create a training plan.	Record what you intend to teach, any required resources, preferred learning styles and experience.
During presentation, demonstration and assessment	With regard to the training plan.	Make notes for any reminders that may be needed to fully complete training or demonstration. Evaluate your training session to help improve any future presentations or demonstrations.
	With regard to any agreed standards and performance outcomes.	Record your assessment of the colleague's competence against the performance criteria. Record any questions and answers that have been used during the assessment to ensure that the full range of abilities/outcomes have been covered.
When the presentation, demonstration or task is complete	With regard to feedback.	Record any constructive feedback comments that have been given.
	With regard to reassessment or progression.	Create an action plan to show what must be done and set a timescale for completion.
	With regard to management consultation.	Record discussions with line managers so that assessment of colleague's abilities can be used during staff appraisals.
	With regard to continuing professional development.	Develop a training plan for CPD.

Chapter 7 Liaising with Vehicle Product Manufacturers and Suppliers on Technical Matters

This chapter will help you develop an understanding of the skills and procedures needed to liaise with vehicle product manufacturers and suppliers on technical matters. It will give an overview of effective techniques that can be used when dealing with products or services offered by motor vehicle parts suppliers and any personal, organisational and legal requirements needed to conduct professional transactions. It will also assist you in developing an efficient approach to competent liaising and communication techniques.

Contents

Information sources

In order for you to effectively communicate with vehicle product manufacturers and suppliers, you will need to have access to accurate information sources. This information can then be used to help you liaise with suppliers in order to obtain parts, technical information, resolve issues and comply with warranty guidelines. Sources of information may include:

Table 7.1 Technical information sources

Verbal information from the driver/customer	Vehicle identification numbers
Service and repair history	Warranty information
Vehicle handbook	Technical data manuals
Workshop manuals/Wiring diagrams	Safety recall sheets
Manufacturer specific information	Information bulletins
Technical helplines	Vehicle registration numbers
Trim codes	Parts suppliers/catalogues and cross-reference
Jobcards	Component type, description and original part number
Engine codes	On vehicle labels/stickers
Transmission codes	Paint codes

Remember that no matter which information or data source you use, it is important to evaluate how useful and reliable it will be when liaising with vehicle product manufacturers and suppliers.

Automotive Master Technician

Effective communication with vehicle parts manufacturers and suppliers

It is important to develop a good rapport with vehicle parts manufacturers and suppliers. You are working in a service industry that relies heavily on an effective supply chain which is able to provide a source of goods, services and information that support your business. The relationship between organisations should be complementary, meaning that a two-way alliance can be refined over a period of time which is mutually beneficial.

Communication types that could be used when dealing with manufacturers and suppliers include:

Face to face - this could be when visiting the parts department of a main agent for example, or dealing with the delivery of stock or components to your premises.

Telephone - this could be when contacting the supplier for parts or information.

Email - this method could be used when placing a large order, which contains many different components, to ensure that the list is comprehensive and no parts are missed out.

Fax (facsimile) - this could be when requesting copies of paperwork, invoices or information for which there is no original electronic source.

Communication techniques and styles

When dealing with manufacturers and parts suppliers, remember to always adopt a professional approach. Many of your interactions will be established and routine, but there may be occasions where difficulties have arisen which need resolving. It should be remembered that no matter what issues are encountered, your relationship with the manufacturer or supplier is mutually beneficial to both organisations when supporting business.

Many of the skills required to communicate effectively may come naturally, but others might require consideration, especially when you are busy or pressurised with working commitments. At these times it is possible to strain or damage your working relationship and this can create barriers to smooth transactions.
Whenever you contact a manufacturer or supplier, ensure that your greeting is polite. Remember that in most circumstances, although they may be providing you with goods or services, they are also your customer to some extent as some of the exchange or transactions will be two-way.
It is often too easy to be caught up in your own situation, and not consider others. Always make sure that you try to recognise any issues or difficulties that the manufacturer or supplier may be experiencing and try to make allowances. It is important to understand the viewpoint of others as some requests may be considered unreasonable if circumstances have created problems beyond their control.
If difficulties arise when undertaking your transactions, ensure that your responses are appropriate and professional. Remain calm, as frustration can often cloud judgements and reduce your effectiveness to communicate. If necessary, go away and decide on a plan of action before proceeding to the next available option.

Effective listening when liaising with vehicle manufactures and product suppliers

Many of us think that communication only involves discussing our needs with the other person, but good communication will also involve sufficient listening skills which enable an in-depth understanding. Effective listening will help to build a relationship with the manufacturer or supplier and also reduce the number of misunderstandings or mistakes that are made. There are a number of techniques that should be used when listening which include:

Advanced Light Vehicle Technology

Stop talking and listen. When somebody else is talking listen to what they are saying, and try not to interrupt, talk over them, or finish their sentences for them. Once the other person has finished talking you may need to clarify certain points to ensure you have understood their meanings accurately.

Prepare yourself for listening. Make sure you put other things out of mind because we are all very easily distracted by other subconscious thoughts such as, 'I wonder what time it is?' or 'what's for lunch?' Relax and focus on the speaker, so that you can concentrate on the messages that are being communicated.

Put the other person at ease. No matter whether you are face to face, or on the telephone, put the speaker at ease.
Help the speaker to feel free to speak and remember their needs and concerns. If you are face to face, nod and use gestures or words that encourage them to continue. Maintain eye contact but don't stare, this will show you are listening and understanding what is being said.
It is more difficult to do this over the telephone as the other person is unable to see your facial expressions or gestures, although many of us still **gesticulate** with our hands as if this will help the other person understand our point of view. It is important while communicating on the telephone, that the other person knows you are listening and paying attention, otherwise they may become discouraged and loose interest as well. Without interrupting, you will need to make encouraging or sympathetic noises which show you are in agreement and understand what has been said, but remember to wait for your turn to talk.

Gesticulate - to make or use gestures, especially in an animated or excited manner, with or instead of speech.

Prevent distractions. Ensure you focus on what is being said and don't doodle, shuffle papers, look out the window, pick at your fingernails or similar. Avoid unnecessary interruptions. These behaviours disrupt the listening process and send messages to the speaker that you are bored or distracted. It is important that the other person doesn't think you have lost interest in what they are saying.

Show empathy. Empathy is the identification with, and understanding of another person's situation, feelings, and motives and the attribution of your own feelings to a situation. Try to understand the other person's point of view and look at issues from their perspective. Let go of preconceived ideas of what things should be like and try to have an open mind. If the other person says something that you disagree with, then wait and construct an argument to counter what is said but bear in mind their views and opinions.

Show patience. A lull or even a long pause, does not necessarily mean that the other person has finished. Be patient and let the other person continue at their own pace, sometimes it takes time to formulate what to say and how to say it. Make sure that you never interrupt or finish a sentence for someone.

Avoid any personal prejudice. Try to be impartial and don't become irritated. Never let the person's habits or mannerisms distract you from what they are really saying. Everybody has a different way of speaking, some people are for example more nervous or shy than others, some have regional accents or make excessive arm movements, some people like to pace whilst talking, others like to be still. Focus on what is being said and try to ignore styles of delivery.

Listen to the style of communication. Listen carefully to how the other person is talking; volume and tone both add to what someone is saying. You may not always be face to face, but everybody will use pitch, tone and volume of voice in certain situations, let these help you to understand the emphasis of what is being said. Tone of voice is often subconsciously picked up, and if the other person is becoming agitated, you might have the tendency to become defensive. If you are able to read the situation from the tone being used, this might help you ease the situation before it escalates and avoid confrontation.

Don't just listen to the words that are being said, you need to get the whole picture, not just isolated bits and pieces. One of the most difficult aspects when listening is the ability to link together pieces of information to reveal the other persons ideas.

If you are face to face, wait and watch for non-verbal communication. Gestures, facial expressions, and eye-movements can all be important. We don't just listen with our ears but also with our eyes, watch and pick up the additional information being disseminated via non-verbal communication. Remember that non-verbal communication is two-way and your gestures and expressions can enhance a conversation or create issues and barriers. Try to be aware of your own reactions when conducting transactions with manufacturers or suppliers. (*See active listening, Chapter 8*).

Figure 7.1 Liaising with product manufacturers and suppliers

Dealing with situations involving emotions

As human beings our behaviours and actions are driven by emotions. These emotions can often provide a barrier to effective communication and therefore make interpersonal exchanges difficult. The main emotions that may cause issue when dealing with others include:

- Anger
- Apprehension
- Confusion

- Frustration
- Nervousness

When dealing with anger, be patient. Give the other person a couple seconds to cool down; no one likes to be spoken to when they are angry. If they are angry at you, apologise and try to work out the problem. If they don't want to, it's probably because they are angry, so don't take it personally.
Don't be pushy, especially if the other person is angry with you. Don't continually apologise, because chances are, they will get even more annoyed and that can make the situation even worse.
If all else fails, go away and talk to them later when they feel better. If they are really angry, it may take time for them to get back to normal.

When dealing with apprehension, it is important to note that the other person may not display any outward signs. Most people experience some form of apprehension when dealing with someone for the first time and this will often be felt as internal anxiety. When conducting exchanges with somebody new, bear in mind that if you feel anxious, they probably do too. Try your best to put the other person at ease as this will reduce tension and help you to both feel more comfortable. Smile, be open, use a calm tone of voice and very soon both parties will feel more relaxed.

Advanced Light Vehicle Technology

A common barrier to effective communication is confusion. This will often occur if the people involved have either too little or too much information. It is important to recognise confusion, and not mistake the emotion for being awkward. Look for signs of misunderstanding and try to gather the facts as to why the confusion has occurred. It is sometimes useful to repeat the source of confusion back to the other person, but make sure that you do this in a way that is not condescending or makes them feel as if they are being mocked. Once the source of the misunderstanding has been identified, you should go back over the key points or issues, making sure that you have provided adequate information to clear up the problem.

Figure 7.2 Dealing with emotions

Frustration is a common emotion that can be experienced by both parties when communication breaks down. Any issues that create barriers when dealing with others can provide a feeling of frustration that rapidly escalates. Unfortunately, frustration is a self-perpetuating emotion, which once experienced can be very difficult to diffuse. It is important to recognise any signs of frustration from either side, and if necessary, postpone the current liaison until another time. This may not always be appropriate when conducting a transaction with a manufacturer or supplier, but a short break in your dealings might be enough to come up with a different approach that works.

Depending on the background of any meeting or transaction, it's not unusual for one or more of the parties involved to feel nervous. Nerves can make any interaction between individuals difficult, which in turn will create barriers to effective communication. It is important to remember that most people will naturally experience nervousness as an initial emotion, and the sooner they can be smoothed over, the better a transaction will take place. Always begin any new interaction with the expectation of nervousness and this will allow you to plan a calming approach from the start.

Conducting transactions

Liaising with manufacturers and product suppliers is an important part of any business transaction. Many of these exchanges will be technical in nature and require the supply of accurate, factual information. Always remember that people working in the manufacturing and product supply industry will possibly not have a workshop based occupational background, and may therefore not use the same type of technical language as you. Ensure that when you discuss your requirements that you provide clear and simplified explanations of any data or terminology with appropriate tone of voice and pace. This will reduce the possibility of crossed-communication and misunderstanding.
The information you provide should be factual, and to avoid ambiguity, there must be a way of distinguishing facts from non-facts.

Appropriate times to contact manufacturers and suppliers

Most of the contact that you have with manufacturers and suppliers will be routine transactions, dealing with the purchase of replacement parts and other resources. There will however be other occasions that fall outside these normal exchanges. Examples of other requirements include:

Product up-dates - when a component or product has been up-dated, it may be necessary to contact the supplier for information regarding special fitment requirements or the correct procedure for the disposal of controlled waste.

Automotive Master Technician

Product clarification - sometimes when a product has been modified it may look different than the original component. In order to ensure that the correct component has been supplied, it may sometimes be necessary to contact the supplier for confirmation. This is especially true when it comes to aftermarket replacement components and how they compare to the original equipment manufacturer OEM part.

Product development - working in the vehicle maintenance and repair industry puts you in a unique position of experience and knowledge when it comes to the failure of components. Although the manufacturer will have thoroughly tested its products and should also have a rigorous quality control system in place, feedback from Master Technicians can be vital to manufacturers when improving design and reliability. If design flaws can be highlighted, then the manufacturer can develop the product further or change its complete conception if required.

Technical clarification - it is common that components are developed, improved or changed during the lifespan of a vehicle. Also within the same model range, there may be a number of alternatives depending on date of manufacture. This means that there will often be occasions when you will need to contact the manufacturer for technical clarification. This could include, cross-referencing, suitability for use, fitment and operating tolerances.

Product health and safety policies - many products or components may be subject to certain health and safety issues. The most common of these is the Control of Substances Hazardous to Health (COSHH 2002 regulations, *see Chapter 1*). If this is the case, you may need to contact the manufacturer or product supplier for a safety data sheet to ensure that you comply with the regulations following a risk assessment. Many new vehicle components and products may contain substances considered hazardous or controlled waste. Some sensors from SRS systems may contain mercury for example. If a component or a product is suspected of containing a hazardous substance, it may be necessary to contact the manufacturer or supplier to find out how it should be disposed of. It may be that the manufacturer will accept it back for recycling or it might have to be treated as controlled waste under the Environmental Protection Act 1990 EPA (*see Chapter 1*). By doing this you will be fulfilling you obligations to the duty of care.

Warranty claims - when a new component is supplied by a manufacturer, it is often protected by a warranty that will cover the reasonable costs involved with the replacement in the event of premature failure. This warranty is an assurance as to the quality of the components supplied, but is subject to certain conditions having been met. The actual conditions of the warranty may vary from manufacturer to manufacturer, so it might be necessary for you to contact the supplier to confirm certain details. These could include:

- Component details - part numbers, fitment type, size, colour or date of manufacture.
- Vehicle details - make, model, VIN number, engine number, colour code, year of manufacturer, date of first registration, mileage, repair or service history.
- Customer details - name, address and confirmation that they are the warranty holder (ensuring that these details do not contravene the Data Protection Act, *see Chapter 8*).

These details will allow the manufacturer or supplier to establish any limits of liability.

Figure 7.3 Dealing with manufacturers and suppliers

Advanced Light Vehicle Technology

Limits of authority

Depending on the size of organisation for which you work, your job role and the amount of responsibility it carries will vary. Lines of communication are normally established which will dictate limits of authority that fall within your responsibility. Figure 7.4 shows an example of an organisational chart for a medium to large company. The levels of authority can be judged by your position on a similar organisational chart constructed for your own company. In this way you should be able to ascertain your limits of authority within an organisation (but this should always be confirmed with your manager).

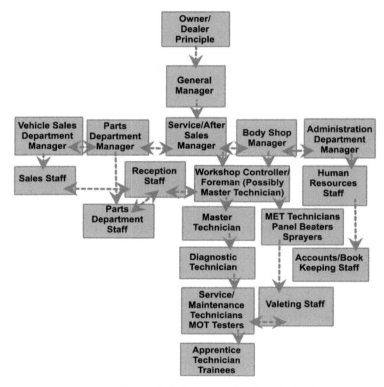

Figure 7.4 An organisational chart

Job roles and requirements

The following section gives examples of the job roles and requirements as shown in Figure 7.4. You should remember that the roles shown are for illustration purposes and may vary depending on company size and organisational requirements.

Dealer principle

The dealer principle is normally in overall charge of the organisation, and in some cases may even be the owner. Most dealer principles will have worked their way up through the motor trade, following the technical, sales or administration route and will therefore have a great deal of knowledge regarding the day to day running of a garage. Their main role will often involve managing and coordinating the company to ensure the efficient running of the organisation.

A dealer principle will have strong leadership skills and be very organised. They will be focused on customer care and also on the financial stability of the company, ensuring the correct balance between profit and loss.

Automotive Master Technician

General manager

If the dealer principle is the owner of the garage, it is not uncommon to have a general manager in charge of the various departments within the organisation. The general manager may perform some of the roles and responsibilities described in the dealer principle section previously, but will also be very focused on liaising with departments and staff to ensure the smooth and efficient running of the company.

Their role will often involve building positive attitudes and good morale with the workforce, devising marketing strategies and effectively dealing with customer complaints.

Service/After sales manager

The service/after sales manager is often in charge of the maintenance and repair workshop. Their duties include dealing with most of the requirements of customers following the purchase of a car, and needs to ensure a high level of customer satisfaction. They will need to ensure the effective running of the after sales department, including the workshops, while making sure that all targets are met for performance, efficiency and customer satisfaction.

Workshop controller/foreman

The workshop controller or foreman will normally be in charge of the practical side of the maintenance and repair workshop. It is their job to organise, motivate and oversee the repairs conducted on customer's vehicles, helping to ensure that a high quality standard of work is maintained. They will often report to the after sales manager and may be the Master Technician.

Master technician

The Master Technician is the most experienced and highly skilled in diagnosis, maintenance and repairs. They will be fully qualified, having considerable motor trade background and may be a member of the industry professional body. The role of a Master Technician can have many different features and may include the management of the workshop as a foreman.

They are often responsible for undertaking the most challenging and complex diagnosis for which there is no prescribed method. Therefore they must have a systematic and logical approach to their work, backed up by a comprehensive knowledge of vehicle systems.

The Master Technician might also be the workshop contact for customers with repair issues. Good customer service skills are therefore required.

They will often have to liaise with product manufacturers and suppliers on technical matters and may have the responsibility for training and up-skilling workshop staff including apprentices.

Diagnostic technician

The role of a diagnostic technician lies somewhere between a service technician and a Master Technician. They will often be qualified to level three and undertake complex diagnosis following prescribed methods. This role will normally require customer facing skills in order to correctly ascertain the issues requiring repair.

Service and maintenance technician

A service and maintenance technician will undertake many of the routine servicing and mechanical repair requirements for customer's vehicles. If they are adequately qualified, they may also be nominated testers for the vehicle MOT test scheme, administered by the Driver and Vehicle Standards Agency (DVSA).

Apprentice technician/trainee

An apprentice workshop technician or trainee is usually someone starting out on a vehicle maintenance career in the motor industry. Initially they will undertake basic maintenance and repair tasks at work and attend college or a training provider to learn the skills required to become a technician. These skills are then put into practice and over the period of their training, and the apprentice will gradually become competent. To gain the appropriate qualifications, the apprentice will be assessed in both the college/training provider, on their skills and underpinning knowledge, and in the workplace when conducting tasks to a prescribed level of competence. Qualifications help demonstrate that apprentices/trainees are able to do certain tasks to a nationally recognised standard.

Parts department manager

The parts department manager is often in charge of a small team of parts/store staff. Their duties include dealing with the efficient running of the parts department for both stock control and supply. They will need to ensure the effective running of the parts department, making sure that all targets are met for sales performance, reliability and profit.

Parts department staff

The role of staff within the parts department is to order, locate and sell vehicle components and consumables. These sales can be made internally to the company's service department workshop and body shop, or they may sell to external customers from the trade and general public. They will often be required to give advice, either face to face or over the phone; taking orders or raising invoices for parts sold. They will also be responsible for maintaining the correct levels of stock within the parts department.

Vehicle sales department manager

The vehicle sales department manager is often in charge of a small team of new and second-hand vehicle sales staff. Their duties include dealing with the efficient running of the vehicle sales department for both stock control and supply. Devising marketing strategies and setting sales targets for their team. These targets can then be used to create incentives and bonus schemes. They will need to compile and analyse sales figures and deal with customer feedback, to ensure performance, consistency and profit.

Vehicle sales staff

Depending on the type and size of organisation, sales staff may be involved with both new and second-hand vehicle transactions. They must be very customer focused as they will be greeting customers, providing information and literature, discussing their particular needs and then advising them on the most suitable type of vehicle. Once a type of vehicle is chosen, they will often arrange a test drive and discuss suitable finance arrangements.

Body shop manager

The body shop manager is normally in charge of a small team who carry out repairs to vehicle bodywork; these may be cosmetic or accident damage. They are often responsible for producing estimates and liaising with insurance companies following an accident. It is their job to ensure that repairs are conducted to the highest standards efficiently and following appropriate quality assurance that meet all requirements for safety and customer satisfaction. They will also undertake transactions with parts and material suppliers in order to maintain an uninterrupted work flow.

Automotive Master Technician

MET technicians, panel beaters and paint sprayers

The job roles of MET (mechanical and electrical trim) technicians, panel beaters and paint sprayers are often highly skilled individuals in their own particular disciplines. They work together and coordinate their roles to ensure that repairs are conducted in an effective manner.

An MET technician is responsible for identifying damaged mechanical, electrical and trim components that need to be removed and replaced during a vehicle bodywork repair. They will also ensure that the vehicle has been stripped out in such a way that any bodywork components that require access for the panel beaters and paint sprayers are available. Once the repairs have been completed, the MET technician will then refit and re-trim the vehicle before it is cleaned, valeted and returned to the customer.

A panel beater/technician is responsible for repairing or replacing of damaged vehicle body panels. They specialise in assessing the level of damage and making a judgement as to whether the components or panels are repairable. If components or panels are repairable, they are then straightened using accepted methods and secured using fixing, welding or bonding techniques that do not compromise the structure or safety of the vehicle.

Once the panel work is complete, it is the job of a sprayer to prepare the bodywork for painting. The paint is then mixed and matched to the vehicle, and the paint sprayer will then skilfully paint the car. Once painted, the car is carefully polished before being returned to the MET technician for re-trimming. When this work is all complete, the car will be sent to the valeting department final cleaning.

Valeting staff

Valeting staff may work with a number of different departments within the same organisation. Their main role is to clean and prepare vehicles for customers. This could be brand-new vehicles, or cars that have been through the workshops or body shop for repair. It is sometimes the case in a large organisation, that valeting staff take on other job roles and responsibilities. These could include the collection and delivery of cars, and organising courtesy cars. No matter which role they fulfil, it is always important that they project a professional image to the customer through their work.

Reception staff

Reception staff will often work in conjunction with a number of departments within a medium to large automotive organisation, although the majority of their time will normally be focused on supporting the after sales and service departments. They are the first point of contact for visitors to the garage and are therefore highly skilled in customer service. It is not uncommon for reception staff to also have specialist knowledge of motor vehicles which will allow them to correctly advise and support customers with their enquiries. Their roles will often involve:

- Greeting customers and providing service information and estimates.
- Booking in and accepting customer's vehicles for repair (this may also include liaising with the service department for workshop loading schedules).
- Directing customer enquiries to the correct department or person.
- Tracking the progress of vehicle repairs and informing the customer of any delays or extra costs involved.
- Contacting the customer upon completion of work and arranging collection.
- Discussing repairs with customers, invoicing and accepting payments.

Administration department manager

The administration department manager is often in charge of a small team who carry out much of the organisational and clerical work behind the scenes; and this may include responsibility as the direct line manager of reception/customer service staff. It is their responsibility to ensure that the administration needs of the company are effectively undertaken and comply with any organisational or legal requirements.

Advanced Light Vehicle Technology

Human resources staff

Human resources staff are responsible for the employment needs of an organisation. They recruit staff, make and administer any company policies regarding the workforce, which may include benefits and safety programmes. Human resources will work with employees to forecast any employment needs, determining and monitoring any training requirements. They are also able to give advice on development, duties, working conditions, wages and opportunities for progression.

Accounts and bookkeeping staff

The staff in the accounts department are responsible for the financial organisation of the company. In conjunction with the human resources department they will often oversee employee payroll and tax contributions. All invoicing, for both purchase and sales will be organised and accounted for, keeping ledgers of all business and financial transactions. It will be the responsibility of the accounts department to track the financial health of the organisation and liaise with the appropriate management and departments to ensure profitability.

Workplace procedures for gaining up-to-date technical information

Technical information is a vital component to any automotive repair. A large amount of technical information can be sourced locally or in-house, but some complex or specific information will have to come from the manufacturer. Examples of locally sourced information include:

- Verbal information from the driver/customer - as long as they are technically competent.
- Service and repair history - although this may require liaising with a previous vehicle repairer.
- Vehicle specific data - manufacturer/make, model, registration number, VIN, paint and engine/transmission codes.
- On-board data - this may include dashboard displays and on-board diagnosis such as, DTC's, PID lists/live data and freeze-frame information.
- Technical data and workshop manuals - these could be in paper based or electronic format.
- On-vehicle stickers and labels - these may include safety, environmental or technical data tolerances.

There are also technical information types that require you to liaise with manufacturers and product suppliers. Examples of manufacturer and product supplier sourced information include:
- Technical specifications - these may be for the vehicle itself or specific component or parts.
- Repair processes - procedures that ensure the vehicle is maintained and repaired correctly in accordance with legislation and warranty clauses (*see Block Exemption*).
- Warranty information - this may include updates, product recalls and bulletins.
- Part supplier catalogued and cross-reference - giving component descriptions and original part numbers.

Remember, when liaising with manufacturers and product suppliers for technical information, build a working relationship that will allow good two-way communication. Although in many cases you may class yourself as the customer, it is important to remember that you might also be considered a business competitor. Therefore some information may be withheld. Although certain material must be released due to Block Exemption regulations, **intellectual property rights** will also create restrictions to the flow of information.

Intellectual property rights - the ability for a person or company to have exclusive rights of its plans, ideas, or other intangible assets without the worry of competition, at least for a specific period of time. These rights can include copyrights, patents, trademarks and trade secrets. These rights may be enforced by law in a court. The reasoning for intellectual property is to encourage innovation without the fear that a competitor will steal the idea and take credit for it.

Automotive Master Technician

Specialist training courses

Another method that can be used to help promote knowledge of technical products, strategies and processes is the opportunity to undertake take specialist training courses provided by vehicle manufacturers and product/equipment suppliers. These courses or seminars are not only good for continuing professional development, but will often come with a large amount of technical information and data. Another opportunity that might present itself during seminars or training courses is networking. Networking can help form contacts and relationships which may then be exploited as access to technical information and processes long after the training has been completed.

Figure 7.5 Specialist training

EU Motor Vehicle Block Exemption Regulation (EU) 461/2010 (MVBER)

The motor vehicle sector benefits from a Block Exemption which helps promote competition rules and agreements for the distribution and servicing of motor vehicles in the European Union.

Before Block Exemption, motorists were required to have their vehicle servicing and repairs conducted within the main dealer network so as not to risk invalidating the vehicle's warranty. The EC Block Exemption Regulation 1400/2002 came into force in October 2003 and allowed motorists more flexibility in selecting where they can get their car serviced.

A revised MVBER ((EU) 461/2010) and guidelines came into force on 1 June 2010 which applies to the repair and maintenance services only. The MVBER is valid until 2023, though the guidelines can be reviewed at any time by the European Commission.

Because of this legislation, maintenance and service work does not have to be done by the main dealer as long as the garage uses Original Equipment Manufacturer (OEM) quality parts, and are recorded as such. The garage must also follow the manufacturer laid down service schedules.

The regulations help safeguard free competition in aftermarket parts, maintenance and services and will benefit independent retailers, repair organisations and motorists by reducing the cost of servicing. Competition allows/promotes better labour rates and helps to reduce the cost of parts.

The rules aim to improve access for the independent repair and servicing sector to the technical information needed for repairs and make it easier to use alternative spare parts.

The MVBER is directly enforced in the UK by the Office of Fair Trading, (OFT) who has responsibility for the enforcement of competition rules. The OFT can be approached for advice on any competition related complaint and can also advise on any cross-border issues.

Reporting systems

The flow of information within a company will require a method of reporting and recording data so that it continues to support organisational processes and provides a history and **audit trail**. This data can then be used in the future and accessed by all requiring it. Most organisations will have company policies and procedures in place for accessing the recording and reporting systems. All data both technical and non-technical may be stored using paper based systems or computer based.

Audit trail - a chronological sequence of records that contain documentary evidence of operation, procedures or events.

Paper based systems

Traditionally most information has been stored as hard-copy, meaning that it is paper based written or printed material. This provides a tangible method of recording and accessing information. The disadvantage if this type of system is that it can be time consuming to prepare and access; it also takes up a large amount of storage space. The advantages of paper based technical information however is that it is often readily portable and therefore good when used in a workshop situation. Data can be accessed using technical and workshop manuals and taken to the vehicle when conducting repairs and diagnosis. It is also less susceptible to the often dirty working environment and less likely to be contaminated or damaged than electronic computer based storage systems.

Computer based systems

Many organisations are now using electronic methods for storing and accessing technical information. Computer stations may be placed in the reception, office and workshop, all having access to the same data. The main advantages of computer based systems are the amount of data and information that can be stored and accessed quickly without the need for large amounts of space. The sharing of information between departments on an intranet, or access to information on a wider network over the Internet can now be exploited. As long as company policy's for the conventions used when recording and storing information are followed, this can be a very efficient use of resources.

Since the introduction of mobile technology, such as portable PC's and tablets, rapid access to information in workshop situations has improved. There are now some organisations that conduct all of their recording and technical data exchanges using tablet PC's. This may include:

- The booking of customers cars for maintenance and repairs (computer based systems are also able to provide electronically generated reminders and workshop loading).
- Workshop loading and job allocation.
- Electronic job cards for both the setting of work instructions and the recording of work undertaken.
- Electronic logging of work time (time spent on and off the job).
- Access to technical data and illustrations.
- Electronic ordering and recording of components and consumables required/used.
- Self-generating service schedules and history.
- Electronic communication with suppliers and customers.
- Invoicing and payment transactions.

In order for any reporting system to work effectively, whether paper based or computer based, it is important that all staff are aware of any requirements to comply with data protection (*see Chapter 8*) and copyright policies to ensure that appropriate legislation is observed. These should also fall in line with any company policies and procedures.

Automotive Master Technician

Recording information and making suitable recommendations

At all stages of liaising or communicating with product suppliers or manufacturers, you should record information and where appropriate, make suitable recommendations. Some of this recorded information will help provide a smooth process to any transactions conducted, but much of it could be used to help provide an audit trail for future reference. The Table below shows some examples of how to do this.

Table 7.2 Recording information and making suitable recommendations/giving feedback

Stage	Information	Records/Recommendations
Before you start	Technical process or data needs analysis.	Record a description of customer/repairing technician information needs.
	Parts or component requirements.	Record a list of component parts and descriptions.
	Warranty conditions.	List vehicle service history and warranty information.
During parts supplier or manufacturer liaisons	Part numbers, costs and fitting instructions.	Ensure that part numbers and any cross-reference codes are recorded, making a full parts list for components and consumables needed. Make careful note of any special requirements such as fitment details, health and safety or waste disposal.
	Technical data and advice.	Record any technical data that will be required to conduct the repairs.
When the transaction is complete	Records to form an audit trail.	Store all records of transactions, such as invoices for future reference. These may be required for warranty/guarantee claims etc.
	Technical information storage.	Any technical data received should be stored in computer based or hard copy format to form a technical library.

Chapter 8 Diagnostic Consultations with Customers in Motor Vehicle Environments

This chapter will help you develop an understanding of customer service within motor vehicle environments. It will give an overview of effective techniques that can be used when dealing with diagnostic consultations and how your actions will impact on the business and company image. You will be expected to work in a professional manner at all times and provide good customer service within legal and organisational requirements. It will also assist you in developing an efficient approach to customer service techniques.

Contents

Information sources

In order for you to effectively conduct consultations with customers, you will need to have access to accurate information sources. This information can then be used to help you discuss vehicle faults in a non-technical manner, resolve issues and provide good customer service. Sources of information may include:

Table 8.1 Technical information sources

Verbal information from the driver/customer	Vehicle identification numbers
Service and repair history	Warranty information
Vehicle handbook	Technical data manuals
Workshop manuals/Wiring diagrams	Safety recall sheets
Manufacturer specific information	Information bulletins
Technical helplines	Vehicle registration numbers
Trim codes	Parts suppliers/catalogues and cross-reference
Jobcards	Component type, description and original part number
Engine codes	On vehicle labels/stickers
Transmission codes	Paint codes

Remember that no matter which information or data source you use, it is important to evaluate how useful and reliable it will be when conducting consultations with customers in motor vehicle environments.

Automotive Master Technician

Organisational requirements

To promote customer service, every good organisation should have a method or policy in place which describes how customers should be treated. Many Master Technician roles will involve dealing with customers, however depending on your position within the company, the amount that you deal with customers will vary. You should be aware of the customer service requirements of your organisation. You also need to find out the limitations of your authority when dealing with customers (what you are allowed to say, do or offer).

These policies should include:

- How the customer is greeted, either in person or over the telephone.
- The promotion of the goods or services offered by your garage.
- How to record vehicle symptoms and faults.
- How to discuss issues and repairs in a non-technical manner.
- How to give estimates for work to be conducted and explain the limitations of vehicle and component warranties.
- How to book vehicles in for work to be conducted.
- How to accept customer vehicles and what information needs to be gathered at the time of acceptance.
- How to conduct a pre-work inspection (preferably with the customer present).
- How to gain agreement for the work to be conducted and obtain signatures for customer service agreements.
- How to keep the customer aware of progress.
- How to inform the customer if work, time or costs are likely to exceed those agreed.
- How to conduct a post-work inspection to ensure customer satisfaction.
- How to inform the customer that the work has been completed.
- How to explain invoicing, including payment methods, and obtain customer signatures when necessary.
- How to explain any further work that needs to be conducted, guarantees or warranties.
- How to accomplish the final handover and return of the vehicle to the customer.
- How to conduct follow-up enquiries to ensure customer satisfaction.

Principles of customer communication and care

Many customers will make up their minds about your organisation within the first few seconds and, as a result, first impressions are important. Good communication and showing that you care about your customers are key to creating a good impression of your garage and the work you do. If you can manage to create a **rapport** with your customer when you first speak to them, you are well on the way to gaining their respect and trust and ensuring their future business with your company.

Good communication involves more than just the words you use. It involves the ability to listen and use positive body language, your attitude to your work and having a professional appearance – both your own appearance and the appearance of the working environment. All of these will combine to create the impression that the customer will take away with them.

Presenting a professional image

It often helps to create a sense of identity. Customers will instantly be able to recognise members of staff. In turn, it may help staff to feel a sense of pride and belonging. A **dress code** can also help foster teamwork. If no company dress code is required, make sure that your appearance is presentable, as this will make a difference to how people react to and interact with you.

Advanced Light Vehicle Technology

Taking care of your appearance is not the only thing you need to do. Consider the impact of your behaviour too. You need to be confident in your actions and behaviour, without being overbearing. When speaking to customers, consider what you want to say. Remember that it's not only what you say but how you say it that will create an impression with the customer. Be positive with your choice of words. Nobody wants to hear what cannot be done, so say what you can do and offer options or alternatives.

Figure 8.1 Presenting a professional image

Dress code – a company's policy regarding the style and type of clothing that you are required to wear.

Rapport – a sense of being comfortable with someone whether or not you know him or her well.

Active listening – paying careful attention to understand more than just the words being spoken.

Listening skills

Make sure that you show interest by actively listening to your customers and not just waiting for your turn to speak. When dealing with customers, you should use a ratio of approximately 80:20 of listening to speaking. Actively listening will help you get things right first time. It will prevent misunderstandings occurring and potential complaints being made.

Active listening involves using your eyes as well as your ears. You need to listen for content as well as underlying emotions. Respond to feelings, sometimes the true meaning is in the emotion not the words used. Tune in to your customer's body language: what are the non-verbal clues telling you?

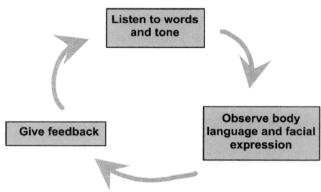

Figure 8.2 Listening skills

Active listening involves:

- Giving your customer your full attention: Make sure you maintain enough eye contact to develop a rapport. Make encouraging noises, for example, say 'OK' especially on the telephone to indicate you are listening. Customers will take this as an acknowledgement that you have heard them.
- Managing your reactions: You are listening, so try not to interrupt. The person speaking the most should be your customer.
- Looking for signs of the true meaning behind the words: Does the customer's facial expression match the words they are using? Does their tone of voice tell you anything about what they are really feeling?
- Giving feedback: Confirm that you have understood by nodding or using phrases like 'I see', 'I understand', 'Oh yes' and 'OK'.

Automotive Master Technician

- Asking questions: If you do not understand, ask questions to find out more information.
- Listening to tone of voice: You can still be an active listener on the telephone. It is not only what your customers say, it is how they say it. Listen carefully for changes in tone of voice and how each word is stressed. Tone is how things are said, for example, loudly, softly, impatiently, quickly, with sensitivity, with respect.
- Reading **body language**: This is an important factor in understanding the difference between hearing and actively listening. The signals given out through body language can be positive or negative.

Body language

It is important to combine words used with body language in order to get the best understanding. This means you should not take one in isolation from the other. For example, it is often said that if a person has their arms tightly crossed, this shows that they are angry, confused or unhappy with the situation. In fact, he or she may just be comfortable in that position or they may be cold.

Figure 8.2 Body language

Body language – non-verbal communication through conscious or unconscious gestures and movements.

You should always strive to use positive body language in order to show your customers that you are willing to help and are respectful of their wishes. This will also give them a positive impression of the organisation. Some examples of positive and negative body language are listed in Table 8.2.

Table 8.2 Positive and negative body language

Positive body language	Negative body language
Smiling with your mouth and eyes	Pursed lips
Leaning forward (but not too much)	Invading personal space
Maintaining eye contact	No eye contact/looking around
Relaxed facial expression	Frowning
Head nodding occasionally	Staring
Head leaning to one side	Yawning
Steady breathing	Rapid breathing
Gesturing with your hands while speaking	Pointing with your hands or fingers
Raised eyebrows while smiling	Tapping your fingers

Advanced Light Vehicle Technology

Gestures are important when used with words because they act to empathise speech. Gestures include movements of the head and moving hands to emphasise what you are saying. If you understand the gestures you tend to use and can recognise them in other people, you can use them successfully to help achieve customer satisfaction.

Positive or supportive gestures will help create **empathy**. For example, your customer will want to know you are listening. You can show that you are by leaning your head to one side and nodding occasionally. Make sure your eyes also really show you are listening. Do not grin from ear to ear, but the occasional smile will help with rapport. This will mean you and your customer have empathy, and you both understand each other's feelings.

Gesture – a movement of part of the body, especially a hand or the head, to express an idea or meaning.

Empathy – understanding and sharing the feelings of others.

Dealing with customers' questions and comments

In your work, you may need to communicate with customers face to face, on the telephone, by electronic means, or using a combination of these methods. Whatever the form of communication, you will need to be respectful, helpful and professional. You can achieve this by being polite and confident and making it very clear what you can and cannot do for your customer.

Figure 8.3 Dealing with customers

Communicating with customers in a professional way involves many different factors in order to create a positive impression. These factors include:

- Appearance (both of you and your surrounding).
- Location.
- Appropriate behaviour.
- Product or service knowledge.

Automotive Master Technician

Your appearance

When dealing with customers face to face, your appearance and that of your surroundings need to convey the right professional impression. You can achieve this by following your organisation's dress code and keeping your working environment clean and tidy. If your job role as a Master Technician, involves some form of minor management, it may be your responsibility to ensure that other staff also keep their appearance and surroundings tidy and professional. You will also find it easier to locate the information that you need to help customers if you are well-organised and keep your workspace/area free of clutter.

Even if you are not dealing with a customer face to face, your colleagues can see you. You need to create the right impression with them too by setting a good example. A tidy appearance and keeping your working area clean and tidy shows your colleagues that you care about your work. This will lead to your colleagues trusting in you and your abilities and will help you to gain their respect.

Hang on please, I'm sure I've got the information here somewhere.

Figure 8.4 Being organised

Selecting an appropriate location to meet customers

Another factor that can impact on how you deal with customers while conducting diagnostic consultations, especially on technical matters is your choice of location. It is important that you select the most appropriate area depending on the type of customer that you are dealing with and their needs or experience. Choices of location could include:

- Office
- Reception
- Workshop
- Showroom

You will need to ensure that the location chosen is free from distractions and supports the type of diagnostic technical consultation taking place. For example a workshop may be appropriate if a demonstration or an example needs to be used to explain how systems work or what has gone wrong, but remember that this environment may be intimidating for some. You will therefore need to assess how customers react and change the location if not appropriate.

Road testing

It is also a good idea, when conducting consultations to do with diagnosis, to offer the opportunity for the customer to road test the car with you. This way they are able to demonstrate or recreate the issues for which they have brought the vehicle into the garage. A one-to-one situation such as this will allow you to experience the problems first hand and explain any processes or possible remedies in a non-technical manner.

Figure 8.5 Road testing

The way you conduct yourself

Effective handling of questions and comments depends to a large extent on your behaviour towards your customers. Equally you will be affected by your customers' behaviour towards you. If a customer behaves in an angry or disrespectful manner towards you, you may be tempted to react in a confrontational manner. Remember to keep your responses professional at all times.

Behaviour refers to everything you do and say. People will draw conclusions about you and your organisation based on your behaviour towards them. You should always behave in a manner that shows you care.

For example, you might find that you hear customers saying similar things many times during the course of your work, and this could get boring. You may show your boredom by your voice becoming flat or by getting easily distracted and your listening skills might suffer. Your customer will pick up on your boredom, and may feel discouraged about your willingness to do the work/diagnosis needed. Remember that to your customer the problem with their vehicle is not boring, and it is part of your job to be interested in finding the solutions that will help them.

Show your customer that their custom is important to you by answering their queries as quickly as you can in a non-technical manner. If you have to research information or seek help from others, do so. Never guess the answer. If the answer is taking longer to find than you anticipated, keep your customer informed of progress.

Product and service knowledge

Ensure you know where to access information about all the products and services you deal with and keep this information up to date. This will help you to make sure that your responses to queries and requests are accurate. It is also important to keep your technical knowledge up to date to ensure that you are able to conduct diagnostic consultations effectively, (*see Chapter 7*).
Know the limits to your authority.
Do not make promises which cannot be kept.

Using questions to check understanding

To avoid misunderstandings and making assumptions, you will sometimes need to check that you have understood what the customer is telling you. Good questioning techniques will also assist you in the initiation of any diagnostic procedure and explanation. You will ensure you have fully understood your customer by:

- Listening actively.
- Asking the right questions.

- Repeating information back.
- Summarising.

Automotive Master Technician

Try using checking phrases such as:

'I just want to make sure that I heard what you said … Did you mean?'

'To be sure I have understood correctly, let me repeat back what you need … Is that right?'

Sometimes you may feel confused by what the customer has said. Repeating back key words you think you have heard could help clarify things. For example: 'Did you say that the warning light on the dashboard only comes on when the car starts to move?'

It is also important to check that a customer has understood what you have said. If they haven't heard you, in their mind you haven't said it. For example: 'Let me clarify, the cost of scanning your car for diagnostic trouble codes will be £39, excluding any other diagnosis'.

Figure 8.6 Using questions to check understanding

Positive and negative language

As with body language, there are also positive and negative words. Choosing the right words will be critical to your success. Customers need to trust you and know you are genuine in what you say and that you care. If you choose negative words you may well create an unwanted situation which could have been avoided with a more positive approach. Say what you can do rather than what you cannot do, but be honest. Table 8.3 lists examples of positive and negative words and phrases.

Table 8.3 Positive and negative words and phrases

Positive words and phrases	Negative words and phrases
Yes	No
How may I help you?	What do you want?
I	They
Definitely	Unlikely
I will find out	I don't know
I will	I can't
Always	Never
I'll do it this morning	I'll do that when I can
I'll sort that out	It's not my fault

Tone of voice

Your customers will also develop an impression of you based on your tone of voice. An appropriate tone of voice is an important tool in customer service because it helps you to bring emotion into putting your message across. If your tone is clear and strong, you come across as confident. If it is halting and softly spoken, you may appear timid and lacking in knowledge. Someone who speaks with a flat tone (that is, with no rise and fall in their voice) may come across as boring and, again, possibly lacking in knowledge and confidence.

Advanced Light Vehicle Technology

Some of the emotions people bring into the tone of voice include:

- Boredom
- Happiness

- Sadness
- Anger

- Frustration
- Worry

You should always aim to speak clearly with warmth and energy. Smiling while you speak will automatically bring warmth to your tone of voice.

Getting the words right is only one part of effective communication. Your tone of voice and the way you choose to emphasise certain words matter too. After all, 'it's not what you say, it's the way you say it'. This phrase has been in use for a long time, but it still holds true.

Being professional on the telephone

Why is dealing with customers on the telephone different from dealing with them face to face? The most obvious answer is that you cannot see each other. So you are both unable to observe the other person's body language. When conducting diagnostic consultations, gestures can be very important when trying to impart an explanation on both sides. We will all use our hands to help emphasise what we are trying to describe and with no visual support, it is possible to misunderstand what is being asked or said. The process of conducting consultations over the telephone will need a slightly different approach than those conducted face to face.

Because customers cannot see what you're doing, never say, 'Hold on please' and leave the customer waiting. Explain exactly what you're going to do and how long it will be. You need to keep them informed of the actions you are taking.

Secondly, there is such an emphasis in today's world of answering the telephone speedily, so when it rings you might be tempted to grab each call quickly. If you do this, you might speak far too quickly. Before picking up the phone, first take a deep breath and then answer the phone calmly, using the agreed company greeting. Put a smile in your voice and be polite and courteous. Remember to slow down and ensure that all of the necessary information is conveyed, and ask them for confirmation that they have understood what you have said.

Thirdly, the customer will feel that time spent waiting (for example, for the call to be answered or for you to find out information) is time spent doing nothing. What might just be a few seconds to you may feel like minutes to a customer who is literally hanging on the other end of the line. Customers get impatient and frustrated more quickly than they would do in a face-to-face situation. If finding information is going to take a while, note down the customer's number and say you will ring them back. Make sure you ring them back when you said you would with the required information.

Being professional using the written word

Effective communication takes place when information is fully understood after it has been passed on and received. With the written word (letters, memos, emails etc.), the customer cannot ask you questions on the spot as you will not be there. It is therefore very important to use words that your customer will understand and to get it right first time. Otherwise, costly errors could be made, your organisation's reputation will suffer and you will waste your customer's time.

Automotive Master Technician

When using the written word:

- Think carefully what you need to say before you write.
- Answer all questions and comments fully.
- Be concise and keep to the point.
- Do not waffle.

- Summarise the key points and any actions to be taken.
- Avoid jargon and specialist technical terms.
- Check your communication for spelling and grammar.

Obtaining and providing relevant information to customers

The exchange of information between a garage and its customers is always necessary in order for a business to function correctly. The more information that is available, the more effective and efficient the customer service will be. If a good exchange of information is available, especially about vehicle symptoms, diagnostic consultations can be more thorough, (*see Chapter 1; the interrogation room*).

Customer needs

So that you can provide your customer with the type and level of service they require, you will need to gather certain information. To assess their requirements the information you gather needs to be relevant and sufficient. The types of information needed can be broken down into three main areas, as described in Table 8.4.

Table 8.4 Types of information

Type	Information
Personal	Name Address Contact details Are they the owner/driver? Payment methods/terms
Vehicle	Make Model Registration number Body style Engine size/number Vehicle identification number (VIN) Milometer reading Any pre-work damage
Work to be conducted	Service history Warranty information Symptoms Issues Faults Diagnostic data Technical data Legal requirements

Once you have gathered this information, you should be in a position to clarify both customer and vehicle needs. Always make sure that you refer to vehicle data and operating procedures when offering advice or guidance, and make sure that any information gathered during this process is kept in accordance with the Data Protection Act.

Advanced Light Vehicle Technology

Giving clear, non-technical explanations

Make sure that you provide customers with accurate, current and relevant advice or information, in a form that the customer will understand.

You may come across some situations where customers find it hard to understand what you're telling them. Vehicle operation, diagnosis and repairs are a very technical subject. You must be careful to avoid confusing your customer with technical information or **jargon**. If your customer does not understand fully what is to be done with their car, they may be cautious about having the work conducted or how much it is going to cost. Your customer might not always tell you that they do not understand as they may not want to appear foolish, so listening and watching for clues is vital.

Jargon – specialised technical terminology that others may find hard to understand.

Clues could include:

- A customer who stays quiet when you're expecting a response.
- A customer who gives a response which doesn't fit with the question you have asked.
- A customer speaking very loudly.
- A customer who fidgets, frowns and looks puzzled.

If a customer displays any of these clues, it is important to rethink your explanation and break it down into simpler terms without being **condescending**.

Giving clear explanations on the phone

You will not be able to see puzzled faces when talking to customers on the phone. So a key consideration is to try quickly to make sure that what you're saying is making sense. Listen out for verbal signs from customers that they do not understand you. Even silence may be a clue that they have not understood what you have said.

It is very important to speak clearly and slowly, without being **patronising**. Don't be in a hurry. Rushing through what you have to say can come across as quite threatening and will only make your explanations even more difficult to understand. Use silence to give customers thinking time before moving on to your next point. Ways to help customers understand what you're saying include:

- Slow down your voice.
- Keep a smile in your voice.
- Avoid using technical terms or jargon.
- Use silence to allow your customer time to understand and respond.
- Use language that is appropriate for the individual customer.
- Check with your customer that they have understood what is being said.
- Encourage the customer to ask questions and seek clarification during your conversation.

Condescending - looking down on someone in a manner that you are somehow superior.

Patronising – treating someone as if you are superior.

Automotive Master Technician

Fulfilling customer expectations

Providing a high quality service involves fulfilling (ideally, exceeding) **customer expectations** within agreed time frames. Poor customer service is often the result of employees' attitudes and behaviour. For example, staff who are helpful, are willing to take responsibility and have the right level of knowledge to help customers will be on their way to providing the correct level of customer service. As a Master Technician it will be your responsibility to deal with customers on advanced technical matters and oversee others in their dealings with customers. You, your colleagues and the systems and processes that support customer service in your organisation are what make or break the customer service experience. You have an enormous part to play in ensuring that customers are dealt with properly and to everyone's satisfaction.

Each customer will have their own ideas of what they expect from your organisation. These customer expectations are based on a number of factors including:

- Advertising and marketing.
- Word of mouth from friends or family.
- Past experiences with your organisation, products or services.

- Reputation of the organisation and its staff.
- Previous experiences with other garages.

The chart in Figure 8.7 gives examples of typical customer expectations.

Customer expectations – what the customer thinks should happen and how they think they should be treated when asking for, or receiving, customer service.

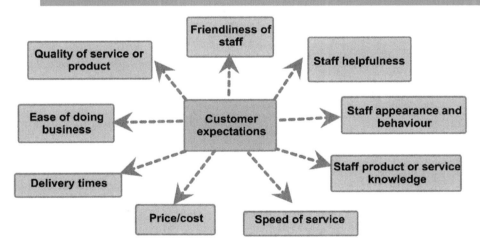

Figure 8.7 Customer expectations

If you provide a good level of customer service and fulfil their expectations, they will go away happy.
If a customer receives poor service, they tend not only to go away unhappy, but will often tell others of their poor experience with your company. It can take a long time to build a good reputation and only moments to destroy it.

Advanced Light Vehicle Technology

Offering to rescan a car 'free of charge' following a repair is not only good customer service, but also helps to reduce 'come-backs'.

Customer reactions

An automotive location can be a daunting or confusing environment for many customers. If consultations are not handled in an appropriate manner, with good customer service/care skills, then reactions may lead to exaggerated emotions including:

- Anger
- Apprehension
- Confusion
- Frustration
- Nervousness

To help you deal with these customer reactions effectively, make sure you:

- Recognise the importance of arguments and viewpoints of others.
- Respond appropriately and effectively, (without becoming defensive or apportioning blame).
- Distinguish facts from non-facts.
- Provide factual information.
- Avoid confrontations.
- Be empathetic.
- Above all, be polite.

Some examples of methods that will help you fulfil or exceed customer service expectations are shown in Table 8.5.

Table 8.5 Methods of fulfilling customer expectations

Expectation	Methods that can be used to fulfil expectation
Friendliness of staff	Greeting customers: Ensure that all staff are trained in the companies approved methods for greeting customers when they enter the garage. Answer your phone promptly and courteously: Make sure that someone is available to answer the phone when a customer calls your business.
Delivery times	Don't make promises unless you will keep them: Reliability is one of the keys to any good relationship, and good customer service is no exception. If you say, 'Your car will be ready on Tuesday', make sure it is ready on Tuesday. Otherwise, don't say it. The same rule applies to appointments, deadlines, etc. Think before you give any promise – because nothing annoys customers more than a broken one.
Ease of doing business	Listen to your customers: Is there anything more frustrating than telling someone what you want or what your problem is and then discovering that that person hasn't been paying attention and needs to have it explained again? Don't use sales pitches, product babble or technical jargon: Let your customer talk and show him or her that you are listening by making the appropriate responses, such as suggesting how to solve the problem.

Automotive Master Technician

Table 8.5 Methods of fulfilling customer expectations

Expectation	Methods that can be used to fulfil expectation
Deal with complaints	Listen attentively to the complaint: No one likes hearing complaints, and many people develop a reflex response to complaints, saying, 'You can't please all the people all the time'. Maybe not, but if you give the complaint your attention, you may be able to please this one person this one time – and your business will then reap the benefits of good customer service.
Staff helpfulness	Be helpful, even if there's no immediate profit in it: Occasionally it's good to give something away for free (especially if its second hand). Garages have often helped a customer out by fitting a small item that gets them back on the road – and charged them nothing! The result is that the customer comes back when they next need a service or repair, and they will tell all their friends and family about their good experience.
Staff product or service knowledge	Staff should be trained to be always helpful, courteous, and knowledgeable. This could be done 'in house' or by professional trainers, *see Chapters 6 and 9*. Talk to your colleagues about good customer service and what it is (and isn't) regularly. Most importantly, every member of staff should have enough information and power to make those small customer-pleasing decisions, so they never have to say, 'I don't know, but so-and-so will be back at ...'
Staff behaviour	Take the extra step: For example, if someone walks into your workshop and asks you to help them find the reception or their car, don't just say, 'It's round the front'. Lead the customer to it. They may not say so to you, but people notice when people make an extra effort and will tell other people.
Quality of service price/cost	Throw in something extra: Whether it's a coupon for a future discount, additional information on how to get the most from their car, or a genuine smile, people love to get more than they thought they were getting. And don't think that a gesture has to be large to be effective.

Dealing with complaints

Your organisation may have a process in place for dealing with customer complaints. For example: when to say 'sorry', when to give refunds, what information needs to be recorded and how to reach a satisfactory conclusion. Depending on your own organisation's practices, you may have full authority to give a refund or a sum of money as a gesture of **goodwill**. Alternatively, you may have to refer to someone else for permission to do so.

Many organisations actively encourage people to contact them with complaints and comments. This is because a complaint is a form of feedback; it tells an organisation about specific problems which need attention. Once the organisation knows that something has gone wrong, it can take action to protect its relationship with the customer. A complaint could be about a member of staff, a faulty product or poor communication – literally anything which has caused the customer to feel dissatisfied.

Complaints procedure

Many organisations have a complaints policy in place which describes in detail what a customer needs to do in order to make a complaint. This shows a willingness to help and to use the information to try and improve customer service in the future. For example, an organisation may describe the procedure in its literature or on a 'how to' page on their website.

Typically the complaints procedure will include answers to frequently asked questions such as:

*   How does the complaints process work?
*   Who will reply and when?
*   What if I have not had a reply?
*   What if I don't like the reply and what happens next?
*   Can I complain to an outside body?

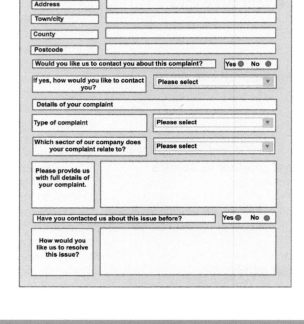

Figure 8.8 An online complaints form

Goodwill – the positive reputation of a business that is not linked to income or money.

Feedback – Information given about the performance or quality of service of an individual or organisation.

Clearly, you need to know how your company's complaints procedures operate, because if a complaint is handled badly, a difficult situation will quickly escalate into something much worse. Make sure you know how to record details of any complaint made so that it can be followed up appropriately.

If there is no complaints procedure where you work, the flow chart in Figure 8.9 may help you to understand what role you need to take. You may need to seek guidance from an appropriate person to help resolve some of the questions.

The company's systems and procedures are there to protect you, your customer and organisation. Try not to think of any system that helps you sort out problems as 'something else to worry about'. It is there to help you and provide you with a framework dealing with customer problems.

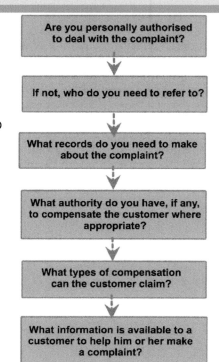

Figure 8.9 Your role in the complaints procedure

Automotive Master Technician

Obtaining customer feedback

Your organisation may have a method or system in place for measuring the effectiveness of how you deal with customers. These methods and systems will vary but they will involve obtaining **feedback** and analysing and using the feedback.

GDS- top-class repairs, first class service

How happy were you with your car service? Tick excellent, good, fair, poor for each statement.

	Excellent	Good	Fair	Poor
The speed with which you attended to when booking the appointment	☐	☐	☐	☐
The explanation of the work to be done	☐	☐	☐	☐
The availability of parts and accessories	☐	☐	☐	☐
How well you were kept informed of any changes to work agreed	☐	☐	☐	☐
The wait time when collecting your car	☐	☐	☐	☐
Overall value for money for the service or repair	☐	☐	☐	☐
How valued and respected you felt when dealing with our staff	☐	☐	☐	☐

Figure 8.10 A feedback form

Feedback is information given about things that you or your organisation do. Sometimes customers will give it without prompting, for example sending a thank you letter or a letter of complaint. On other occasions, customers may give feedback as a result of a request from you or your organisation. You may also receive feedback from colleagues who have observed your work or from a line manager or supervisor as part of the organisation's performance and appraisal system.

Every organisation will have its own way of obtaining feedback, depending on:

• The size of the organisation.
• How sophisticated the organisation's systems and procedures are.
• Whether funds are available to undertake research.
• Whether the organisation wants to listen to customers.
• The organisation's willingness and ability to put any changes necessary (as a result of feedback) into effect.

While more and more people are complaining or mentioning when they would like something to be done differently, there is still a huge silent majority who do not give any feedback. When the customer does not complain, it is very easy to assume that everything is fine and you are doing everything right. The organisation might believe its products and services are exactly what the customers want, but this might not be the case at all. If your organisation does not actively seek feedback, it runs the risk of making assumptions on behalf of its customers. Instead of giving feedback, customers might simply be walking away and finding what they want elsewhere.

Remember: it takes a long time to build a good reputation and a very short time to lose it.
There are many ways you and your organisation can set about obtaining feedback, including:

- Questionnaires.
- Direct mailings.
- Telephone surveys.
- Comments/suggestion boxes.

Company products and services

No matter what size of company or organisation you work for, there will be standards in place that specify how work is conducted and how products are sold. As a Master Technician, you may be involved in the setting of these standards or ensuring that they are met. These standards can be broken down into three different categories as shown in Table 8.6.

Advanced Light Vehicle Technology

Table 8.6 Service standard categories

Service standards	Example
National: standards set by nationally recognised organisations that seek to assure a high quality of service provided by a garage.	Kitemark for Garage Services scheme: The Kitemark scheme for Garage Services is a voluntary, **third party** scheme that ensures that the standards of PAS 80 are met and maintained. The PAS 80 specification has been developed by the British Standards Institution (BSI) in conjunction with respected members of the automotive industry. It defines standards for customer service, ensures that technical and service standards are maintained and provides a quality framework for garages, including fast-fit outlets, that service and repair automotive vehicles. The specification includes a framework of performance measures that focus on aspects of service quality which have been identified as important to customers. Holding a garage services Kitemark licence shows your customers and competitors that you are serious about delivering a quality service. The reputation of the Kitemark ensures that your customers come back – reassured by your Kitemark status that you are committed to customer service, fair trading and safety. The Kitemark scheme for garage services covers the critical elements in delivering a quality service, including: • customer service • customer satisfaction • customer facilities • staff competencies • technical inspection
Manufacturer: standards set by motor vehicle manufactures to ensure that work and products are of a quality expected by the original equipment manufacturer.	If you work for a dealership, selling and repairing just one make of car, many manufacturers place requirements and stipulations on the garage to ensure that a certain level of service and corporate image is maintained. If you work for an independent garage, many manufacturers require that you use approved repair principles and parts that are of the same quality as those of the original equipment manufacturer (OEM). The use of inferior parts or working practices can jeopardise warranties and guarantees and may also make the vehicle unsafe. (*See Chapter 7, Block Exemption*).
Organisational: standards set by an individual garage to show the level of service expected from their staff.	Many organisations have a set of guidelines for working practices within the garage. These guidelines are designed to promote a professional approach to the services provided. They are focused mainly on attracting and maintaining a solid customer base. These could include: • A dress code and corporate image • Customer care standards • The level of training or qualification required to conduct certain repairs • Quality control procedures

Third party - Someone other than the principals directly involved in a transaction or agreement.

Automotive Master Technician

The British Standards Institution (BSI) is the national standards body of the UK, with a globally recognised reputation for independence, integrity and innovation in the production of standards that promote best practice. It develops and sells standards and in many different fields of business and industry.

In order for you to provide a good customer service experience, you need to be fully aware of the range and type of services offered by your organisation. Table 8.7 gives typical examples of services offered by many garages.

Table 8.7 Types of services offered by garages

Service type	Description
Servicing 	Servicing includes inspection and general maintenance designed to ensure that a customer's car is kept operating within legal requirements and with optimum performance and reliability.
Repair 	Repair work normally includes the rectification of mechanical faults. These faults could have occurred in the form of a breakdown, or maybe as the result of wear and tear.
Warranty	Warranty work is normally conducted when a component has prematurely failed within a manufacturer's guarantee period. The cost of repair for both parts and labour is normally covered by the vehicle manufacturer, as long as the car has been maintained in accordance with the manufacturer's instructions. Warranty work is normally conducted by a garage at a reduced labour rate.
MOT testing 	MOT testing is an annual inspection of a vehicle to ensure that testable items meet a minimum required legal standard. These standards are developed and regulated by the Driver and Vehicle Standards Agency (DVSA). The MOT test is a visual inspection of safety critical items, but should not be considered a certificate of roadworthiness.

Table 8.7 Types of services offered by garages

Service type	Description
Fitment of accessories/ enhancements 	Many garages will undertake the fitment of additional vehicle equipment and enhancements on behalf of a customer. Examples could include: • in-car entertainment • body styling • telecommunication • interior trim
Diagnostic 	Most garages offer a diagnostic service for advanced vehicle systems. This normally requires a large amount of investment on behalf of the garage in the way of equipment and knowledge. This specialist area is becoming a necessity for many garages in order to conduct normal maintenance and repair.

Resolving customer problems

When a customer comes to a garage to book in their vehicle, there are two main types of work that could be done. These are **proactive** and **reactive**.

- Proactive work normally involves scheduled maintenance, including servicing and annual MOT testing.
- Reactive maintenance is normally the result of a breakdown or mechanical failure.

Proactive – taking action before something has happened.

Reactive – taking action as a result of a problem.

Adapting effective technical language

When conducting technical consultations with customers, there are a number of important stages which will ensure that a diagnosis, repair and description of the task completed, work in a mutually beneficial manner. The chart shown in Figure 8.11 describes how to conduct technical consultations with customers.

Automotive Master Technician

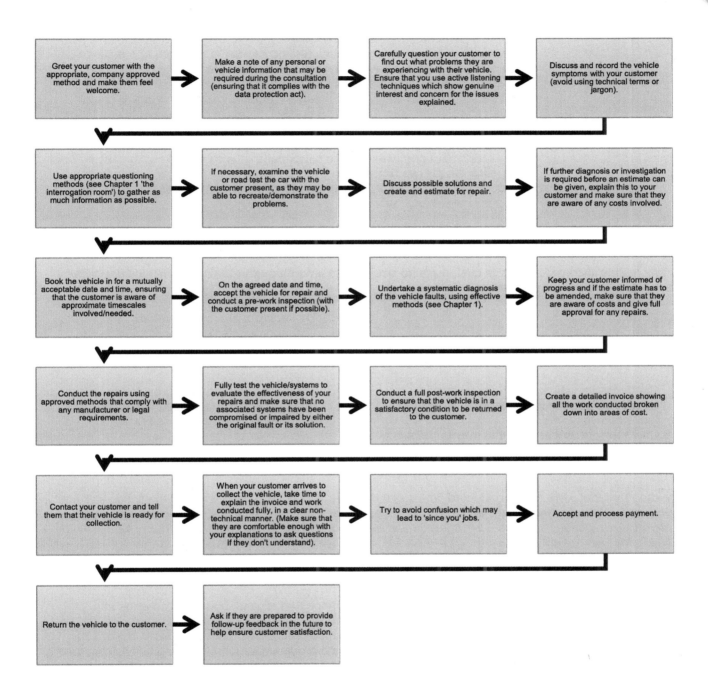

Figure 8.11 Conducting technical consultations with customers

Advanced Light Vehicle Technology

Notes

A common issue facing garages following repairs to customer's vehicles is the 'since you' job. This often arises because a customer is unaware of the scope or range of the repairs that have taken place. For example 'since you fixed my windscreen wipers, my exhaust has fallen off'.

Some systems are so completely unrelated, that the misunderstanding can be easily clarified, but others require careful description in order to justify any further repair work or expense.

To help prevent this kind of issue, ensure that you fully explain the repair or invoice to your customer upon collection, in a manner that is clear and non-technical. Try to avoid using the phrase or terminology that the car is 'all fixed', as you may be describing the actual repair, but the customer hears or assumes that the entire vehicle is now without fault.

Questioning techniques

To help any technical consultation with customers go smoothly, appropriate questioning techniques should be employed. Make sure that any questions asked are:

- Polite.
- Clear and concise.
- Use appropriate tone of voice.
- Use appropriate pace of communication.

When questioning customers on diagnostic matters, make sure you:

- Clarify the vehicle faults and symptoms.
- Clarify any sequence of events leading up to fault and symptoms.
- Clarify any recent repairs that may be relevant.
- Evaluate all information provided by customer.
- Identify possible connection between information given and possible faults.

Agreeing and undertaking work for customers

It is very important that work to be conducted on a customer's car is agreed in advance, and the scope and cost is fully understood. When agreeing work with a customer you are entering into a service contract which could be written or verbal. Any changes to the work agreed must be authorised by the customer before any further action is taken.

Extent and nature of the work to be undertaken

Garages often have a set pricing scheme for scheduled maintenance and repairs, and the processes and timescales can normally be accurately predicted. Take care when you are describing service schedules and costs to a customer, as you must take into account any extra maintenance procedures that occur within a particular service interval due to time and mileage. These extra maintenance procedures, such as timing belt replacement, brake fluid change, gearbox oil change, air bag replacement, etc., are normally charged as extra costs.

It can be more difficult to predict costs and timescales for reactive maintenance and repairs. Because of the nature of breakdowns and mechanical failures, many tasks will require some form of diagnosis before an estimate can be given. If diagnosis is required, many garages operate a policy of conducting some investigation work for a set price so that an estimate can be created to give the customer an idea of the costs and time involved. Some garages have a diagnostic labour rate that is different from the standard charge.

The difference between an estimate and a quote

When pricing a job for a customer, make sure you are clear whether you are giving the cost of repairs as a quotation or an estimate.

Automotive Master Technician

- A quotation is normally a fixed price which can be legally binding.
- An estimate is normally an approximate price that may be subject to change if the situation develops further during the repair process.

It is vitally important to keep the customer informed of progress and costs throughout the entire diagnostic and repair procedures, so that they are able to give their authorisation for work to be conducted.

The terms and conditions of acceptance

When accepting a vehicle for repair, the garage is entering into a contract with the customer. This contract may be verbal and informal or it may be formalised and the customer will have to sign to say that they agree that the work can be undertaken and that they will pay upon completion. At this point, the terms and conditions of payment are sometimes agreed, including the methods by which the customer may pay. This is often known as the **customer service agreement** and, no matter whether it is formal or informal, it will usually be legally binding. As work is conducted, it is common for situations to develop or change. Any changes will affect the contract that has been agreed with your customer. The contract will need to be updated so that the customer is aware of any changes to timescales and costs.

Customer service agreement – a contract between the organisation and the customer that agrees the level of service to be provided and what work will be undertaken.

Factors affecting timescales and costs

A number of factors impact on the timescales and costs involved with the repair of the customer's car. Some of these will be fixed and you can explain them to the customer at the time when you estimate and book in a job. Other factors will change as work is being conducted and the customer will need to be kept informed. Examples are shown in Table 8.8.

Table 8.8 Factors affecting timescales and costs

Factor	Reason and effect
Availability of equipment	During the diagnosis of a customer's car, you may discover that a specialist piece of equipment is required to conduct the repair. This piece of specialist equipment may need to be borrowed or hired from the manufacturer, adding extra cost and time to the job.
Availability of technicians	Staffing levels within a workshop will have an effect on how and when a repair can be conducted. Many workshops have technicians who specialise in certain types of diagnosis and repair. Depending on the task, there may only be one or two technicians with the skills required to conduct the work needed. Technicians may be on holiday or on sick leave and this will affect how long the job may take.
Workshop loading systems	Many garage workshops operate a booking system in order to make sure that all of the technicians, equipment and space are utilised in an efficient manner. This is known as a workshop loading system. If an unexpected problem occurs during the repair of a customer's car, this may mean that a technician or piece of equipment becomes unavailable. This will affect the workshop loading plan and the timescale of the job will also be affected.

How to access costing and work completion time information

No matter what job the customer books their car into your garage for, they will want to know roughly how much it will cost to repair. There are often set repair times and costs for scheduled maintenance, but other types of maintenance and repair will need to be estimated. Manufacturers supply given times for the inspection, replacement and repair of most vehicle components. You can access these repair times using manuals or computer-based equipment. Repair times are normally shown to two decimal places, with 0.1 of an hour equalling six minutes. The image in Figure 8.12 shows some example repair times.

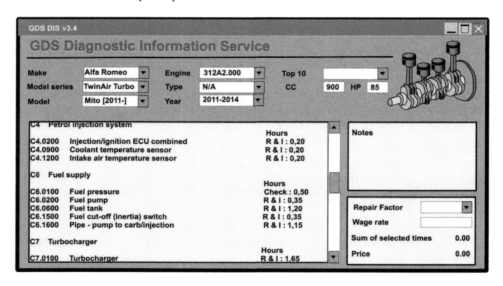

Figure 8.12 Electronic technical information

Vehicle information systems, servicing and repair requirements

During servicing and repair you will need to have access to information. Table 8.9 gives some examples of information that you may need and where you can find it.

Table 8.9 Information needed during service and repair

Information type	Information source
Technical data, including diagnostics	Vehicle identification numbers (VIN) Technical data manuals Workshop manuals Wiring diagrams Safety recall sheets Technical help lines Advice from other technicians/colleagues Parts suppliers/catalogues Diagnostic trouble codes Oscilloscope waveforms
Servicing to manufacturer requirements/standards	Manufacturer's literature Manufacturer's technical help line
Repair/operating procedures	Vehicle handbook Workshop manuals Manufacturer's technical help line

Table 8.9 Information needed during service and repair

Information type	Information source
MOT standards/requirements	The Driver and Vehicle Standards Agency (DVSA) provides a range of licensing, testing and enforcement services with the aim of improving the roadworthiness standards of vehicles. DVSA oversees the MOT scheme for quality and standards
Quality controls – interim and final	Company policy and standards Service or maintenance schedules Compliance with BSI Kitemark standards Involvement with Automotive Technician Accreditation (ATA)
Requirements for cleanliness of vehicle on return to customer	Company policy and standards Post-work inspection sheets
Handover procedures	Company policy and standards Service history and handbook

Pre-work and post-work check sheets

It is good practice to perform pre-checks before you carry out any service, maintenance or repair work. These checks usually include a visual inspection of the vehicle's exterior and interior for any damage or cosmetic faults. Many garages use pre-check sheets to identify and record any damage. By completing this inspection before you begin work, you can get the customer to sign the check sheet and confirm the condition of the vehicle prior to the repair. This will help to avoid any potential conflict after the repair if the customer claims they didn't know about any existing damage. Ideally, the customer should be present while the check is completed, so that any queries can be cleared up immediately. A typical process involved in carrying out pre-work checks is shown in Figure 8.13.

To meet the customer's expectations, you should carry out all service and repairs so that they conform to a high standard. You also need to return the vehicle in a clean and acceptable condition following any work carried out. You will need to complete a post-work check sheet by referring to the notes written on the original pre-work check sheet and the road test checklist. This will make sure that you haven't missed anything and that you send the vehicle back to the customer in a roadworthy condition.

Consumer legislation

Your organisation will have rules in place in order to comply with legislation. Some legislation applies to all workplaces, for example laws relating to data protection and health and safety. Some rules are specifically designed to protect consumers, for example laws relating to the description and quality of goods that are sold.

Knowing about the legislation will help you understand the reason for the rules that you have to follow in your workplace.

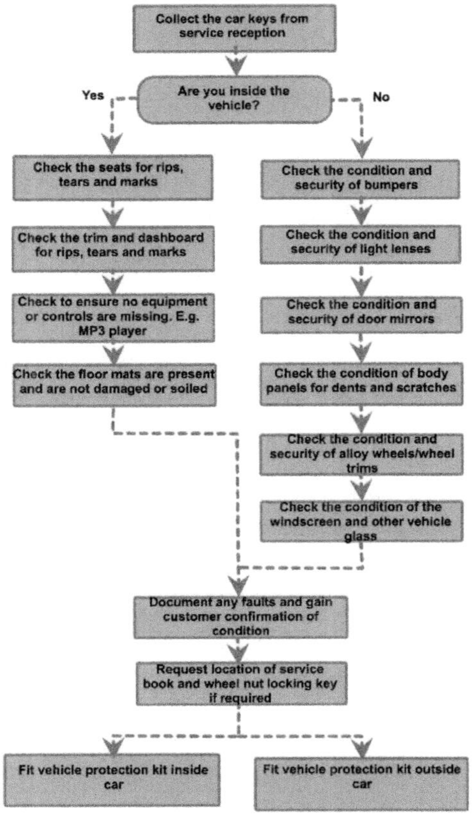

Figure 8.13 Pre-work flow chart

Advanced Light Vehicle Technology

Trade Descriptions Act 1968

This act states that traders must not describe something falsely that is on sale and must not make false claims about services or facilities. It is a criminal offence to falsely describe something on sale. This applies to any description a garage might make, such as an advertisement in your service reception, or a verbal description given by one of the members of staff.

You need to be careful when you describe the services that your garage provides because if the description is reckless as well as false, you are breaking the Trade Descriptions Act and can be prosecuted under criminal consumer law.

A customer has three years in which to take any legal action under this act.

Consumer Protection Act 1987

This act states that only safe goods should be put on sale. Under the 28th rule, it also prohibits misleading price indications. When displaying goods or that are on sale, both the previous price and the sale price of the goods must be stated. The goods must have been on sale at the previous price for at least 28 consecutive days in the last six months at the same branch.

The Sale of Goods Act 1979 (as amended)

The Sale of Goods Act lays down several conditions that all goods sold by a trader must meet. The goods must be of a **merchantable** and of a satisfactory quality. They must be **as described** and **fit for purpose**.

To be of a satisfactory quality, goods must have nothing wrong with them (unless any defect was pointed out at the time of sale) and should last a reasonable time. This means that the act does not give a customer any rights if the fault was obvious or pointed out when the customer bought the product. The act covers the appearance and finish of the goods, as well as their safety and their durability.

Merchantable – suitable for purchase or sale.

As described – this refers to any advertisement or verbal description made by the trader. For example, if an engine oil is described as fully synthetic then it must be so.

Fit for purpose – good enough to do the job it was designed to do. This covers not only the obvious use or purpose of an item, but also anything you say the item will do when trying to sell the product.

If any product bought by the customer does not meet any of the conditions set out in the Sale of Goods Act, the customer is entitled to a full refund. You cannot expect them to accept a repair, replacement or credit note if they don't want to. The customer does not have any rights to an automatic refund if he or she has had a change of mind, made a mistake and bought the wrong product or been told about the fault before the purchase was made.

Supply of Goods and Services Act 1982

This law covers work done and the products supplied by tradesmen and professionals. The supply of goods and services act contains the following points:

- A tradesman or professional has a duty of care towards the customer and his or her property.

Automotive Master Technician

- Any price stated or agreed with the customer must be honoured. Where you and your customer have not agreed a price, the customer does not have to accept an outrageous bill. All the customer has to pay is what he or she considers is reasonable. A reasonable charge is considered to be what other similar garages in the same geographical area would make for the same job.
- The work must be done to a reasonable standard and at a reasonable cost (if not otherwise agreed in advance).

The Data Protection Act 1998

Most garages store information about their customers, either in written or electronic formats. Types of information stored can include:

- Names
- Addresses
- Contact details
- Vehicle details
- Service history
- Financial and payment history

This information is known as data, and there are two types of personal data. Straightforward personal data includes information such as:

- Names
- Addresses
- Banking details

Sensitive personal data includes information such as:

- Racial or ethnic origin
- Political opinions
- Religion
- Membership of a trade union
- Health
- Criminal record

People want to know that information about them is being kept private and that it will be used appropriately. Customers may have concerns about the information being misused or falling into the wrong hands. They might have concerns about:

- Who can access the information?
- How accurate is it?
- Is it being copied?
- Is it being stored without their permission?

The Data Protection Act governs the use of personal data in the United Kingdom. The act does not stop organisations storing information about people; it just ensures that they follow rules. Some of the rules include:

- Data should only be used for the reasons given and not disclosed to unauthorised people.
- Data cannot be sold or given away without authority to do so.
- You must only hold enough detail to allow you to do the job for which the data is intended.
- Make sure the information you request from the customer is needed for a genuine reason.
- You must keep the data away from people who are not authorised to access it. Leaving information out on a desk or not password protected may mean it could be misused.
- Data should not be kept for longer than is necessary, although the act does not state how long that should be.

Customers' rights

Under the Data Protection Act 1998, people have rights concerning the data kept about them, which include:

- Right of subject access: Anyone can request to see the personal data held about him or her.
- Right of correction: You are obliged to correct any mistakes in data held once these have been pointed out.

- Right to prevent distress: This right prevents the use of information if it will be likely to cause a person distress.
- Right to prevent direct marketing: People can prevent their data being used in attempts to sell them things (for example, by direct mail or cold calling).
- Right to prevent automatic decisions: People can specify that they do not want a data user to make automated decisions about them (for example: computerised credit scoring of a loan application).
- Right of complaint to the Information Commissioner: People can ask for the use of their personal data to be reviewed by the Information Commissioner, who can enforce a ruling using the act. The commissioner may inspect the garage's computer to help the investigation.
- Right to compensation: People can use the law to get compensation for damage caused if personal data about them is inaccurate, lost or disclosed.

Health and safety

In your working environment there may well be a number of potential hazards to yourself, your customers and your colleagues. Some examples of hazards are:

- Chemical substances.
- Dust and fumes.
- Excessive noise.
- Moving vehicles.
- Moving parts in machinery.

- Electricity.
- Extremes of heat and cold.
- Uneven floors.
- Exposed wiring.
- Cabling over the floor or ground.

The Health and Safety at Work Act 1974 (HASAWA) covers the responsibilities that employers have to their employees and also to customers who are on their premises. An employer needs to take steps as far as is **reasonably practicable** to put in place measures to control health and safety risks. (For more information on the Health and Safety at Work Act, *see Chapter 1*).

Figure 8.14 Health and safety responsibilities

Reasonably practicable – can be carried out without incurring excessive effort or expense.

Automotive Master Technician

Discrimination

It is important to treat people fairly and equally regardless of who they are, where they live or how much you like or dislike them. In other words, while you should treat people as individuals, what makes one person different from another does not mean he or she should have any advantage or disadvantage over anybody else in relation to the customer service delivery. All your customers should be treated fairly, regardless of their age, gender, race, sexual orientation, disability, gender reassignment, religion or beliefs.

The Equality Act 2010

The Equality Act 2010 came into force on 1st October 2010 and replaces the Disability Discrimination Act 1995 (DDA). It is one of the most important pieces of legislation you need to know, regarding disability discrimination and equal rights. People who access your goods, facilities or services are protected from discrimination because of certain 'protected characteristics'. These are:

- Age.
- Disability.
- Gender reassignment.
- Marital or civil partnership status, pregnancy and maternity.

- Race – this includes ethnic or national origins, colour and nationality.
- Religion or belief.
- Sex.
- Sexual orientation.

Definition of disability

Disability has a broad meaning. It is defined as a physical or mental **impairment** that has a **substantial** and long-term adverse effect on the ability to carry out normal day-to-day activities.

Substantial - more than minor or trivial.

Impairment - covers long-term medical conditions such as asthma or diabetes and fluctuating or progressive conditions such as rheumatoid arthritis and motor neurone disease. Mental impairment includes learning difficulties such as dyslexia, Down's syndrome and autism.

Some people are automatically protected as disabled people by the act, including those with cancer, multiple sclerosis and HIV/AIDS.

Helping disabled people

Disabled people must not be treated less favourably than others because of their disability. Businesses have an obligation to make reasonable adjustments to help disabled people access their goods, facilities and services. Some organisations build ramps to ensure that wheelchair users can access their premises or provide information in Braille for visually impaired customers.

What is a reasonable adjustment? This will depend on a number of circumstances, including cost. The Equality Act 2010 requires that service providers must think ahead and take steps to put right things that may stop disabled people using their goods, facilities or services. Your organisation should not wait until a disabled person experiences difficulties using a service, as it may then be too late to make the necessary adjustment. This could mean making:

- Reasonable changes in the way things are done: for example, changing practices, policies or procedures which could put disabled people at a substantial disadvantage', such as amending a 'no dogs' policy.

Advanced Light Vehicle Technology

- Reasonable changes to premises: for example, changing the structure of the building to improve access, such as fitting handrails alongside steps.
- Providing auxiliary aids and services: for example, providing information in large print or an induction loop for customers with hearing aids.

To find out more information about the Equality Act 2010, visit the Home Office Equalities website and the Directgov website.
http://www.direct.gov.uk/en/DisabledPeople/RightsAndObligations/DisabilityRights/DG_4001068

The Equalities and Human Rights Commission (EHRC) is the statutory body responsible for protecting, enforcing and promoting equality. We recommend that you visit their website for guidance and information, including equality matters within different industry sectors. You can also use this website to keep yourself up to date with changes to equal opportunities legislation.

Automotive Master Technician

Recording information and making suitable recommendations

At all stages of your diagnostic consultation with customers on technical matters, you should record information and where appropriate, make suitable recommendations. Some of this recorded information will help provide a smooth process to explain any work to be conducted and help to create estimates. It will also help you to comply with any organisational or legal requirements. The Table below shows some examples of how to do this.

Table 8.10 Recording information and making suitable recommendations

Stage	Information	Records/Recommendations
Before you start.	Customer and vehicle details.	Enough customer and vehicle information to enable you to conduct the work for which the vehicle has been booked in (held in accordance with the Data Protection Act 1998)
	Customer needs and symptoms caused by vehicle faults.	Information about why the vehicle has been brought to the garage to enable you to produce an estimate for the cost of repair.
	Warranty/service repair history.	Any service information that might have an impact on the technical diagnosis or any possible warranty claim.
During the diagnosis and repair.	Record of work undertaken, parts and processes used.	A complete record of any parts and diagnostic/repair processes which can be used to update any estimate or customer service agreement, so that the customer is kept fully informed of costs and timescales.
When the diagnosis and repair is complete.	Invoicing.	Complete a fully detailed invoice that can be used to explain the diagnosis and repair in a non-technical manner. Inform the customer if the vehicle will need to be returned for any further work. Advise the customer of any other issues noticed during the repair.
	Service and repair history.	Complete any service or repair history as required, in order to put in place any warranty or guarantees. Store any diagnostic information to help with future diagnostic strategies.

Chapter 9 Fundamental Management Principles in the Automotive Industry

This chapter will help you develop an understanding of fundamental management principles in the automotive industry. It will give an overview of effective leadership techniques that can be used when supervising colleagues and workloads within your organisation. You will be expected to work in a professional manner and motivate staff, leading by example. It will also assist you in developing an effective management style which promotes moral and a productive business.

Contents

Roles and responsibilities

One of the duties of a Master Technician, is often to perform some sort of management role within an organisation. Fundamental management skills will be required in order to undertake these responsibilities. Not only will you need the self-discipline to organise your own work, but also that of other people. It might take some time to develop these skills as not everyone is a born leader, but with careful planning and practice you will eventually become a successful manager.

Motivation

Motivation has a key role to play in any managerial success. This is both personal motivation and the encouragement and motivation of others.
Motivation is the reason that someone has for acting in a particular way and is normally the desire or willingness for someone to do something.
Before you can motivate and inspire others, you will need to understand your own personal desires and goals. Your personal goals will have an impact on your success as a manager and if you can understand what is driving you, it will help you understand the desires and goals of your colleagues and staff which can lead to mutual job satisfaction. It can be a useful tool to create a list of your personal goals which you are able to review and revise at different times in your career in order to set targets for achievement. These could be broken down into a number of different categories including:

- Personal happiness.
- Job satisfaction.
- Financial security.
- Promotional aspirations.
- Friendship.

- Qualifications and life-long learning.
- Lifestyle.
- Professional attitude.
- Kudos.

Automotive Master Technician

A way of archiving these goals can sometimes be to create an action plan, in which you can set yourself targets. Target setting can be difficult so it is a good idea to use the SMART method.
This is where your targets are:
S - Specific
M - Measurable
A - Achievable
R - Realistic
T - Timed (given a target date to aim for)

SMART target setting will allow you to break your goals down into smaller sections, making them easier to achieve and helping you to judge your own progress. The achievement of goals will bring benefits such as, motivational and aspirational ideals.

Motivational needs of others

Once you have discovered your own goals and aspirations, you will need to understand the motivational needs of others. For many people, personal requirements fall into categories known as the hierarchy of motivational needs; a theory first proposed by Abraham Maslow in 1943 in a paper called 'A theory of human motivation'. Maslow's theory consists of two parts, the classification of human needs, and a consideration of how the classes relate to each other. The classification of needs can be summarised in a diagram as shown below in Figure 9.1.

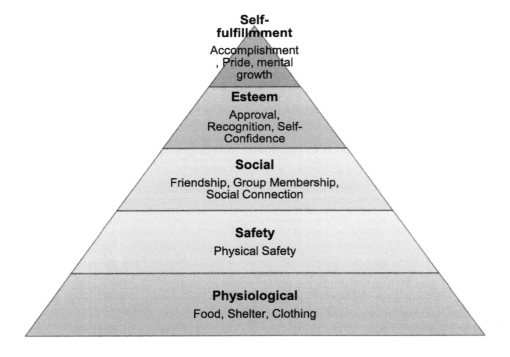

Figure 9.1 Motivational needs pyramid

How the hierarchy works

Starting at the bottom of the hierarchy pyramid, a person will initially seek to satisfy basic needs (e.g. food, shelter). Once the physical needs have been satisfied, they are no longer a motivator. You then move up to the next level. The next level indicates safety needs at work. These could include physical safety, e.g. protective clothing, as well as protection against unemployment, loss of income through sickness etc.

Advanced Light Vehicle Technology

When these have been satisfied, the next level recognises social needs, showing that most people want to belong to a group. These social needs can include the need for love and belonging, e.g. working with colleagues who support you at work, teamwork, communication and a supportive home life.

The next level up describes a person's needs about being given recognition for a job well done, helping to build esteem. This reflects the fact that many people seek the esteem and respect of others; a promotion at work might achieve this.

Self-actualisation is about how people think about themselves. This is often measured by the extent of success and/or challenge at work.

When it comes to managing others, Maslow's hierarchy of motivational needs can help you can find out which level your colleague has reached. You can then provide incentives or rewards that will help them progress personally which will also be beneficial to the organisation.

Maslow's model of motivational needs however, cannot be solely relied on. There are a number of issues that this categorisation fails to take into account:

Everybody is different and therefore individual behaviour will often respond to several needs, not just one.
The same needs, such as socialising, may cause quite different behaviour in different people.
It can often be difficult to decide when a level within the pyramid has actually been fully achieved and therefore satisfied.
The theory ignores the patient behaviour of some people, who will tolerate low-pay for the promise of future benefits for example.

Some people will require **incentives** in order for them to progress, and others will not. Whichever type of person you are dealing with, most people can be motivated by the promise of certain rewards such as bonus or promotion leading to career progression. These incentives can often play on an individuals, needs wants and desires.

Incentives - something that tends to incite an action or greater effort as a reward offered for greater productivity.

Personal development

It is important to also have your own targets and rewards in order for you to have motivation. These targets or ambitions could be personal but achieved via your work life. Most people gain promotion through a driven and focused attitude and this can be enhanced by creating your own personal development plan (PDP).

A PDP will often consist of an action plan developed by analysing your values and setting goals to gain an awareness of your own personal development within a career, education, self-improvement or relationship context. Once developed, a personal development plan can help you reflect on your progress and realise your goals through accomplishment. An example of a personal development plan is shown in the next image.

Automotive Master Technician

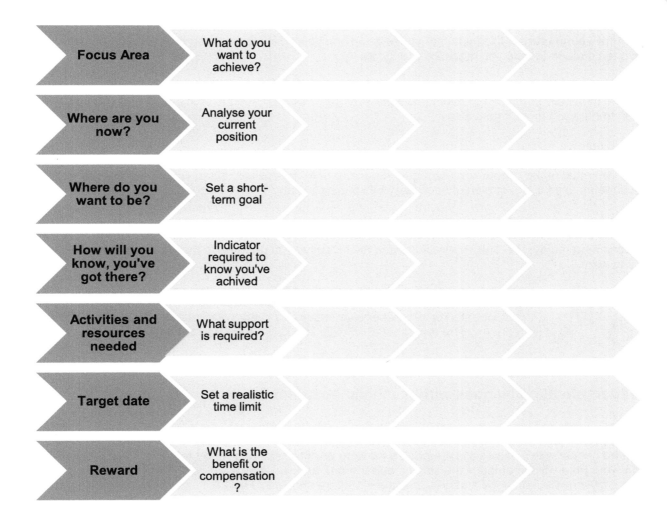

Figure 9.2 Personal development plan

Figure 9.2 shows an example of a PDP and the explanation below will help you understand how to use it and complete the Table.

Step 1 Focus Area
Choose 1 to 3 areas of your personal development to focus on. Too many and you'll never get anywhere, too few and your thought process may be too narrow.

Step 2 Where are you now?
You need to know where you're starting from so you can measure your progress as you work through your personal development plan. The better understanding of your starting point that you have, the easier it will be to focus on your achievements.

Step 3 Where do you want to be?
If by the end of the day you could have resolved certain development problems, what would that be like?
What would you feel?
What would you see?
By knowing what you want to achieve you can increase your chances of achieving it.

Step 4 How will you know you've got there?

What **quantitative measure** will tell you that you're achieving your goal? Even if you give yourself a simple 1-5 score it is often enough to help you measure your goals.

Step 5 Activities and resources needed
What practical steps are you going to take to achieve your goal? Do you need to study, do research or gain a qualification to help you to your next step?

Step 6 Target date?
It is important to plan your time carefully for specific activities, like you would a regular appointment. Take small steps and if you must make changes, re-schedule for a later date. Be realistic about your commitments and make some adjustments to your life, or plan, if necessary by keeping a diary.

Step 7 Reward
All work and no play, makes Jack a dull boy. Give yourself rewards as you achieve your mini-goals to help make the whole process more enjoyable. This may just provide the incentive you need to do the work that will make the big changes you want.

Quantitative measure - data that is used to describe a type of information that can be counted or expressed numerically. This type of data can be statistically analysed and may be represented visually in graphs, histograms, tables and charts.

Team leadership of colleagues within vehicle operations

If you work for a medium to large organisation, it is common for the company to arrange its staff into different departments, *see Chapter 7*. The reason for this is that many businesses are arranged by a matrix management system, where people with similar skills are pooled for work assignments and report to a line manager. As a Master Technician, your area will encompass the service department, and you may even be the line manager. This type of management style has advantages and disadvantages and are shown in Table 9.1.

Table 9.1 Matrix management

The advantages of a matrix management style include:	The disadvantages of a matrix management style include:
Individuals can be chosen according to the specific needs of the task/job.	A conflict of loyalty between different departments and a culture of 'us and them' can develop.
Department managers are directly responsible for completing work within a specific deadline and budget.	Costs can be increased if more managers (i.e. department managers) are created through the use of separate departments.

Structuring teams

Once a departmental team can be identified, a structure within that team can be defined. If done correctly this will lead to harmony and productivity. Individuals will have different personalities and skill-sets, which if correctly managed, can contribute to effective team working.
Once a good team has been established, proactive management is then often able to match work or tasks which play to an individual's strengths. This can be a productive method, leading to a very motivated and efficient team. A downside to team structuring is that if someone leaves or joins the group, the dynamics can be disrupted, manning the whole process must be reorganised.

Automotive Master Technician

No matter what the format of the team is, good communication within the department is vital so that everybody understands the reasons behind work loading, team organisation and hierarchy. To promote harmony, it may fall to you within your management role to provide discipline and structure, but also encouragement in the form of praise when appropriate.

Fundamental leadership principles

A good manager is able to provide leadership qualities which will influence, inspire and motivate others. In order to do this you will need to recognise any limitations within your own leadership style. The ability to lead effectively is based around a number of key skills. Ideally you will want colleagues and staff to follow you because of the trust and respect that you have earned, not because they have been told to.

Some people will possess natural leadership skills, while others will develop these over time by recognising the requirements of a good manager.

One of the most important skills of a leader is the ability to make decisions. To do this you will need to have a focus or target aim. Once the goal has been set, you can make the choices based on your decisions, but remember that every option will involve others and have an effect on them. Some of your decisions may occasionally conflict with your management style.

Problem solving is another key leadership skill. Problems come in all shapes and sizes, from those on a strategic level, to those about managing your team, and everything else in between. A good leader will not be afraid of problems, but instead face them head on and try to turn them into opportunities for progress. Many problems do not have to be solved alone, and with your management skills you will often be able to rely on the support of others, gaining information and providing solutions.

In order to be a good leader, you will also require comprehensive planning skills, although not everything always goes to plan. You will need to realise that the goals or outcomes you have chosen during your planning are likely to change, and adapt quickly and efficiently to new situations, balancing opportunity and risk.

Another good leadership skill is to be a strong **facilitator**. For this you will need to provide the necessary support for your team that will allow them to achieve their goals. Support can be as simple as encouragement or may require that you provide the resources needed to complete a task. To be a facilitator, you will need to identify barriers or obstacles and provide solutions.

A mistake that leaders and managers often make is to try and do everything themselves. A good leader will **delegate** responsibility to team members and then facilitate the requirements of the team to enable success. Delegation will also allow you to find out more about the strengths and weaknesses of your team, and therefore you will be in a better position to make informed decisions about their individual roles and responsibilities.

Facilitator - a person that is responsible for leading or coordinating the work of a group.

Delegate - to commit powers, functions or targets to another person.

Autocratic - being dictatorial or domineering.

Management styles

Management achieves objectives and outcomes by engaging others to find solutions.

Management styles can be categorised by varying degrees, ranging from the **autocratic**, through to a hands-off approach.

The main management styles can be split into three groups:

- Authoritarian or autocratic.
- Participative or democratic.
- Hands-off.

Management styles will inevitably vary from manager to manager, and also by the task being undertaken and by the personalities and capabilities of those being managed.

Authoritarian or autocratic management style

Although this style of management may seem very domineering, it can be a highly effective and well suited to specific situations that require absolute clarity on objectives, outcomes and methods. It can also be necessary when time is scarce or team members are averse or reluctant to participate.

Participative or democratic management style

A democratic management style involves the participation of team members in decision making. The manager acts as the facilitator, but will play the key role in shaping both the context and the final decision, as well as resolving any differences.

The main advantage of this management approach is it can make use of skills that might not be held by the manager themselves. It will also build the commitment of your team members through their involvement in the project or task.

A disadvantage to this method can be the time involved with this process, and it is therefore not suitable for all management situations, especially those that need a quick decision. The advantages of this style however, is that it will often lead to better strategies or plans, and also in the way the team works together towards mutual goals.

Hands-off management style

This approach places a great deal of authority and trust in your team members, relying on them to both understand and deliver on your set objectives. The risks involved in this management style include:
If team member's mistake or misunderstand what they are working towards, it can lead to large degree of variation in the overall outcome, and in some cases, to underperformance.

The gains involved in this management style include:
A manager who can successfully use this style will normally allow team members to get on with their work. This can lead to possible boosts in productivity, but will need to ensure the manager's door is 'always open' for consultation and discussion.

All of these management styles can be employed to maintain channels of communication and discipline, but they can also be used to motivate and inspire. You will need to plan, organise and deploy the described management strategies and styles in order to provide good leadership to your teams, achieving maximum performance and harmony.

Figure 9.3 Hands off management style

Measuring performance

In order to judge the effectiveness of your management, you will need to measure your own performance as a leader. This way you can alter or develop your management style to suit the common goals of the organisation. There are seven key factors in measuring your own performance which include the following:

1. Make sure that you are aware of the organisations goals and define them into tangible targets.
2. Define specific indicators of your success as a manager, making sure that you take into account any outside factors that are beyond your control.
3. Review your progress periodically, this could be on a monthly, biannual or yearly basis.

Automotive Master Technician

4. Evaluate your ability to balance your personal and work life. Try to keep the two separate and make sure that they don't have a detrimental effect on each other.

5. Get feedback from your colleagues as to how they feel about your management style. (This may have to come via a line manager, as sometimes your colleagues may feel uncomfortable addressing any issues directly with you).

6. Discuss your leadership qualities with your line manager to help identify strengths and weaknesses.

7. Observe your colleagues moral and staff retention rates. A company with a high staff turnover might be experiencing management inefficiencies.

Effective teamwork

It is important to understand teamwork; a successful organisation will be built on teamwork. The key objectives of teamwork are to improve performance through increased productivity and competitiveness; to improve the industry's image by changing its culture, developing people and engaging them with other individuals, businesses, organisations and industry associations.

The participation and management of teamwork is an important tool in achieving a company's objectives.

Teams are groups of people, with skills that complement each other, who are committed to a common purpose and share the responsibility for the achievement of their goals. Many teams will develop a distinct identity, particularly within different departments, working together in a co-ordinated manner which can sometimes require careful management.

A successful team is characterised by a team spirit based around trust, mutual respect, helpfulness and friendliness.

Simply putting people together as a group within a department does not necessarily mean that they will function effectively as a team or make appropriate decisions and this is where careful management is involved. Effective leadership will ensure that the team works well together and that they have any information and resources required to carry out the task.

Fundamental management that creates teamwork results from:

- A team whose involvement, size and resources match the task.
- Leadership and attention to team-building.
- Commitment by team members to understand and recognise the common goals to create a shared vision.
- A team that has common ownership of the task and joint responsibility for its achievement.
- Co-ordinated effort and planned sharing of tasks evenly across the team (but with the ability to delegate where individual skills can be utilised to produce the best results).
- An open exchange of information within the team.

Effective teamwork can be undermined by a variety of problems, including:

- Disorganisation, poor communication, misunderstandings or inadequate procedures for problem-solving.
- Team functioning can be weakened by barriers faced by individual members within the team, as well as by difficulties linked to the task.

A good and effective team leader looks to their team for answers and welcomes suggestions for alternative courses of action. He or she needs to be able to deal with conflict constructively through the processes of mediation or negotiation. There will be occasions when the leader must be prepared to take difficult decisions and be willing to explain the reasons why they have been taken.

Figure 9.4 Effective teamwork

Advanced Light Vehicle Technology

Team meetings

It can often be a good idea to hold team meetings as these provide face to face contact at regular intervals. They can be valuable for both social and business purposes. The social value of meetings should not be underestimated, trust, respect, team identity and familiarity with one another's ways of working can all be developed. Team meetings could include:

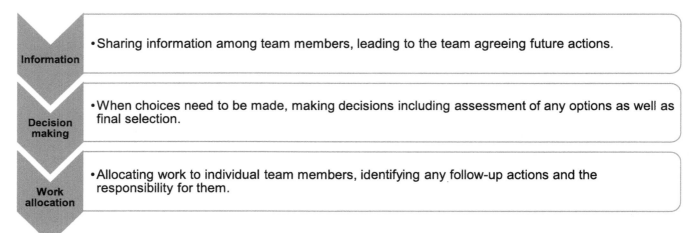

Information • Sharing information among team members, leading to the team agreeing future actions.

Decision making • When choices need to be made, making decisions including assessment of any options as well as final selection.

Work allocation • Allocating work to individual team members, identifying any follow-up actions and the responsibility for them.

Figure 9.5 Team meetings agenda

Chairing a meeting is a key skill for any manager or team leader. It will be your responsibility to make sure that there is a set agenda, that all of the items that need discussion are covered and keeping it on time so that it doesn't drift and is properly concluded. If regular meetings are held, then issues can be discussed and will often not have time to escalate; try not to fall into the trap, however, of having meetings just for the sake of it.

Teamwork models

The teamwork grid in Table 9.2 has been laid out to help teams to assess themselves for effectiveness. To use the table, first photocopy it, and choose a column. Then find the cell which will best describe your current teamwork practice, and mark it so that it can be identified. Repeat this for the rest of the columns and when complete the markings can be joined to show a profile of your effectiveness. The rows are labelled 1 to 5 to indicate the capability of your teams.

- Peaks on the grid show where your team management is well developed.
- Troughs show where there is room for improvement.
- A good balance shows that your team is well managed.

This table can then be used during team meetings to help give insights into teamwork and promote good practice. If you compare individual members' profiles with one another it can reveal areas of agreement and disagreement about how well the team is working together.
Updating profiles periodically can be used to measure changes in teamwork practices over time and by concentrating on the columns with low effectiveness, improvements can be encouraged.

Automotive Master Technician

Table 9.2 Teamwork models

Effectiveness level	Team identity	Common vision	Ability to communicate	Participation and collaboration	Issue negotiation and resolution	Reflection and self-assessment
5	The team takes ownership of any problems and accepts joint responsibility for its achievement.	The team has a shared vision and set of objectives. The objectives have been developed together and are regularly reviewed.	Team members actively share their knowledge and ideas around the whole team.	The collective expertise of the team provides full participation due to familiarity, honesty and mutual trust.	Different viewpoints are encouraged as a source of differentiation which spur on the team's creative problem-solving.	The team regularly reviews members' roles and their impact on the team and acts on the outcome.
4	The team recognises that its members have both individual and team goals. Tasks are distributed accordingly.	The team has developed for itself both a common/shared vision and clear outcomes.	Team members communicate information and knowledge freely around the team.	All team members are given opportunities to contribute, construct and expand on the suggestions of others.	Any sources of disagreement are addressed openly and are resolved directly through constructive negotiation.	The impact on the team for each member's different role and responsibility is clearly recognised and discussed.
3	The focus of the team is on the tasks that individual members need to solve.	The whole team is working to a set of common objectives.	Team members communicate information when others need it.	The team seeks ideas, proposals and solutions from all of its members.	Team members are prepared to explain their underlying suppositions and negotiate realistic options.	Different roles and responsibilities are discussed from time to time.
2	Team members only take a limited interest in issues that lie outside their own immediate area of responsibility.	Members of the team are clear about their own objectives but the team has no shared vision.	Individuals tend to be protective of their own information and share it reluctantly, only in response to specific requests.	Team members prefer to work alone and give priority to their own concerns over those of the team.	Contentious issues are avoided or ignored completely; conflict is dealt with only at surface level.	The team acknowledges its members have several roles and responsibilities but they are not reviewed.
1	Team members concern themselves only with their own responsibilities.	Team members, both individually and collectively, lack a common vision and clear objectives.	Information is passed to team members strictly on a 'need to know' basis.	A lack of trust or power struggles reduce participation and collaboration.	Conflicting opinions remain unaddressed.	The team places no value on considering and negotiating how they work together.

Effective communication within vehicle operations

You will spend a large proportion of your time communicating, either directly or indirectly. Methods of direct communication are normally verbal, and indirect tend to be written. The type of communication, the flow and the amount needs to be correct in order for it to be effective. Too little and information is lost. Too much and the entire system can grind to a halt. There is a strong relationship between good communication and effective management. The benefits of good communication include, improved team morale, elevated performance of colleagues, increased business performance and better customer satisfaction and retention.

Channels of communication

Within any organisation, no matter how large or small there will be channels of communication. These can be vertical or horizontal.
Vertical communication is upwards and downwards; i.e.
From managers to supervisors and then down to the staff.
Feedback from staff to supervisors and back to management (an organisation with good communication will always encourage feedback from the staff).

Horizontal communication is sideways; i.e.
Across the organisations departments, service, sales and parts for example.

No matter whether the communication is vertical or horizontal, it is important that it is two-way.

Formal and informal communication

Every company will have a flow of communication that is either formal or informal. The most common type is informal face-to-face and goes on constantly among colleagues and customers. The smaller the organisation, the more informal communication tends to be. Informal communication may also include, texting (SMS), post-it notes or email.

A large proportion of formal communication tends to be routine and will normally be conducted in a written format. This can be paper based, but many organisations are beginning to adopt electronic methods in preference due to their speed, flexibility and limited storage requirements. Routine communication may include examples such as:

- Job cards
- Invoices
- Requisitions
- Service sheets
- Estimates

Non-routine communication includes things that happen infrequently or are a one off occurrence. These may include examples such as:

- Memos
- Special reports
- Notices

Each method of communication will have advantages and disadvantages when it comes to communicating the type of information required and whether it is intended for individual or group communication. Examples of some types of communication are listed below and it will be up to you to choose the most appropriate form for your specific needs, both formal and informal.

- Verbal
- Written
- Signs and notices
- Memos
- Reports
- Telephone
- E-mail
- Notice boards
- Newsletters
- Websites/intranet
- SMS text messaging
- Letters

It is important when managing a company, department or area, that staff and colleagues are kept informed of the organisations goals and strategies. When appropriate, company strategies should be **disseminated** amongst the staff. This way everyone is able to contribute to a shared vision. It is vital therefore that communication is two-way with feedback from others actively encouraged. Allowing colleagues to provide input to the decision making process may provide alternative ideas or methods which can improve processes and productivity. Contribution from colleagues can also raise morale by inclusion, making them feel part of a team (*see teamwork*). The biggest barrier to this is poor communication, and your methods should be reviewed to ensure that they are effective and the best type for any given situation.

Automotive Master Technician

Disseminate - to disperse, spread out and share widely, especially when giving information.

Organisational changes

Change is inevitable, in fact it is a requirement for progress and growth. Without change an organisation will become stagnant, characterised by a lack of development or progressive movement. In business, 'standing still' is 'going backwards'.

Too many companies become comfortable within their own sector and are not proactive in their approach to business.

There are many driving factors for change, including the examples show in Table 9.3.

Table 9.3 Organisational changes

Driving factor for change	Example
Technical	The automotive industry is a highly technical environment. Vehicles are becoming more and more advanced and the growth of technology is **exponential**. Without investment of time effort and finance, many garages will soon be left behind with the scope and type of work that they can undertake.
Legal	Legislation is constantly being introduced and updated. This can be to do with business, health and safety, environmental protection or vehicle law. Legal changes are often beyond the control of an organisation and your company will have to adapt in order to comply with the requirements.
Economic	Every business is finding that they are having to do more for less. In order for a company to be competitive, it may have to review its staffing and work policies, making cuts that are required from a commercial point of view.
Political	Most politics are driven by government and will often have an effect on legislation. There are, however often internal politics in an organisation where individuals use strategy and tactics to obtain a position of power or control. Some internal politics will drive change within a company that affects all staff.
Social	Social change is connected to the life, welfare and relations of human beings in a community. If either the local community around the organisation changes or the internal structure of a company changes, adaption is needed in order to maintain harmony.
Demographic	Statistical data about a population showing average age, income and education is known as **demographics**. In order for a business to be viable, it must take into account the services and products that it supplies relative to its local demographic. If the society or community in which the organisation trades changes, then so will its needs. It will therefore be necessary for the company to adjust or modify accordingly.
Business practices	How an organisation conducts its business has an impact on its success. Business practice will constantly evolve to keep up with legislation, economics, technology and demographics. New ideas drive business practice change and change drives adaptions in business practice as a perpetual cycle.
Structural	Reorganisation of a company structure, staff or departments, is a periodical process. The larger the organisation, the more regularly this may happen. Staff turnover can sometimes be a contributing factor to this form of change, while it is often in an effort to increase efficiency.
Financial	Commercial success is strongly related to cash flow. **Turnover** is about how much work is being done and the money it brings into an organisation, but may not necessarily be profitable. A company might have a large turnover but a small or negative profit margin. In order to ensure that a company remains profitable, it will have to regularly review its finances and make adjustments where required.

Advanced Light Vehicle Technology

If handled correctly, change provides opportunities including:

- Financial gains.
- Expansion.
- Increased productivity.

- Job security.
- Improved facilities and resources.
- Customer satisfaction and reputation.

Exponential - something that builds on itself with a snowball effect.

Demographic - a statistic characterising human populations (or segments of human populations broken down by age or sex or income etc.).

Turnover - the amount of money taken by a business in a particular period.

Planning for change

Any change in an organisation will affect its employees. A successful business plans for and manages change, which will assess and identify the development, and ease any transition helping to reduce barriers. Depending on your position within the company, the timescale for targets will vary. In a supervisory position, you may be dealing with short-term change, or in a more senior role, you could be called upon to facilitate long-term changes.
Factors that could affect change include the requirements for:

- New equipment and tooling.
- New skills.
- New business practices.
- Additional resources.

- Additional expense.
- Increased workload.
- New customer demands and expectations.

A **SWOT** analysis is a good tool for helping you assess the required changes that will improve the business while identifying any obstacles that will need to be overcome.
It is a structured planning method used to evaluate the **Strengths, Weaknesses, Opportunities**, and **Threats** involved in a project or in a business venture.
To create a **SWOT** analysis you will need to identify a target for the business venture or project and identify any internal or external factors that are positive or negative to achieving the desired outcomes.
Setting the objectives should be done after the **SWOT** analysis has been performed. This will allow achievable goals or objectives to be targeted for the organisation.
The sections of a **SWOT** analysis are grouped as follows:

Strengths- these are characteristics of the business or project that give it an advantage over others.
Weaknesses- these are characteristics that place the team at a disadvantage relative to others.
Opportunities- these are elements that the changes created by a project could exploit to its advantage.
Threats- these are elements in the environment that could cause trouble for the business or project.

When using this method you should consider whether the objective is attainable, given the **SWOTs**. If the objective is not attainable you will need to choose a different objective and repeat the process.
In order for this method to be successful you will need to ask yourself questions that generate meaningful information for each category (strengths, weaknesses, opportunities, and threats) helping to find the competitive advantage.

Another aspect that must be considered when using a **SWOT** analysis are any internal or external factors that could affect the outcome.

Internal factors – The strengths and weaknesses internal to the organisation.

External factors – The opportunities and threats presented by the external environment to the organisation.

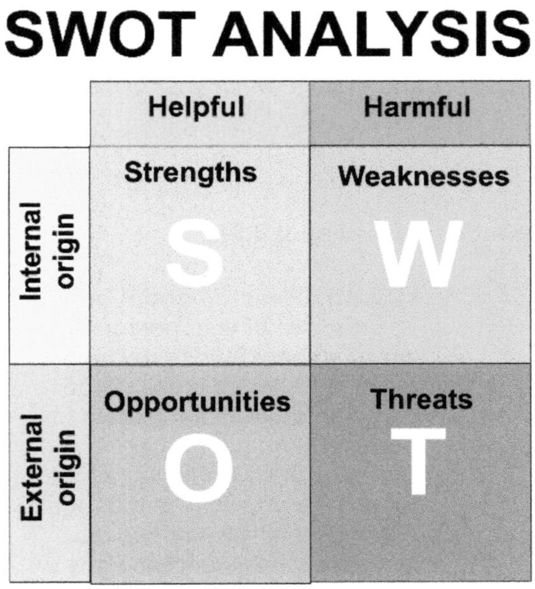

SWOT ANALYSIS

	Helpful	Harmful
Internal origin	**Strengths** S	**Weaknesses** W
External origin	**Opportunities** O	**Threats** T

Figure 9.6 A SWOT analysis matrix

The internal factors can be viewed as strengths or weaknesses depending upon their effect on the organisations' objectives. What may represent strengths with respect to one objective may be weaknesses for another objective.

The factors affecting this can include: personnel, finance, manufacturing capabilities and resources.

The external factors can include: technical, legal economic, political, social and demographic (*see Table 9.3*).

The planning and results are often presented in the form of a matrix or chart as shown in Figure 9.6.

It is important that everyone knows about any upcoming changes, when appropriate, and that the information is disseminated appropriately.

If this is not handled correctly, rumours can cause misunderstanding and bad feeling. Where necessary, you will need to identify and implement any training or support for people that will be affected by the change.

Advanced Light Vehicle Technology

Recording information and making suitable recommendations

At all stages of your fundamental management process, you should record information and where appropriate, make suitable recommendations. Some of this recorded information will help provide an efficient and productive department within your organisation. It can also help promote good teamwork and morale within your workforce. The Table below shows some examples of how to do this.

Table 9.4 Recording information and making suitable recommendations

Stage	Information	Records/Recommendations
Before you start	Personal development plan PDP with SMART targets.	Create a list of your personal goals which you are able to review and revise at different times in your career in order to set targets for achievement.
	Plan for change.	Create a SWOT analysis as a tool for helping you assess the required changes that will improve the business while identifying any obstacles that will need to be overcome.
On-going during working activities or tasks	Teamwork and progress monitoring.	Hold minuted team meetings to provide face to face contact at regular intervals with colleagues. These can help develop, trust, respect, team identity and familiarity with one another's ways of working.
Future strategic planning	Company and staff targets.	Evaluate and reflect on current company and staff position. Hold regular recorded staff appraisals, which set individual targets with achievable deadlines. Action plan for both short-term and long term change

Automotive Master Technician

Common Acronyms/Abbreviations

A - Amperes
A/C - Air Conditioning
A/F - Air Fuel Ratio
A/T - Automatic Transmission
AAC - Auxiliary Air Control Valve
AAT - Ambient Air Temperature
ABD - Automatic Brake Differential
ABS - Antilock Brake System
ABV - Air Bypass Valve
AC - Alternating Current
ACC - Automatic Climate Control
ACC - Air Conditioning Clutch
ACR - Air Conditioning Relay
ACR4 - Air Conditioning Refrigerant, Recovery, Recycling, Recharging
ACV - Air Control Valve
ADU - Analogue-Digital Unit
AEV - All Electric Vehicle
AFC - Air Flow Control
AFL - Advanced Front Lighting System
AFM - Air Flow Meter
AFR - Air Fuel Ratio
AFS - Air Flow Sensor
AGM - Absorbed Glass Matt
Ah - Amp Hours
AIR - Secondary Air Injection System
AIS - Automatic Idle Speed
ALC - Automatic Level Control
AM - Amplitude Modulation
API - American Petroleum Institute
APS - Atmospheric Pressure Sensor
ARC - Automatic Ride Control
ARS - Automatic Restraint System
ASARC - Air Suspension Automatic Ride Control
ATC - Automatic Temperature Control
ATDC - After Top Dead Centre
ATF - Automatic Transmission Fluid
ATS - Air Temperature Sensor
AVO - Amps Volts Ohms
AWD - All Wheel Drive
AWG - American Wire Gage
AYC - Active Yaw Control
B/MAP - Barometric/Manifold Absolute Pressure
BARO - Barometric Pressure
BCM - Body Control Module
BCM - Battery Control Module
BDC - Bottom Dead Centre
BEV - Battery Electric Vehicle
BHP - Brake Horsepower
BOB - Breakout Box
BP - Barometric Pressure
BPP - Brake Pedal Position Switch
BTDC - Before Top Dead Centre
BTS - Battery Temperature Sensor

Btu - British thermal unit
BUS N - Bus Negative
BUS P - Bus Positive
C - Celsius
CA - Cranking Amps
CAN - Controller Area Network
CANP - **EVAP** Canister Purge Solenoid
CAS - Crank Angle Sensor
CBW - Clutch by Wire
CC - Catalytic Converter
CC - Climate Control
CC - Cruise Control
CC - Cubic Centimetres
CCA - Cold Cranking Amps
CD - Compact Disc
CDI - Capacitor Discharge Ignition
CFC - Chlorofluorocarbons
CFI - Continuous Fuel Injection
CI - Compression Ignition
CKP - Crankshaft Position Sensor
CL - Closed Loop
CLC - Converter Lockup Clutch
CLV - Calculated Load Value
CMP - Camshaft Position Sensor
CNG - Compressed Natural Gas
CO - Carbon Monoxide
CO2 - Carbon Dioxide
COC - Conventional Oxidation Catalyst
COP - Coil on Plug Electronic Ignition
COSHH - Control of Substances Hazardous to Health
CP - Crankshaft Position Sensor
CP - Canister Purge (GM)
CPP - Clutch Pedal Position
CPU - Central Processing Unit
CRC - Cyclic Redundancy Check
CRD - Common Rail Diesel
CRS - Common Rail System
CTP - Closed Throttle Position
CTS - Coolant Temperature Sensor
CV - Constant Velocity
CVT - Continuously Variable Transmission
DBW - Drive by Wire
DC - Duty Cycle
DC - Direct Current
DCS - Dual Clutch System
DI - Distributor Ignition (System)
DI - Direct Ignition
DIS - Direct Ignition (Waste Spark)
DIS - Distributor-less Ignition System
DMF - Dual Mass Flywheel
DMM - Digital Multimeter
DLC - Data Link Connector (OBD)
DOHC - Dual Overhead Cam
DPF - Diesel Particulate Filter

Advanced Light Vehicle Technology

DRL - Daytime Running Lights
DTC - Diagnostic Trouble Code
DVD- Digitally Versatile Disc
EAIR - Electronic Secondary Air Injection
EBCM - Electronic Brake Control Module
EBP - Exhaust Back Pressure
EBD- Electronic Brake Force Distribution
ECC - Electronic Climate Control
ECM - Engine/Electronic Control Module
ECS - Emission Control System
ECT - Engine Coolant Temperature
ECU - Electronic Control Unit
EDC- Electronic Diesel Control
EECS - Evaporative Emission Control System
EEGR - Electronic EGR (Solenoid)
EEPROM - Electronically Erasable Programmable Read Only Memory
EFI - Electronic Fuel Injection
EFT - Engine Fuel Temperature
EGO - Exhaust Gas Oxygen Sensor
EGR - Exhaust Gas Recirculation
EGRT - Exhaust Gas Recirculation Temperature
EMF - Electromotive Force (voltage)
EMI - Electromagnetic Interference
EOBD - European On Board Diagnostics
EOP - Engine Oil Pressure
EOT - Engine Oil Temperature
EPA - Environmental Protection Act
EPB- Electronic Parking Brake
EPROM - Erasable Programmable Read Only Memory
EPS- Electronic Power Assisted Steering
ESP- Electronic Stability Programme
ESS - Engine Start-Stop
EVAP - Evaporative Emissions System
EVAP CP - Evaporative Canister Purge
FM- Frequency Modulation
FOT - Fixed Orifice Tube
FSD- Full Scale Deflection
FT - Fuel Trim
FWD - Front Wheel Drive
GDI - Gasoline Direct Injection
GND - Electrical Ground Connection
GPS- Global Positioning System
GWP - Global Warming Potential
H – Hydrogen
HASAWA- Health and Safety at Work Act
H2O - Water
HC - Hydrocarbons
HCA- Hot Cranking Amps
HDI- High Pressure Direct Injection
HEGO - Heated Exhaust Gas Oxygen Sensor
HFC- Hydrogen Fuel Cell

HFC- Hydro-fluoro Carbon
HFO- Hydro-fluoro Olefin
Hg – Mercury
HICE- Hydrogen Internal Combustion Engine
HID- High Intensity Discharge (lighting)
HO2S - Heated Oxygen Sensor
hp - Horsepower
HSE- Health and Safety Executive
HT - High Tension
HUD - Head up Display
HVAC - Heating Ventilation and Air Conditioning
Hz - Hertz
I/O - Input / Output
IA - Intake Air
IAC - Idle Air Control (motor or solenoid)
IAT - Intake Air Temperature
IC - Integrated Circuit
IC - Ignition Control
ICE- In Car Entertainment
ICE – Internal Combustion Engine
ICM - Ignition Control Module
IFS - Inertia Fuel Switch
IGBT- Insulated Gate Bipolar Transistor
IGN - Ignition
IGN ADV - Ignition Advance
IGN GND - Ignition Ground
IPR - Injector Pressure Regulator
ISC - Idle Speed Control
ISO - International Standard of Organisation
KAM - Keep Alive Memory
Kg/cm2 - Kilograms/ Cubic Centimetres
KHz - Kilohertz
Km - Kilometres
KPA - Kilopascal
KPI- Kingpin Inclination
KS - Knock Sensor
KWP - Keyword Protocol
l - Litres
LCD - Liquid Crystal Display
LED - Light Emitting Diode
LHD - Left Hand Drive
Li-ion- Lithium ion
LOOP - Engine Operating Loop Status
LOS - Limited Operating Strategy
LPG - Liquefied Petroleum Gas
LSD- Limited Slip Differential
LTFT - Long Term Fuel Trim
LWB - Long Wheel Base
M/T - Manual Transmission
MAC - Mobile Air Conditioning
MAF - Mass Air Flow Sensor
MAP - Manifold Absolute Pressure Sensor
MAT - Manifold Air Temperature

Automotive Master Technician

MCM- Motor Control Module
MEF- Methane Equivalency Factor
MF- Maintenance Free
MFI - Multiport Fuel Injection
MIL - Malfunction Indicator Lamp
MPG - Miles per Gallon
MPH - Miles per Hour
mS or ms - Millisecond
mV or mv - Milivolt
N - Nitrogen
NCAPS - Non-Contact Angular Position Sensor
NCRPS - Non-Contact Rotary Position Sensor
NGV - Natural Gas Vehicles
Ni-MH- Nickel Metal Hydride
Nm - Newton Meters
NOx - Oxides of Nitrogen
NPN- Negative Positive Negative
NTC - Negative Temperature Coefficient
O2 - Oxygen
OBD I - On Board Diagnostics Version I
OBD II - On Board Diagnostics Version II
OC - Oxidation Catalytic Converter
OD - Overdrive
OD - Outside Diameter
ODP - Ozone Depletion Potential
OE - Original Equipment
OEM - Original Equipment Manufacturer
OFN - Oxygen Free Nitrogen
OHC - Overhead Cam Engine
OHV - Overhead Valve
OL - Open Loop
OS - Oxygen Sensor
P/N - Part Number
PAG - Polyalkylene Glycol
PAIR - Pulsed Secondary Air Injection
PATS - Passive Anti-Theft System
PCB - Printed Circuit Board
PCM - Powertrain Control Module
PCV - Positive Crankcase Ventilation
Pd-Potential Difference (volts)
PEF- Propane Equivalency Factor
PEM- Proton Exchange Membrane
PFI - Port Fuel Injection
PGM-FI - Programmed Gas Management Fuel Injection
PID - Parameter Identification Location
PKE - Passive Keyless Entry
PNP- Positive Negative Positive
POT - Potentiometer
PPE- Personal Protective Equipment
PPM - Parts Per Million
PPS - Accelerator Pedal Position Sensor
PROM - Programmable Read-Only Memory
PSI - Pounds per Square Inch

PTC - Positive Temperature Coefficient Resistor
PTO - Power Take Off (4WD Option)
PUWER- Provision and Use of Work Equipment Regulations
PWM - Pulse Width Modulation
RAM - Random Access Memory
RBS - Regenerative Braking system
RCM- Reserve Capacity Minutes
RDS - Radio Data System
REF - Reference
RFI - Radio Frequency Interference
RHD - Right Hand Drive
RIDDOR- Reporting of Injuries Diseases and Dangerous Occurrence Regulations
RKE - Remote Keyless Entry
RMS - Recovery Management Station
ROM - Read Only Memory
RON - Research Octane Number
RTV - Room Temperature Vulcanizing
RWD - Rear Wheel Drive
SAE –Society of Automotive Engineers (Viscosity Grade)
SAI- Swivel Axis Inclination
SCR- Selective Catalytic Regeneration
SCS - Sick Car Syndrome
SFI - Sequential Fuel Injection
SI- Spark Ignition
SIPS - Side Impact Protections System
SOC- State of Charge
SOHC - Single Overhead Cam
SPFI - Single Point Fuel Injection (throttle body)
SRI - Service Reminder Indicator
SRS - Supplementary Restraint System (air bag)
SRT - System Readiness Test
STFT - Short-Term Fuel Trim
SWB - Short Wheel Base
SWL- Safe Working Load
TAC - Throttle Actuator Control
TACH - Tachometer
TBI - Throttle Body Injection
TC - Turbocharger
TCC - Torque Converter Clutch
TCM - Transmission Control Module
TCS - Traction Control System
TD - Turbo Diesel
TDC - Top Dead Centre
TDI - Turbo Direct Injection
TOOT- Toe Out On Turns
TP - Throttle Position
TPM - Tyre Pressure Monitor
TPP - Throttle Position Potentiometer
TPS - Throttle Position Sensor
TSB - Technical Service Bulletin
TV - Throttle Valve

Advanced Light Vehicle Technology

TXV- Thermal Expansion Valve
UART - Universal Asynchronous Receiver-Transmitter
UJ- Universal Joint
USB - Universal Serial Bus
UV - Ultraviolet
V - Volts
VAC - Vacuum
VAF - Vane Airflow Meter
VDP- Variable Diameter Pulley
VDU- Visual Display Unit
VIN - Vehicle Identification Number
VPE- Vehicle Protection Equipment
VSS - Vehicle Speed Sensor
W/B - Wheelbase
WOT - Wide Open Throttle
WSS - Wheel Speed Sensor
YRS - Yaw Rate Sensor

Index

Lightning Source UK Ltd.
Milton Keynes UK
UKOW07f1650260916

283836UK00012B/120/P

9 780992 949228